HYDROLOGY OF ARTIFICIAL AND CONTROLLED EXPERIMENTS

Edited by **Jiu-Fu Liu** and **Wei-Zu Gu**

Hydrology of Artificial and Controlled Experiments
http://dx.doi.org/10.5772/62972
Edited by Jiu-Fu Liu and Wei-Zu Gu

Contributors

Wolfgang Schaaf, Werner Gerwin, Christoph Hinz, Markus Zaplata, Reinhard Hüttl, Sanyuan Jiang, Qiande Zhu, Seifeddine Jomaa, Lihu Yang, Xianfang Song, Yunfeng Qiao, Wei-Zu Gu, Yoshitaka Komatsu, Till H. M. Volkmann, Peter Troch, Luke Pangle, Katerina Dontsova, Nate Abramson, John Adams, Greg Barron-Gafford, David Breshears, Aaron Bugaj, Jon Chorover, Stephen DeLong, Matej Durcik, Ty Ferre, Ciaran Harman, Edward Hunt, Travis Huxman, Minseok Kim, Raina Maier, Antonio Meira, Laura Meredith, Russell Monson, Guo-Yue Niu, Jon Pelletier, Craig Rasmussen, Joaquin Ruiz, Scott Saleska, Marcel Schaap, Michael Sibayan, Markus Tuller, Joost Van Haren, Yadi Wang, Xubin Zeng, Aditi Sengupta, Michael Pohlmann, Alejandro Cueva, Nicola Pastore, Jan Frouz, James V. Bonta, Martin J. Shipitalo, Lloyd Owens, Beibei Zhou

First published in London, United Kingdom, 2018 by IntechOpen
IntechOpen is the global imprint of INTECHOPEN LIMITED, registered in England and Wales, registration number: 11086078, The Shard, 25th floor, 32 London Bridge Street
London, SE19SG – United Kingdom
Printed in Croatia

British Library Cataloguing-in-Publication Data
A catalogue record for this book is available from the British Library

Additional hard copies can be obtained from orders@intechopen.com

Hydrology of Artificial and Controlled Experiments, Edited by Jiu-Fu Liu and Wei-Zu Gu
p. cm.
Print ISBN 978-1-78923-558-6
Online ISBN 978-1-78923-559-3

For EU product safety concerns:
IN TECH d.o.o., Prolaz Marije Krucifikse Kozulić 3, 51000 Rijeka, Croatia,
info@intechopen.com or visit our website at intechopen.com.

Meet the editors

Jiu-Fu Liu received his BSc and MSc degrees in Hydrology from the Hohai University, Nanjing, China. He has been a research professor at the Nanjing Hydraulic Research Institute (NHRI) of the Chinese Ministry of Water Resources since 2003. He engages himself in the development of scientific hydrology and hydroinformatics and in the renovation of hydrological experiments. He has presided the National Key R&D Program of China and several key projects of the National Natural Science Foundation. He is a winner of two second prizes of the National Scientific and Technological Progress Award, four first prizes, and one second prize of Scientific and Technological Progress Award in provincial/ministerial levels. He has published c.a. 30 academic articles. He became the director of the Institute of Hydrology and Water Resources of NHRI in 2007 and the deputy secretary-general of Hydrological Professional Committee, Chinese Hydraulic Engineering Society, and so on.

Wei-Zu Gu graduated from the Jiao-Tung University, Shanghai, China, in 1952. He has been a research professor at the Nanjing Hydraulic Research Institute of the Chinese Ministry of Water Resources since 1982 and an adjunct professor of Hefei Polytechnical University (1987–1996), Heilongjiang University (1990–1996), Hohai University (2003–2006), Xi'an University of Science and Technology (1994–2014), and Nanjing Center of Chinese Geological Survey (2003–2009). He dedicated himself to the Chinese field of basin studies for more than 60 years since 1953. He was awarded by the Chinese Ministry of Science and Technology (2009) as the outstanding contributor of the field work. He is an editor/coeditor of three books and c.a. 50 published articles. His current research focuses on the renovation of watershed hydrological experimental systems and the applications of isotope hydrology.

Contents

Foreword

Controlled experiments in hydrology are rare. The best ones have helped redefine the field: John Hewlett's trough hillslope experiment at Coweeta in the late 1950s and early 1960s that changed how we conceptualize hillslope drainage, Alan Rodhe's Gardsjon roof experiment in Sweden in the 1980s that helped inform new understanding of hillslope storage and release and its control on concentration-discharge relations, and the Biosphere-2 LEO hillslope led by Peter Troch in the past decade that is opening up new and fundamental understanding of coevolution of hillslope climate, vegetation, and solute weathering. Of course, these experiments allow boundary conditions to be constrained and defined (something rare in field experiments in natural systems) and allow one to study events outside the narrow range of conditions one might be able to "see" in a typical field experiment. As models flourish and usurp much basic field work, controlled experiments and artificial soil blocks, hillslopes, and mini-catchments are increasingly needed to discover new process behaviors and responses and to quantify model performance.

It is against this backdrop that the very useful *Hydrology of Artificial and Controlled Experiments* comes along. Its editors, Jiu-Fu Liu and Wei-Zu Gu from the National Hydraulic Research Institute in Nanjing, are the forefront of such study in China—both linked to the Chuzhou Hydrological Laboratory and the famous Hydrohill experimental hillslope—the single greatest public work efforts in artificial hillslope hydrology. I served on the Biosphere-2 "LEO planning committee" in the years before its construction and can say that many-a-planning-discussion was influenced by what had been done at Hydrohill in the decades prior.

Liu and Gu have done an outstanding job of assembling an impressive list of authors and studies from around the world—from lysimeter studies in Japan to mine cover sites in Germany and laboratory slopes and runoff plots in China along with chapters on LEO and Hydrohill. The book *Hydrology of Artificial and Controlled Experiments* is a comprehensive overview of studies of artificial and controlled experiments today and a useful benchmark for students and practitioners alike.

Jeffrey J. McDonnell
Global Institute for Water Security
University of Saskatchewan, Saskatoon, Canada

Preface

The river basin, watershed, or catchment is central to many of the concepts in hydrology [1], and scientific hydrology was found in two basin studies in the Seine river basin during the end of the seventeenth century as suggested by UNESCO/WMO/IAHS [2]. However, basin studies developed slowly until the end of the nineteenth century when public demands accelerated. Since the early twentieth century, a multitude of basin studies have appeared, including the Wagon Wheel Gap experiment of the USA begun in 1910 in two forested basins, the Valdai Branch of the State Hydrological Institute (for field experimental investigations) of the former USSR in 1933, Coweeta Hydrologic Laboratory of the USA set up since 1934, Harz Mountains experiment of Germany begun in 1948, Alrance experiment of France started in 1950, Bluebrook Runoff Experiment of China established in 1953, and so on. This can be regarded as the first stage of hydrological experimentation. A period of rapid worldwide development resulted from the representative and experimental basin (EB) program provided by the first International Hydrology Decade (UNESCO/IHD) since 1965, with an estimated 3000 basin studies conducted over the world during the decade. It can be regarded as the second stage of development. Since this flourishes it has been going on for more than five decades up to now, what faced in the present is a changing nature of great transition with anthropogenic perturbation, replumbing of the hydrologic cycle, and natural climate oscillations. In 2006, a paper coauthored by 12 scientists resulted from a CUAHSI vision workshop pointed out that "Yet, most field experiments and observations in watershed science to date, remain largely descriptive. Many of these field studies have not set out to seek fundamental truth or understanding (nor test any formal theory or hypothesis per se)" and "We should instead focus on the development of systematic measurement programs that are specifically targeted to the generation of tests of new theories" [3]. We have to thus aim at the substantial progress in hydrologic science toward "a new unified hydrologic theory" as Sivapalan has suggested [4]. May we believe that the third stage of the transition of hydrological experiments of reforming is irresistible, perhaps it has begun already.

Werner Heisenberg had warned that "what we observe is not nature herself, but nature exposed to our method of questioning." Perhaps, there is a misunderstanding in hydrological experiments to place hopes on natural watershed for trying to lift her veil of complexities. In fact even in small scale most natural watershed will keep doggedly her own character of complexity with some degree of organization, it follows that the manipulation experiments very likely can play their unique role to making some changes for it. "Nothing ventured, nothing gained," the venture for watershed experimentation may be using some kind of measures to manipulate a part of the nature for what we want her to expose.

Actually, since John Hewlett's trough hillslope experiment at Coweeta in the late 1950s, a lot of development has been achieved including controlled slope, artificial catchment, and even that in phytotron. The goal of this book is to stimulate the approach of manipulation in promoting watershed hydrological experimentation. Hopefully, it can demonstrate that the controlled and artificial experiments are a promising way of useful and effective generation of tests of new theories.

This book is organized on the basis of nine different manipulation types from six countries, including field lysimeter, field runoff plot, field manipulated EB, field artificial catchment, laboratory pedon (rock), laboratory lysimeter, laboratory hillslope, phytotron artificial catchment, and artificial river segment (Table 1).

In completing this book, first and foremost, we would like to thank the contributors for their interests, expertise, and insights that have made this book possible. We would like to thank all the peer reviewers who donated their valuable time. We are grateful to Jian-Yun Zhang, Jeffrey McDonnell, Henry Lin, Werner Gerwin, Peter Troch, Jan Frouz, James V. Bonta, and Martin J. Shipitalo for their encouragements and support during the planning of this book. Thanks also go to Dajana Pemac, Danijela Duric, Ana Pantar, and Maja Bozicevic, Publishing Managers of IntechOpen, for initiating this book and for their patience and editorial efforts in bringing this book to completion.

Jiu-Fu Liu and Wei-Zu Gu
Institute of Hydrology and Water Resources
Nanjing Hydraulic Research Institutes
Ministry of Water Resources, China

References

[1] Rodda JC. Basin Studies. In: Rodda JC (ed.) Facets of Hydrology. London: John Wiley & Sons; 1976. pp. 257-297.

[2] UNESCO/WMO/IAHS. Three Centuries of Scientific Hydrology. Paris: Unesco; 1974.

[3] McDonnell JJ, Sivapalan M, Vache K, Dunn S, Grant G, Haggerty R, Hinz C, Hooper R, Kirchner J, Roderick ML, Selker J, Weiler M. Moving beyond heterogeneity and process complexity: A new vision for watershed hydrology. Water Resources Research. 2007;43:W07301.

[4] Sivapalan M. Pattern, Process and Function: Elements of a Unified Theory of Hydrology at the Catchment Scale. In: Anderson MG (ed.) Encyclopedia of Hydrological Science. John Wiley & Sons; 2005. pp. 193-219.

Table 1. Facility types

Chapter (from country)	Type													Main monitoring items					
	Field EB	Field EB (M)	Field plot	Field lysimeter	Field artificial catchment	Lab. pedon	Lab. rock	Lab. lysimeter	Lab. slope	Lab. artificial catchment	Lab. river	Rainfall simulator	Phytotron	Water	Solute	Gas	Geo-media	Heat	Soil ero
1 (USA)	☆			☆										☆	☆				
5 (China)	☆		☆											☆	☆				☆
11,12 (China)	☆		☆	☆										☆	☆				☆
		☆												☆	☆		☆		
7 (Japan)			☆									☆		☆	☆				
8 (China)			☆									☆		☆	☆				☆
5 (China)								☆	☆			☆		☆	☆				☆
								☆				☆	☆	☆	☆	☆	☆		
											☆	☆		☆					
6 (China)						☆		☆				☆		☆	☆			☆	☆
9 (Italy)							☆							☆	☆			☆	
10 (China)								☆	☆			☆		☆	☆				☆
3 (Germany)			☆		☆									☆	☆		☆		
4 (Czech Republic)					☆									☆	☆		☆		
11, 12 (China)					☆							☆		☆	☆		☆		
2 (USA)										☆		☆	☆	☆	☆	☆	☆		

EB – experimental watershed; M – manipulated; Lab – laboratory, soil ero – soil erosion

Chapter 1

Experimental Watersheds at Coshocton, Ohio, USA: Experiences and Establishing New Experimental Watersheds

James V. Bonta, Martin J. Shipitalo and Lloyd Owens

Additional information is available at the end of the chapter

http://dx.doi.org/10.5772/intechopen.73596

Abstract

The North Appalachian Experimental Watershed (NAEW) in Ohio was established in 1935 to improve economical and physical sustainability in agriculture. The objectives were to test management practices on small watersheds, investigate scaling of runoff and erosion to larger areas, and research ways to extrapolate the results to ungauged areas. The facility was equipped with a permanent infrastructure consisting of runoff stations and rain gauges for watersheds ranging in size from 0.26 to 1854 ha, and 11 large (0.008 ha) monolith lysimeters to investigate small-scale water balances, all in an area greater than 2000 ha. After about 1970, the NAEW was reduced in size to 425 ha consisting of mostly small watersheds ("test beds") ranging in size from 0.26 to 3.07 ha. The NAEW was in operation for approximately 81 years generating a long record of runoff and other data for various watersheds, and closed in 2015. A wide variety of experiments were conducted on the NAEW with many high-impact accomplishments and addressing emerging issues that founders never envisioned. Nearly, 500 publications came from investigations during the history of the facility, and insights for establishing new experimental watersheds are presented covering site selection, funding, site specificity, extrapolation of results, generation of runoff in different physiographic regions, collaboration, off-site investigations, and instrumentation. The research on water quality was added to the research objectives in the 1970s, including nutrients (nitrogen and phosphorus) and pesticides in surface runoff and subsurface flow.

Keywords: experimental watersheds, lysimeter, precipitation measurement, runoff measurement, agriculture, hydrology, water quality

1. Background

In the decades leading up to the mid-1930s in the United States (USA), agricultural enterprises were increasingly physically and economically unsustainable due to soil erosion and flooding. The United States Department of Agriculture (USDA) recognized that there was insufficient underlying science supporting the management of agricultural lands that could be assembled into practical land-management guidance for producers. As a result, a national effort established large and small scale research projects that would test the effectiveness of land-management practices under natural-weather conditions in different regions of the country to minimize agricultural environmental problems nationwide.

Consequently, three large-scale experimental watersheds were established in the USA in the mid-1930s [1]. In 1935, one of the large-scale areas established was the outdoor laboratory for land and water management research at the North Appalachian Experimental Watershed (NAEW, also known as the "Coshocton watersheds") near Coshocton, Ohio, the focus of this chapter. This chapter draws heavily from three prior publications that describe the NAEW. Reference [1] describes the NAEW as part of the three original large-scale experimental watersheds, Ref. [2] concentrates on the NAEW history and capabilities, and Ref. [3] describes the types of NAEW data available.

The purposes of this chapter are to: (1) present the history and design of the NAEW, instrumentation, physical features, unique capabilities, data available, research portfolio, and examples of accomplishments, and (2) discuss challenges likely to be encountered when establishing new experimental watersheds and suggest possible remedies. This chapter summarizes the information given in the NAEW history, research portfolio, and capabilities found in [2]. It differs from the other publications on the NAEW listed above in that it raises challenges and provides guidance for establishing new experimental watersheds based on the research experiences at the NAEW. Some of the information in [2] is reiterated here, and the reader is referred to that publication for more detailed information. As noted in this chapter, the NAEW was unique in data collected and physical features found nowhere else in the USA.

The founding document for the NAEW [4] listed three overall objectives:

1. "To determine the effect of land use and erosion control practices upon the conservation of water for crops and water supply and upon the control of floods under conditions prevailing at the North Appalachian Region [NAR] of the US";

2. "To determine the effect under (1) for small and large areas and to trace variations in this effect from the smallest plot and lysimeters through a series of intermediate watersheds to the largest watershed on the project"; and

3. "To determine the rates and amounts of run-off for precipitation of different amounts and intensities for watersheds typical of the NAR of different configuration, size, shape, topography, cover, underground conditions, land use, and erosion control practices. To furnish data needed for use in the design of erosion control structures and in the design and operation of the Muskingum Watershed Conservancy District and other flood control projects lying within the NAR."

The originators of the experimental watershed program were visionaries as the concepts above are the needs required for ideally researching landscapes to minimize environmental damage and maximize sustainability even today. The objectives above were to investigate, within a region (hill lands of the North Appalachian Region, NAR), conservation practices at the small field scale (areas manageable by the producer—objective 1), to investigate the watershed response when the smaller nested areas (possibly not owned by the same producer) collectively interact at increasingly larger areas (objective 2—scaling), and to generalize/extrapolate the site-specific field results to ungauged areas (todays "modeling"—objective 3—recognizing the site-specificity of monitored watersheds). The founders recognized the temporal and spatial variability of weather and the landscape, the complex nonlinear areal behavior of runoff and erosion processes, the need for measuring watershed responses, and the need for developing guidance for producers in the absence of field data for unmonitored fields and watersheds.

The NAEW was originally operated by the USDA—Soil Conservation Service, Division of Research, and in 1954 became part of the newly created Agricultural Research Service (ARS). NAEW has worked collaboratively throughout the 81-year history with The Ohio State University, especially through its agricultural research center located in Wooster, Ohio. Throughout its history, NAEW scientists have collaborated with university scientists and students worldwide, state and Federal agencies, and persons in the private sector. These collaborations supplied scientific expertise required for specific project objectives and facilitated addition of new capabilities for the NAEW.

Construction of administrative and shop buildings, and instrumentation infrastructure started at the NAEW in about 1935 using workers from the Works Progress Administration (WPA), Civilian Conservation Corps (CCC), and the Civilian Public Service (CPS) program for construction and data collection in the early years. The NAEW was closed in December, 2015, a duration of about 81 years since the construction began. The earliest data records began in about 1937 spanning 78 years of data collection on the NAEW (instrumentation and data to be presented later).

2. Physical setting

The NAEW was chosen for its "representativeness" in the NAR which included southeast Ohio, eastern Pennsylvania, northern Kentucky, and northern West Virginia (**Figure 1**). Physical features considered for "representative" experimental watershed selection included soil types, climate, and other factors. Determining representativeness using physical map overlays at the time was comparable to the use of modern-day geographical information systems (GIS). The selected site was one of 86 candidate sites [2, 4].

The NAEW consisted of agricultural lands in east-central Ohio (Ohio map inset in **Figure 2**) with slopes typically ranging from 18 to 25% and elevations ranging from about 250 to 350 m. About half of the area was in grassland with corn, soybeans, wheat, and forest comprising the remaining area [2]. The latitude of the NAEW is about 40.4° N.

Originally, the NAEW comprised a 1854-ha watershed area with several nested gauged watersheds (Little Mill Creek [LMC] watershed, **Figure 2**, left). This watershed was chosen to address

Figure 1. View of NAEW landscape and administrative buildings.

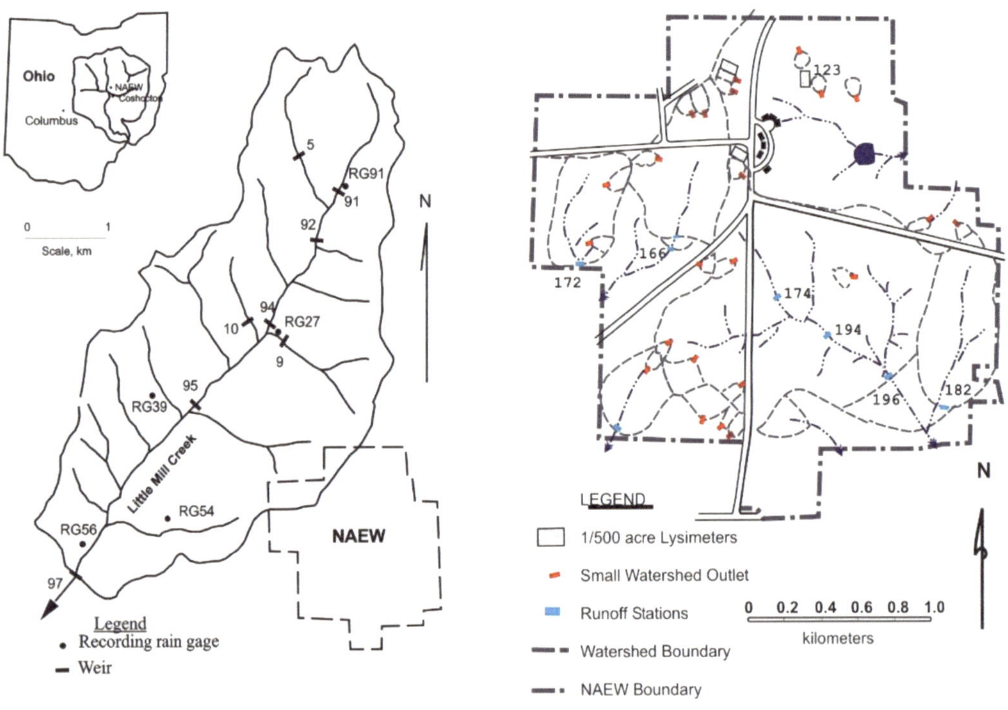

Figure 2. The North Appalachian Experimental Watershed (NAEW) comprises the 1854-ha Little Mill Creek (LMC) watershed (left) and the smaller 425-ha NAEW area (right). Inset shows the location of the NAEW within the state of Ohio.

mainly objective 2—scaling issues. Additionally, part of the NAEW included 425-ha in the southeast area of LMC (**Figure 2**, right). On this area were several small monitored watersheds of the order of 0.4 ha to address mainly objective 1—evaluating impacts of specific practices on a small (producer-managed) areas, where there were no confounding influences of other land-management activities. In approximately 1970, monitoring in the LMC watershed ceased and the NAEW was reduced in size to 425-ha with the largest gauged watershed at 123 ha (**Figure 2**, right).

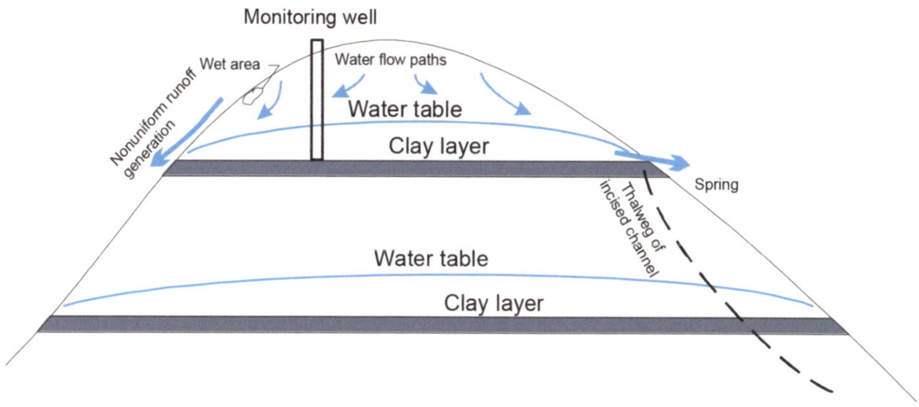

Figure 3. Schematic drawing of perched water table due to geological clay layers in unglaciated sedimentary strata on the NAEW in a hilltop, landscape incision of a stream channel intercepting these water sources, and elements of nonuniform runoff generation.

The average annual air temperature is 10.4°C and the average annual precipitation is 959 mm. Cool air from the northwest and moist air from the south often converges to form storms over the NAEW [2]. Soil during winter often freezes for short periods causing precipitation to immediately runoff. Snowmelt also is a source of runoff.

The geology of the NAEW consists of unglaciated sedimentary strata composed of mostly sandstone and shale, with interbedded strata of coal, clay, and limestone. An underlying anticline, local synclines, and strata slightly dipping to the southeast characterize the structure of geological formations [5], (**Figure 3**).

Soils of the NAEW were developed in residua of weathered sandstone and shale. Three dominant soil types include well-drained sandstone-derived soils (Inceptisols), soils with an argillic horizon derived from shale (Alfisols & Ultisols), and soil between these extremes [2, 5].

Small watersheds were chosen from many swale areas on the landscape. They were characterized by ephemeral areas that shed water only during heavy rain storms and snowmelt, with no incised channel. Larger watersheds have incised channels and drain areas with multiple land-management areas.

3. Serendipitous physical features affecting hydrology and water quality

Using general factors such as geology, soils, weather, etc., to select a "representative" site for the NAEW was necessary. Other hydrologically beneficial features of the NAEW location, however, became apparent as experiments were conducted on the site. For examples:

1. *The imperviousness of geological clay layers underlying coal seams supported perched water tables* [2]. These perched water bodies allowed an index measure of the ground-water impacts of surface land-management treatments. Ground-water impacts were evaluated, where the intersection of geological clay layers intersected the landscape surface forming springs that were monitored beneath treated hilltops [5], (**Figure 3**). Ground water beneath areas as

large as 15 ha have been monitored by using springs because of this favorable geological structure. Impacts of land management on ground water became a significant area of research on the NAEW.

2. *Nonuniform runoff generation* [2]. Due to springs at the ground surface and persistently high soil-profile-water-content areas, runoff is generated nonuniformly on the surface and with time during an event. NAEW measurements of natural-precipitation infiltration showed that water simultaneously emerges from the soil (exfiltration) and infiltrates into it during a runoff event at different locations. Watershed models today are deficient in modeling this runoff-generating process. Superimposed on these physical processes are wide ranging anthropogenic influences on the land surface as watershed areas increase that also help to generate runoff nonuniformly over a landscape.

3. *Interflow process* [2]. Closely related to nonuniform runoff generation is the interflow process in which water moves laterally within the soil profile. This process was apparent on the NAEW and is also not well simulated in watershed models (**Figure 3**).

4. *Natural lysimeter* [2]. A lysimeter (discussed under "Instrumentation" section) is usually considered an isolated block of soil that accounts for the sources and distribution of water in a contained area. It was discovered that a thick clay layer underlying a coal seam outcropped along the periphery of a hilltop enclosed an approximate area of 2.8 ha (known as Urban's Knob). The synclinal structure of the sedimentary bedrock within the hilltop forced all water entering the hilltop to its center where it discharged to a surface spring. Consequently, the source of all water within the hilltop was from precipitation as no ground water flowed from adjacent areas as often occurs in ground-water studies, forming a "natural" lysimeter. The area was instrumented with a network of wells and piezometers, a spring, two watersheds, a rain gauge, and profiles of ceramic suction cup lysimeters to investigate unsaturated flow of water and chemicals.

5. *Macropore flow* [2]. There is significant transport of chemicals and water in larger pores in the soil (particularly holes caused by earthworms), a poorly simulated process in watershed models. This became a significant area of research at the NAEW as explained later.

4. Instrumentation

Instrumentation was planned to achieve the general objectives listed above under natural precipitation and weather conditions—small scale evaluations of treatments, evaluations of watershed responses at larger scales, and "modeling." Generally, instrumentation for measuring watershed responses to treatments was to be permanently available for experiments. This allowed the immediate use of experimental watersheds with a long runoff record to be used in comparisons when evaluating land treatments and reduced the cost of monitoring runoff.

Small watersheds, ranging in size from 0.26 to 3.07 ha, were installed on the smaller 425-ha area (**Figure 2** right) in natural-swale, ephemeral, overland-flow areas on the hillsides where runoff occurs during large intensity rains and snowmelt. These watersheds were used as "test beds" to determine the effectiveness of different land-management treatments. The treatment for an individual watershed was implemented over the entire area so that runoff-response data were not confounded by runoff from other areas with different land managements. Runoff from the smaller watersheds were measured using H flumes ([6], **Figure 4**). More recently, two watersheds were monitored using drop-box weirs which provide better flow measurement in sediment-laden runoff water [7], (**Figure 5**). Because of spatial variability of precipitation, each watershed was instrumented with a weighing-bucket rain gauge. Runoff and precipitation data were historically tabulated with depth and time resolutions of 0.25 mm and 1 min, respectively, and when a change in flow depth or precipitation intensity was apparent. Larger watersheds on the 425-ha area up to 123 ha were monitored using Parshall flumes initially and later short-crested V-notch weirs replaced them [6], (**Figure 6**).

The LMC watershed was instrumented with a network of recording rain gauges and weirs (**Figure 2**, left). Nested watersheds ranged in size from approximately 39 to 1854 ha. As mentioned before, LMC was closed in about 1970 so there is approximately 30 years of runoff and precipitation data available from most of these watersheds and rain gauges. These watersheds were useful for documenting the nonlinearity of runoff ("scaling," **Figure 7**) at Coshocton, and have potential for other investigations such as for regional model parameterization and routing.

Figure 4. H flume and original Coshocton wheel rotating-slot sampler.

Figure 5. Turbulent flow in a NAEW drop-box weir for flow measurement in sediment-laden runoff.

Figure 6. Short-crested V-notch weir replaced the Parshall flume upstream in the view on a larger NAEW watershed.

Figure 7 shows how watersheds in different physiographic and climatological regions in the USA respond to climate as watershed area increases. For the unglaciated watersheds in the Coshocton area, the nonlinearity of watershed area vs. runoff relationship reflects the increase in baseflow to a relatively constant value as more and larger stream channels intersect perched water tables in this region of sedimentary strata (dashed line in **Figure 3**).

http://dx.doi.org/10.5772/intechopen.73596

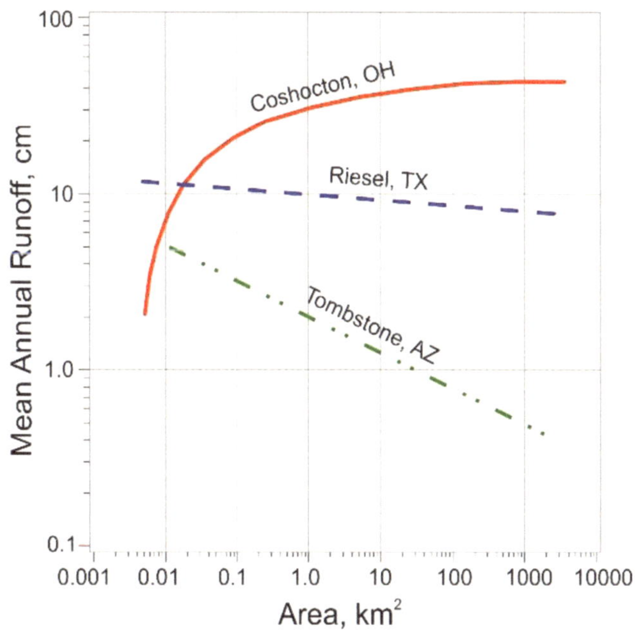

Figure 7. Watershed area versus annual runoff for different physiographic locations in the USA. Graph from [8, 9].

Figure 8. Construction and installation of three lysimeters at the NAEW.

While small watersheds provided data on runoff responses at a small watershed scale, the originators of the experimental watershed program wanted to investigate on a very small scale the water balance on isolated blocks of undisturbed soil ("monolith lysimeters," **Figures 8** and **9**). Eleven lysimeters were installed in the three dominant soil types on the NAEW, four each on two soil types and three on the third soil type. Each lysimeter had a horizontal surface area of ~0.0008 ha (width ~1.8 m and length ~4.3 m), depth was ~2.4 m, and enclosed an undisturbed monolith of the soil profile. The 2.4-m depth included undisturbed surface soil and weather bedrock. Each lysimeter measured percolation (ground-water recharge) from

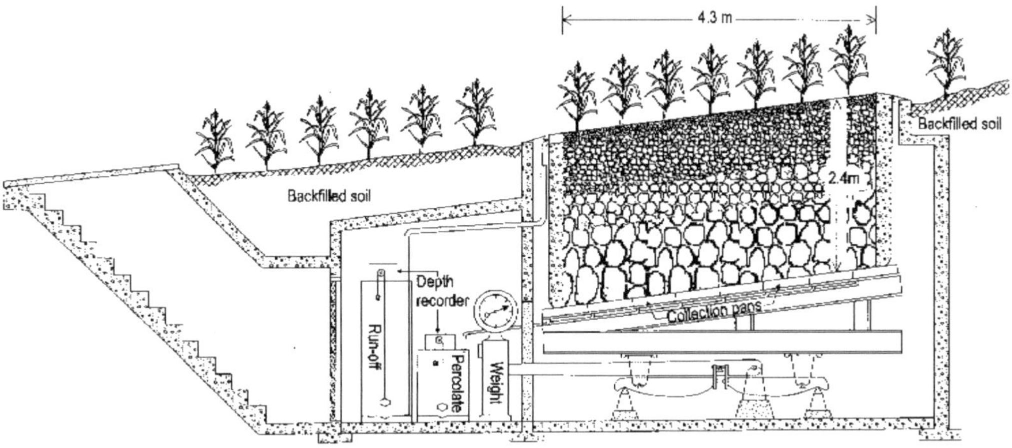

Figure 9. Schematic profile of an underground weighing lysimeter with and undisturbed profile of weathered bedrock near the bottom and soil at the top.

the bottom and runoff from the surface. Additionally, one lysimeter at each set of lysimeters within a soil type was weighed to provide evapotranspiration and ground-level precipitation data.

Recognizing the spatial variability of precipitation over small areas (especially during summer months), precipitation data were monitored by using weighing-bucket rain gauges with orifices placed approximately 1 m above the ground at most small watersheds. Gauges were similarly placed at each of the three sets of lysimeters.

Weather data were measured at a single weather station on the NAEW and included wind speed and direction, air temperature, humidity, solar radiation, evaporation pan, barometric pressure, soil temperature, and precipitation. Since about 1985, data loggers were used to monitor all NAEW data (except precipitation) with a radio-telemetry system. This system allowed more frequently measured weather, runoff, and precipitation data to be recorded. Prior to ~1985, runoff and other charts were hand tabulated. For the entire period, manual measurements were made of some weather elements.

Soil loss from the small experimental watersheds was an original concern; however, no reliable water sampler was available to measure sediment concentration during a runoff event. Consequently, the "Coshocton Wheel" was invented in about 1945 to obtain a flow-weighted composite measurement of the total sediment concentration during runoff events (**Figure 4**). The sampler consisted of a water (runoff)-powered wheel with a rotating slot (no power requirements), and obtained a constant fraction of the total sediment load from the watershed (single sample). The sampler has been used worldwide. When event concentration is multiplied by total runoff, an estimate of event sediment load is obtained for the treatment on the small watershed. Prior to the NAEW invention of the Coshocton Wheel, all runoff and sediment were collected in concrete troughs that were dug out manually to obtain a measurement of sediment yield (**Figure 10**). Note that the flume is full of sediment after the runoff event flowed over the bare, steep soil. This photo demonstrates the need for an automatic water sampler and the desirability of using a drop-box weir that will keep sediment moving through the flume (**Figure 5**).

In the 1970s, water quality beyond sediment concentrations became important [3]. Pesticides, major anions and cations, and nutrient losses, especially nitrogen, from the small watersheds became a concern and various land management treatments were evaluated using these measures as well as runoff volumes and sediment loads, and subsurface flow. The Coshocton Wheel provided the sample needed for laboratory analyses of these constituents and loads were similarly computed as for sediment concentrations.

Occasionally, new watershed and plots sites on and off the NAEW were required such as for coal surface-mine (**Figure 11**), paper-mill byproduct, and manure studies. NAEW scientific and technical personnel provided the needed expertise.

The small NAEW experimental watersheds were managed with a new treatment following an old one on the same watershed. This allowed comparisons of current treatments with previous ones. Occasionally, a paired watershed approach was used.

Figure 10. H flume and downstream sediment trough that catches all sediment from the bare, steep watershed after an extreme event.

Figure 11. NAEW scientists investigated the effects of drastic land disturbances due to surface mining for coal before and during mining, and after reclamation on ground and surface water hydrology and water quality.

5. Data

The basic data types collected include runoff, precipitation, weather, and water-quality data. Tables listing details of runoff, precipitation, lysimeter, weather, land-management, and other data are listed in [3] through about 2009 and are not repeated here. Due to specific project, financial, and personnel constraints during NAEW history, data for some watersheds such as runoff, precipitation and other data were not obtained for the entire 79-year period of monitoring. ARS operation of the NAEW ceased in late 2011, however, data collection continued from seven watersheds as part of a grant-funded project through Dec 2015 when all monitoring operations were discontinued. Through 2015, approximately 2125 station years of runoff data and 1126 station years of precipitation data were collected.

At the time of this writing, the NAEW data are being reviewed, corrected, and uniformly formatted from variety of original formats. The location for the data on the internet is yet to be determined. As part of the NAEW data-review process, a GIS of the NAEW has been developed documenting locations of runoff, precipitation, weather stations, etc. GIS will become part of the NAEW data base posted on the web site. The list of NAEW publications will also be posted.

Even though data collection has ceased, the database is valuable for further investigations of hydrology and water quality. Watershed modeling, in particular, can be studied at large (LMC—scaling, spatial parameterization of watershed models, and hydrology) and small scales (small watersheds—watershed modeling, hydrology, and water quality). The long precipitation and runoff records are valuable because they have experienced a wide array of weather conditions, even during a period of trending climate. Other precipitation, infiltration, soil moisture, ground water, and soil characterization data bases have not been analyzed and are available in hard copy form in the NAEW files. These data would have to be converted from hard copy to electronic form.

6. Examples of accomplishments of the NAEW

The outdoor laboratory of the NAEW has historically addressed the challenges of emerging national issues and addressing stakeholder's needs. Over 500 reports and peer-reviewed publications originated from NAEW research. The NAEW was a world-class facility and examples of accomplishments of the NAEW are [2]:

Crop rotations: The early record of runoff measurements (first ~28 years) documented the benefits of rotating crops and planting on the contour to reduce erosion in agricultural fields in the hill lands of Appalachia.

No-till/Conservation tillage: No-till farming reduces (and can essentially eliminate) soil losses and runoff (**Figure 12**). The USDA funds a national farm program and recommends the no-till practice for improving agricultural lands as a best practice. The NAEW was the first facility in the world to evaluate the water quality benefits of no-till on a watershed basis with experiments beginning in 1964 and continuing through 2011 [8]. The practice allows more frequent harvesting of high value crops, produces yields that are the same or greater than with conventional tillage (especially during droughts), increases soil-carbon storage (resulting in larger soil moisture for crops), and reduces energy needs. The environmental benefits of other types of conservation tillage have also been investigated [9–12].

Grazing: The NAEW developed environmental recommendations for pasture fertilizer application rates based on nitrogen [13, 14], sources of nitrogen fertility for pastures [15], and over-wintering practices on grazing lands [16].

Management-Intensive Grazing (MIG): The NAEW management-intensive grazing (MIG) project investigated the water-resource benefits of frequent rotation of livestock between small paddocks in a pasture for organic and non-organic production, and included impacts on surface and subsurface water quality, animal health, and changes in plant species [17]. Potential benefits of MIG to the producer include extended grazing season, less cost, and more leisure time.

Nutrient movement in stormflow vs. baseflow: Major transport of nutrients can occur in both baseflow and stormflow from mixed agricultural watersheds [18].

Preferential movement of water in soil: Fundamental knowledge from NAEW experiments on the fate of infiltrated water and chemicals in the subsurface through preferred pathways (e.g., earthworm burrows and cracks) has been used by scientists worldwide and in the development of a macropore component of a watershed model [19–21]. Additionally, guidance was developed on manure application in tiled fields [22, 23].

Evaluation of best management practices: A method was published to estimate the variability of chemical concentrations in runoff when there are few water samples and a history of runoff using duration curves [24].

Pesticide transport: Research on herbicide concentrations for weed control on watersheds showed that concentrations in runoff can reach levels of concern, particularly in the first few events after application, and that by reducing application rates by replacing herbicides with short half-life types, concentrations can be reduced [25–27].

Figure 12. Original no-till experiments on steeply sloping experimental watershed.

Climate change: Climate was changing starting in about 1980 at the NAEW. Several studies of precipitation showed that underlying runoff parameters in the "curve number" method of estimating runoff were changing as extreme precipitation events were increasing in magnitude, and that air temperature was trending upward [28]. Watershed data were used for modeling watersheds for climate change impacts [29].

Organic agriculture: The NAEW investigated the organic-agriculture component of a large nationwide study on the effect of climate on corn production (data still being analyzed). Other organic-related research included comparing impacts of continuous and MIG as they underwent transition to organic agriculture.

Precipitation modeling: Several studies on modeling short-time increment precipitation data (of the order of minutes) for modeling purposes have been completed. Studies on parameterization, data quality, seasonal variation, times between storms, climate change, etc., support a model that generates independent storms of any duration [30, 31].

Ground-water recharge: The Glugla method for estimating ground-water recharge was verified by using the NAEW lysimeters [32]. This method is used nationwide in Germany.

Evapotranspiration (ET): Lysimeter data have been useful for investigating ET losses under different management practices. The Glugla method also estimates long-term ET losses. The lysimeters were used in modeling the ET component for verifying the engineering design procedure for alternate and cheaper landfill covers [33].

Watershed modeling: Models of some small watersheds allowed the evaluation of climate change and runoff, and the adequacy of a model-parameterization procedure [34]. NAEW data are currently being used to nationally update a runoff-estimation procedure used worldwide ("curve number" model).

Biofuel removal: It was documented that removal of large amounts of crop residue for ethanol production can negate many soil and water-quality benefits of long-term no till [35, 36].

Urbanization: It was shown that a low level of imperviousness, either close to or far from a stream channel, on a 3-ha watershed had no effect on runoff. It was also shown that minor surface disturbances can increase runoff potential, but that the land surface can recover, using grazing a surrogate for urban land surface disturbance [37].

Winter application of manure: Data from large runoff plots of various sizes and treatments, and experimental watersheds were used to provide guidance for applying manure during the winter [38].

Pathogens from manure applications: The winter-manure application project provided an opportunity to add another dimension to NAEW research—pathogens. The data were used to provide pathogen guidance for winter application of manure.

Instrumentation: NAEW scientists developed and adapted many hydrological and water-quality instruments required for specific research objectives and not commonly available commercially. Examples are the Coshocton Wheel water sampler (invented in ~1945 and used worldwide in remote areas), Coshocton Vane water sampler, large sediment particle runoff sampler, drip-flow meter and sampler, hydraulic studies of the drop-box weir, adaptation of the Coshocton Wheel for the drop-box weir [39], natural-precipitation infiltrometer, worm-burrow infiltrometer, and rainfall simulator for macropore studies.

Filter sock performance: Coshocton data showed that there were limits to use of on-field control of pesticides on watersheds. Netting in the form of tubes filled with materials that can adsorb pesticides and nutrients from surface runoff have been studied and published. The results are useful for contractors that use filter socks in controlling chemical and sediment water quality from construction sites and farm fields [40, 41].

Paper mill byproducts: In collaboration with the State of Ohio and the paper industry, NAEW studies on use of paper-mill sludge (waste product) applied to surface mines showed that the State's upper limit of land application rate was environmentally acceptable [42]. This provided paper mills a cheaper, more environmental beneficial, alternative to landfill disposal, and at the same time provide a good source material for revegetating and controlling erosion when reclaiming surface mines.

Carbon sequestration: The long-term nature of the management practices on the small watersheds including continuous no-till corn with over 40 years of runoff records have enabled numerous investigations into the impact of land management on carbon sequestration (**Figure 13**). Sediment-bound carbon losses from various conservation tillage practices and organic carbon losses in subsurface flow were also measured [43, 44].

Surface mining and reclamation: A landmark study on effects of coal mining and reclamation on surface and ground water in three watersheds (~16 ha) showed the temporary and permanent effects of this drastic land disturbance (**Figure 11**). Watersheds were monitored before, during, and after mining and reclamation. Monitoring could not have been possible without the use of the drop-box weir [7] due to the large sediment-laden flows from areas such as shown in **Figure 5**. Runoff potential increased (large curve number) to a near constant after reclamation regardless of original geology, and erosion can be controlled to near pre-mine levels with the right reclamation practice [45–47]. The results have been used in court cases and in regulations.

Figure 13. Soil carbon increases in soil planted to continuous no-till corn (bottom right) compared with soil from conventionally tilled soil (upper left).

Landfills: Lysimeter data validated an engineering design model for a new type of landfill cover that utilized the ET processes in the soil to minimize water percolating to ground water [33].

Other research: Other research conducted at the NAEW (not exhaustive list) were projects related to water quality of spent foundry sand, dairy wastes, and nursery operations. Frozen soil, rain gauges, soil moisture, soil characterization, etc., have also been topics.

Emerging issues: A significant advantage of the NAEW facility is its long-term data base and the permanent monitoring infrastructure. It has been used for many investigations which were never imagined at the outset. Examples are placement of impervious structures for urbanization studies, evapotranspiration landfill caps, macropore investigations, advanced modeling, organic agriculture, climate change, ground-water recharge studies, spent foundry sand, nursery operations, dairy wastes, filter socks, carbon sequestration, pathogens and estrogens in runoff [48], biofuels, surface mining and reclamation, paper mill sludge, and long-term water-quality response times in natural systems.

7. Insights from NAEW experiences for establishing new experimental watersheds

Many paths can be followed to establish new experimental watersheds to conduct watershed-science research ("outdoor laboratories for water and land-management research") such as the NAEW. Watershed science involves expertise in the biological and physical sciences to solve national problems. Occasionally required expertise can be acquired from university, government, and private-sector partners and stakeholders. The 81-year experience of managing experimental watersheds may be useful for establishing new experimental sites at the scale of the NAEW, and some important considerations from the NAEW experience are listed below.

1. When establishing new experimental watersheds, "representativeness" is important, and newer GIS technology should be used to identify potential sites. However, there will be other research benefits discovered as a facility is managed. In the case of the NAEW, the geological and soil characteristics became important for potentially providing new knowledge on hydrological processes such as nonuniform runoff generation, interflow, macropore flow, perched water tables, etc. Other potential site benefits should be considered in site selection.

2. For an ideal comprehensive watershed-science program, the three original NAEW objectives are required. However, funding may become a problem and pursuit of the three objectives can be spread at more than one location (e.g., one location can perform modeling and another field experimentation). However, having modelers conducting field research is also of value to experience data characteristics and natural variability found in landscapes to help formulate algorithms. It is important to have field data for validation of watershed models.

3. Monitoring a large experimental watershed requires sufficient funding to sustain scientists, support staff, and experimental resources. The value of experimental watersheds is that they can provide an uninterrupted record of runoff, water quality, etc., spanning years, with dry years producing insufficient numbers of runoff events and longer periods of runoff records may be required. Furthermore, *watershed-science research can be considered long-term and high risk* because experiments are subject to weather extremes (e.g., droughts and other project factors that are affected by the weather). It is expensive to maintain such a record, and continuous funding must be maintained—temporary grants will interrupt long-term records after the grant period is completed, and a sufficient record of runoff may not be recorded.

4. It can be difficult to exploit characteristics unique to a site (e.g., at the NAEW—understanding and quantifying interflow, nonuniform runoff generation, etc.) because of funding and a wide range of expertise needed. At the NAEW, some of these features were not fully exploited.

5. **Figure 7** shows the wide variety of watershed behaviors for three experimental watersheds in different physiographic and climatological regions of the USA [49, 50] and generally describes how runoff is generated on landscapes. It is apparent from **Figure 7** that the unglaciated NAEW area will follow a convex-upward curve where smaller overland flow areas do not support baseflow (annual runoff is small). For larger areas, however, baseflow is increasing as incising stream channels drain water from intersected water tables, and the curve approaches an apparent constant. For arid areas (Tombstone, AZ), runoff decreases with area in a log-log manner due to channel transmission losses and isolated storms. For the location at Reisel, TX, the response is nearly flat due to its climate and soil conditions leading to a more uniform generation of runoff. The reader is referred to [49] for more particulars of **Figure 7**. If a network of experimental watersheds is developed, a plot of data in a similar manner may lead to a general characterization of watershed sites under consideration, and may help differentiate proposed sites.

6. Site specificity of experimental watersheds must be expected. Soil and geology are important factors that can affect different responses of two similarly treated small adjacent watersheds subjected to similar precipitation and weather drivers. This variability affects project results, numbers of watersheds needed for experiments, and highlights the need for watershed modeling to extrapolate field data to ungauged areas. Furthermore, the history of an individual watershed is known and quantified with permanently monitored sites. Some watersheds may still be responding to prior treatments when a new treatment is initiated. In the case of new watersheds, the effects of prior treatments may be unknown, yet they may affect interpretations of the data.

7. Seasonal air temperatures and percent of snow in a precipitation record are important for seasonal runoff generation mechanisms that can affect water quality also. At the NAEW, lower winter air temperatures did not always insure frozen soil, and sometimes frozen soil occurred intermittently. Consequently, latitude and climate are important to consider.

8. Instrumentation selection is important in managing experimental watersheds. Two important measurements are runoff and precipitation. For runoff, it is known that large sediment concentrations in runoff can affect the rating curve of H flumes [51], a commonly used flow-measuring device. A weir that has been tested for a wide range of field conditions under large sediment loads (including rocks) is the drop-box weir [7]. Drop-box weirs of any size can be constructed from small runoff/erosion plots to large watersheds with incised channels. It is important to house weirs from freezing weather to prevent damage and maximize the opportunity for good winter runoff records.

9. Precipitation measurement is a persistent problem because the gauge shape and orifice height affect the wind flow around the orifice, resulting in an under catch of precipitation. This is because smaller diameter rain drops and light weight snowflakes are carried with the wind away from the orifice. This error can be as high as a 20% under catch on average during the winter and approximately 2% during the nonwinter months [52]. True ground-level precipitation measurements for individual events can be much higher during events with high wind speeds. Furthermore, often-used tipping bucket rain gauges do not measure snowfall, and under catch of precipitation is complicated by using heaters to melt the snow because precipitation is evaporated [53], and snow intensities will not be measured. In arid areas this may not be a problem. The effects of under estimating precipitation are to under estimate runoff, erosion, and water quality in watershed modeling [49] in a nonlinear manner. Suggestions for improved precipitation measurement include shielded gauges such as the dual fence gauge of the Climate Reference Network [54] that was evaluated by US Forest Service [55]. Their study suggested that the CRN gauge is the best available. A set of dual gauges (one shielded and one not shielded) were not tested in the Forest Service study, but is likely to be a contender and should be investigated [56]. Furthermore, weighing buckets are preferred measuring technology compared with tipping buckets. Advances in other emerging technologies should also be explored.

10. Water quality sampling is important for evaluations of the performance of land treatments. Two types of sampling are possible—composite and discrete sampling. For composite sampling, the same fraction of the flow is sampled for each flow rate during the runoff event and only one sample is obtained for the event. The Coshocton Wheel has been a useful tool for composite sampling of small watersheds [6, 39]. Larger watersheds require smaller fractions of flow sampling to manage the size of a composite sample and for a range of runoff volumes. Commercial samplers and the Coshocton Vane sampler [57] are available for this purpose. An instantaneous sample is pumped for discrete sampling a preselected times or changes in runoff depth, and many samples are obtained for an individual event. This type of sampling is more expensive and may not be as useful as composite sampling unless there is a research objective for this sampling strategy. For evaluations of water quality effects of land treatments, a composite sample is adequate (and may be preferred). For either sampling type, a flow measurement record is required.

11. It is possible that experimental watershed investigations can be affected by a changing climate. Climate was found to be changing at the NAEW.

12. It should be recognized that some investigations will be affected by watersheds with low runoff potential (e.g., some forested sites). It may take many years for such watersheds to provide enough data for evaluations to experience larger infrequent rainfalls.

13. Knowledge of challenges in conducting watershed research in disturbed lands in particular is presented in [58] and is pertinent to selecting new watersheds.

14. It is highly likely that as an emerging issue arises, that an experimental watershed facility would be a likely place for pertinent investigations. The permanent monitoring infrastructure allows for a relatively rapid implementation of a proposed land treatment and monitoring. Funding for such issues is important.

15. Another opportunity for monitoring experimental watersheds is off site from the home site. This was necessary for monitoring watersheds with three different coal seams in the disturbed land (coal mining) project conducted at the NAEW [58], **Figure 11**.

16. It is important for all data to be checked and be made available on the internet as soon as possible after collection.

8. Summary

The North Appalachian Experimental Watershed (NAEW), in east-central Ohio near Coshocton, Ohio, was one of the three large watershed facilities established in 1935 to advance watershed science of agricultural lands to improve their economical and physical sustainability. It was an outdoor laboratory for land and water management research. The original objectives were to test management practices on small watersheds (small swales in the hilltops), investigate scaling of runoff and erosion to larger areas, and provide a way to extrapolate the results to ungauged areas (modeling). The NAEW was in an unglaciated sedimentary geological setting (strata nearly horizontal) and originally spanned an area of approximately 2000 ha. The facility was equipped with a permanent infrastructure consisting of runoff stations and rain gauges for watersheds ranging in size from 0.26 to 1854 ha. After about 1970, the NAEW was reduced to a 425-ha area consisting of mostly small watersheds ("test beds") ranging in size from 0.26 to 3.07 ha but with a few up to 123 ha. The smaller watersheds were equipped with the well-known Coshocton Wheel composite runoff samplers. The NAEW was in operation for approximately 81 years with an approximate 79-year record of runoff and other data for various watersheds, and was closed in 2015. Eleven large monolith lysimeters were also constructed to investigate small scale water balances.

A wide variety of experiments were conducted on the NAEW with many high impact accomplishments (listed in the section titled, Examples of Accomplishments of the NAEW). Many investigations used the facility for emerging national issues that the founders never envisioned (e.g., surface mining impacts, landfill caps, organic agriculture, climate change, filter socks, carbon sequestration, pathogens in runoff, biofuels). Nearly 500 publications were developed from investigations during the 81-year history of the facility.

Experiences in the operation of the facility during the 81 years provide insights for establishing new experimental watersheds in the future. Watershed science involves the expertise in the biological and physical sciences and engineering to solve national problems. Sixteen suggestions for new facilities are presented covering site selection, funding, site specificity, extrapolation of results, generation of runoff in different physiographic regions, collaboration, off-site investigations, and instrumentation. Instrumentation suggestions are particularly important for precipitation because it is a major driver of watershed responses and must be more accurately gauged than commonly measured. Runoff measurements also can be affected by large sediment concentrations using common flow-measuring devices. Watershed modeling will be sensitive to precipitation inputs and validation of runoff amounts in modeling will be affected by runoff measurements.

Acknowledgements

The authors are grateful to all those who worked at the Coshocton watershed throughout its 81-year history.

Author details

James V. Bonta[1]*, Martin J. Shipitalo[2] and Lloyd Owens[2]

*Address all correspondence to: ohky@ymail.com

1 USDA-Agricultural Research Service, Oxford, MS, USA

2 USDA-Agricultural Research Service, Ames, IA, USA

References

[1] Harmel RD, Bonta JV, Richardson CW. The original USDA-ARS experimental watersheds in Texas and Ohio: Contributions from the past and visions for the future. Transactions of the ASABE. 2007;**50**(5):1669-1675

[2] Bonta JV, Owens LB, Shipitalo MJ. Watershed research at the north Appalachian experimental watershed at Coshocton, Ohio. In: Rogers JR, editor. Environmental and Water Resources Milestones in Engineering History. Reston, VA: ASCE/EWRI; 2007. pp. 127-134

[3] Owens LB, Bonta JV, Shipitalo MJ. USDA-ARS North Appalachian Experimental Watershed: 70-year hydrologic, soil erosion, and water quality database. Soil Science Society of America Journal. 2010;**74**(2):619-623

[4] Ramser CE, Krimgold DB. Detailed working plan for watershed studies in the North Appalachian Region relating to water conservation, flood control, and run-off as influenced by land use and methods of erosion control; 1935. WHS#1, November, 1935

[5] Kelley GE, Edwards WM, Harrold LL, McGuinness JL. Soils of the North Appalachian Experimental Watershed. USDA Misc. Publ. #1296; 1975. p. 145

[6] Brakensiek DL, Osborn HB, Rawls WJ. Field Manual for Research in Agricultural Hydrology. U.S. Dept. of Agriculture, Agriculture Handbook No. 224. Washington, DC: U.S. Government Printing Office; 1979. p. 547

[7] Bonta JV, Pierson FB. Design, measurement, and sampling with drop-box weirs. Applied Engineering in Agriculture. 2003;**19**(6):689-700

[8] Harrold LL, Triplett GB, Youker RE. Watershed test of no tillage corn. Journal of Soil and Water Conservation. 1967;**22**(3):98-100

[9] Shipitalo MJ, Edwards WM. Runoff and erosion control with conservation tillage and reduced-input practices on cropped watersheds. Soil and Water Tillage Research. 1998;**46**:1-12

[10] Shipitalo MJ, Owens LB, Bonta JV, Edwards WM. Effect of no-till and extended rotation on nutrient losses in surface runoff. Soil Science Society of America Journal. 2013;**77**:1329-1337

[11] Edwards WM, Shipitalo MJ, Dick WA, Owens LB. Rainfall intensity affects transport of water and chemicals through macropores in no-till soil. Soil Science Society of America Journal. 1992;**56**(1):52-58

[12] Shipitalo MJ, Dick WA, Edwards WM. Conservation tillage and macropore factors that affect water movement and the fate of chemicals. Soil & Tillage Research. 2000;**53**:167-183

[13] Owens LB, Van Keuren RW, Edwards WM. Hydrology and soil loss from a high-fertility, rotational pasture program. Journal of Environmental Quality. 1983;**12**(3):3441-3346

[14] Owens LB, Van Keuren RW, Edwards WM. Nitrogen loss from a high-fertility, rotational pasture program. Journal of Environmental Quality. 1983;**12**(3):346-350

[15] Owens LB, Bonta JV. Reduction of nitrate leaching with haying or grazing and omission of nitrogen fertilizer. Journal of Environmental Quality. 2004;**33**(4):1230-1237

[16] Owens LB, Shipitalo MJ. Runoff quality evaluations of continuous and rotational overwintering systems for beef cows. Agriculture, Ecosystems & Environment. 2009;**129**(4): 482-490

[17] Owens LB, Barker DJ, Loerch SC, Shipitalo MJ, Bonta JV, Sulc RM. Inputs and losses by surface runoff and subsurface leaching for pastures managed by continuous or rotational stocking. Journal of Environmental Quality. 2012;**41**(1):106-113

[18] Owens LB, Edwards WM, Van Keuren RW. Baseflow and stormflow transport of nutrients from mixed agricultural watersheds. Journal of Environmental Quality. 1991;**20**(2): 407-414

[19] Shipitalo MJ, Edwards WM, Dick WA, Owens LB. Initial storm effects on macropore transport of surface-applied chemicals in no-till soil. Soil Science Society of America Journal. 1990;**54**(6):1530-1536

[20] Shipitalo MJ, Butt KR. Occupancy and geometrical properties of *Lumbricus terrestris* L. burrows affecting infiltration. Pedobiologia. 1999;**43**:782-794

[21] Shipitalo MJ, Nuutinen V, Butt KR. Interaction of earthworm burrows and cracks in a clayey, subsurface-drained, soil. Applied Soil Ecology. 2004;**26**:209-217

[22] Shipitalo MJ, Gibbs F. Potential of earthworm burrows to transmit injected animal wastes to tile drains. Soil Science Society of America Journal. 2000;**64**(6):2103-2109

[23] Hoorman JJ, Shipitalo MJ. Subsurface drainage and liquid manure. Journal of Soil and Water Conservation. 2006;**61**(3):94A-97A

[24] Bonta JV, Cleland B. Incorporating natural variability, uncertainty, and risk into water quality evaluations using duration curves. Journal of the American Water Resources Association. 2003;**39**(6):1481-1496

[25] Shipitalo MJ, Edwards WM, Owens LB. Herbicide losses in runoff from conservation-tilled watersheds in a corn-soybean rotation. Soil Science Society of America Journal. 1997;**61**(1):267-272

[26] Shipitalo MJ, Owens LB. Atrazine, deethylatrazine, and deisopropylatrazine in surface runoff from conservation tilled watersheds. Environmental Science & Technology. 2003; **37**(5):944-950

[27] Shipitalo MJ, Malone RW, Owens LB. Impact of glyphosate-tolerant soybean and glufosinate-tolerant corn production on herbicide losses in surface runoff. Journal of Environmental Quality. 2008;**37**:401-408

[28] Bonta JV. Curve number method response to historical climate variability and trends. Transactions of the ASABE. 2015;**58**(2):319-334

[29] Gautam S, Mbonimpa EG, Kumar S, Bonta JV, Lal R. Agricultural policy environmental eXtender model simulation of climate change impacts on runoff from a small no-till watershed. Journal of Soil and Water Conservation. 2015;**70**(2). DOI: 10.2489/jswc.70.2.101

[30] Bonta JV. Stochastic simulation of storm occurrence, depth, duration, and within-storm intensities. Transactions of ASAE. 2004;**47**(5):1573-1584

[31] Bonta JV. Development and utility of huff curves for disaggregating precipitation amounts. Applied Engineering in Agriculture. 2004;**20**(5):641-653

[32] Bonta JV, Müller M. Evaluation of the Glugla method for estimating evapotranspiration and groundwater recharge. Hydrological Sciences Journal. 1999;**44**(5):743-761

[33] Hauser VL, Gimon DM, Bonta JV, Howell TA, Malone RW, Williams JR. Models for hydrologic design of evapotranspiration landfill covers. Environmental Science & Technology. 2005;**39**(18):7226-7233

[34] Mbonimpa EG, Gautam S, Lai L, Kumar S, Bonta JV, Wang X, Rafique R. Combined PEST and trial–error approach to improve APEX calibration. Computers and Electronics in Agriculture. 2015;**114**(2015):296-303. DOI: 10.1016/j.compag.2015.04.014 0168-1699

[35] Blanco-Canqui H, Lal R, Post WM, Owens LB. Changes in long-term no-till corn growth and yield under different rates of stover mulch. Agronomy Journal. 2006;**98**:1128-1136

[36] Blanco-Canqui H, Lal R, Post WM, Izaurralde RC, Owens LB. Rapid changes in soil carbon and structural properties due to stover removal from no-till corn plots. Soil Science. 2006;**171**(6):468-482

[37] Bonta JV. Managing landscape disturbances to increase watershed infiltration. Transactions of the ASABE. 2013;**56**(4):1349-1359

[38] Owens LB, Bonta JV, Shipitalo MJ, Rogers S. Effects of winter manure application in Ohio on the quality of surface runoff. Journal of Environmental Quality. 2011;**40**(1):153-165

[39] Bonta JV. Modification and performance of the Coshocton wheel with the modified drop-box weir. Journal of Soil and Water Conservation. 2002;**57**(6):364-373

[40] Shipitalo MJ, Bonta JV, Dayton EA, Owens LB. Impact of grassed waterways and compost filter socks on the quality of surface runoff from corn fields. Journal of Environmental Quality. 2010;**39**(3):1009-1018

[41] Shipitalo MJ, Bonta JV, Owens LB. Sorbent-amended compost filter socks in grassed waterways reduce nutrient losses in surface runoff from corn fields. Journal of Soil and Water Conservation. 2012;**67**(5):433-441

[42] Shipitalo MJ, Bonta JV. Impact of using paper mill sludge for surface-mine reclamation on runoff water quality and plant growth. Journal of Environmental Quality. 2008;**37**(6): 2351-2359

[43] Owens LB, Malone RW, Hothem DL, Starr GC, Lal R. Sediment carbon concentration and transport from small watershed under various conservation tillage practices. Soil & Tillage Research. 2002;**67**:65-73

[44] Owens LB, Starr GC, Lightell DL. Total organic carbon losses in subsurface flow under two management practices. Journal of Soil and Water Conservation. 2002;**578**(2):74-81

[45] Bonta JV, Amerman CR, Dick WA, Hall GF, Harlukowicz TJ, Razem AC, Smeck NE. Impact of surface coal mining on three Ohio watersheds - physical conditions and ground-water hydrology. Water Resources Bulletin. 1992;**28**(3):577-596

[46] Bonta JV, Amerman CR, Dick WA, Harlukowicz TJ, Razem AC. Impact of surface coal mining on three Ohio watersheds - ground-water chemistry. Water Resources Bulletin. 1992;**28**(3):597-614

[47] Bonta JV, Amerman CR, Harlukowicz TJ, Dick WA. Impact of coal surface mining on three Ohio watersheds - surface-water hydrology. Journal of the American Water Resources Association. 1997;**33**(4):907-917

[48] Shappell NW, Billey LO, Shipitalo MJ. Estrogenic activity and nutrient losses in surface runoff after winter manure application to small watersheds. Science of the Total Environment. 2016;**543**:570-580

[49] Baffaut C, Dabney SM, Smolen MD, Youssef MA, Bonta JV, Chu ML, Guzman JA, Shedekar VS, Jha MK, Arnold JG. Hydrologic and water quality modeling: Spatial and temporal considerations. Transactions of the ASABE. 2015;**58**(6):1661-1680

[50] Glymph LM, Holtan HN. Land treatment in agricultural watershed hydrology research. In: Moore WL, Morgan CW, editors. Effects of Watershed Changes on Streamflows. Austin: University of Texas Press; 1969. pp. 44-68

[51] Gwinn WR. Chute entrances for HS, H, and HL flumes. Journal of Hydraulic Engineering. 1984;**110**(5):587-603

[52] McGuinness JL. A comparison of lysimeter catch and rain gage catch. ARS. 1966;**41**:124

[53] Hanson CL, Zuzel JF, Morris RP. Winter precipitation catch by heated tipping-bucket gages. Transactions of ASAE. 1983;**26**(5):1479-1480. DOI: 10.13031/2013.34154

[54] NOAA, National centers for Environmental Information, National Oceanic and Atmospheric Administration. https://www.ncdc.noaa.gov/crn/ [Accessed: 2017–08-28]

[55] Hansen S, Davies MA, Windshields for Precipitation Gauges and Improved Measurement Techniques for Snowfall, https://www.fs.fed.us/t-d/pubs/htmlpubs/htm02252325/ [Accessed: 2017–08-28]

[56] Hanson CL, Pierson FB, Johnson GL. Dual-gauge system for measuring precipitation: Historical development and use. ASCE Journal of Hydrologic Engineering. 2004;**9**(5): 350-359

[57] Edwards WM, Frank HE, King TE, Gallwitz DR. Runoff Sampling: Coshocton Vane Proportional Sampler. Vol. 50. ARS-NC; 1976. p. 9

[58] Bonta JV. Challenges in conducting hydrologic and water quality research in drastically disturbed watersheds. Journal of Soil and Water Conservation. 2005;**60**(3):121-133

Chapter 2

Controlled Experiments of Hillslope Coevolution at the Biosphere 2 Landscape Evolution Observatory: Toward Prediction of Coupled Hydrological, Biogeochemical, and Ecological Change

Till H. M. Volkmann, Aditi Sengupta, Luke A. Pangle, Katerina Dontsova, Greg A. Barron-Gafford, Ciaran J. Harman, Guo-Yue Niu, Laura K. Meredith, Nate Abramson, Antonio A. Meira Neto, Yadi Wang, John R. Adams, David D. Breshears, Aaron Bugaj, Jon Chorover, Alejandro Cueva, Stephen B. DeLong, Matej Durcik, Ty P. A. Ferre, Edward A. Hunt, Travis E. Huxman, Minseok Kim, Raina M. Maier, Russell K. Monson, Jon D. Pelletier, Michael Pohlmann, Craig Rasmussen, Joaquin Ruiz, Scott R. Saleska, Marcel G. Schaap, Michael Sibayan, Markus Tuller, Joost L. M. van Haren, Xubin Zeng and Peter A. Troch

Additional information is available at the end of the chapter

http://dx.doi.org/10.5772/intechopen.72325

Abstract

Understanding the process interactions and feedbacks among water, porous geological media, microbes, and vascular plants is crucial for improving predictions of the response of Earth's critical zone to future climatic conditions. However, the integrated coevolution of landscapes under change is notoriously difficult to investigate. Laboratory studies are limited in spatial and temporal scale, while field studies lack observational density and control. To bridge the gap between controlled laboratory and uncontrollable field studies, the University of Arizona built a macrocosm experiment of unprecedented scale: the Landscape Evolution Observatory (LEO). LEO comprises three replicated, heavily instrumented, hillslope-scale model landscapes within the environmentally controlled

Biosphere 2 facility. The model landscapes were designed to initially be simple and purely abiotic, enabling scientists to observe each step in the landscapes' evolution as they undergo physical, chemical, and biological changes over many years. This chapter describes the model systems and associated research facilities and illustrates how LEO allows for tracking of multiscale matter and energy fluxes at a level of detail impossible in field experiments. Initial sensor, sampler, and soil coring data are already providing insights into the tight linkages between water flow, weathering, and microbial community development. These interacting processes are anticipated to drive the model systems to increasingly complex states and will be impacted by the introduction of vascular plants and changes in climatic regimes over the years to come. By intensively monitoring the evolutionary trajectory, integrating data with mathematical models, and fostering community-wide collaborations, we envision that emergent landscape structures and functions can be linked, and significant progress can be made toward predicting the coupled hydro-biogeochemical and ecological responses to global change.

Keywords: critical zone, hillslope, model system, water cycle, carbon cycle, energy balance, soil weathering, geomorphology, plant ecology, microbiology, climate change, coevolution, closure relations, coupled-process modeling

1. Introduction

The structure and function of the critical zone—the interface between the solid Earth and its fluid envelopes—are the product of ongoing coevolution of biota, soils, and landforms. Physical and chemical weathering of the parent material lead to soil formation, landscape denudation, and geomorphic alteration. Biological processes modulate these landscape transformations, as microbial and vascular plant communities establish, compete, and adapt while sequestering important elements, such as carbon and nitrogen, for sustaining life cycles. Gravitational, thermal, and chemical gradients controlling fluxes of water, energy, and nutrients mediate the evolution of the critical zone and are in turn determined by its emergent abiotic and biotic characteristics. The complex linkages between these physicochemical and biological processes and the degree of surface and subsurface heterogeneity they create pose a fundamental challenge to Earth scientists across disciplines attempting to predict changes in critical zone behavior [1–3]. Hydrologists, geologists, biologists, and atmospheric scientists have approached critical zone research primarily from disciplinary perspectives [4], thereby overlooking important spatio-temporal process interactions and feedbacks and introducing inherent uncertainty into parameterizations of resource cycling and ecosystem dynamics. Even when interdisciplinary efforts are made (e.g., [5]), the integrated coevolution of landscapes under change remains extremely difficult to investigate. Field studies generally lack the multifaceted observational density and control needed to capture the wide range and variability of relevant processes. Any natural landscape will also have a history of geologic, climatic, and anthropogenic forcing that cannot be fully known, complicating interpretation of present day "snapshot" observations [2]. Laboratory studies, in turn, offer known boundary conditions and a high level of control but are typically limited in the spatial and temporal scales at which they can be performed. Therefore, effects of multiscale process interactions and heterogeneity on system response cannot be captured, and results may not be immediately transferable to scales relevant for prediction.

To bridge the gap between laboratory and field experiments and meet the challenge of understanding and predicting landscape-scale changes in Earth system behavior, the University of Arizona has constructed a new large-scale and community-oriented research facility: the Landscape Evolution Observatory (LEO). LEO consists of three identical artificial landscapes (each with a surface area of 330 m^2) within the environmentally controlled Biosphere 2 facility near Tucson, Arizona, USA. At LEO, experimental manipulation of climate parameters (rainfall, air temperature, relative humidity, and wind speed) is possible at a scale that is infeasible in a typical lab setting, while fluxes of water, solutes, gases, and geological media can be monitored at a detail that is not possible in the field. The threefold replication allows quantifying experimental variability at the landscape level. The primary scientific objectives are to quantify interactions among hydrologic partitioning, geochemical weathering, ecology, microbiology, atmospheric processes, and geomorphic change associated with incipient landscape coevolution. The LEO infrastructure is designed to facilitate investigation of Earth surface processes by rapidly iterating dense experimental measurement with development and validation of coupled computational models.

Scientists working on the LEO project are specifically pursuing an interdisciplinary approach to experimental design by cultivating a collaborative group that includes representation from hydrology, geomorphology, soil geochemistry, atmospheric science, ecology, microbiology, and genomics. Overarching considerations in the design of the landscapes included that (i) interactions between abiotic and biotic process could be effectively studied, (ii) patterns emerging during landscape coevolution across multiple scales could be unambiguously identified, and (iii) results could be linked to ongoing work in natural systems. Eco-hydrological [6, 7], soil erosion, and geochemical [8] modeling were used to aid the design process. The consensual physical model that implements these criteria consists in a zero-order basin shape with convergent topography, an average slope of 10°, and uniform depth and composition of purely abiotic, granular basalt soil with a loamy sand texture. The zero-order basin shape emulates the fundamental geomorphic unit of Earth's uplands, presenting a highly relevant study object that parallels areas under active investigation in the natural environment (e.g., [5, 9]). The convergent topography creates significant spatial variability in slope angle and aspect, thus promoting diversity in net radiation, hydraulic gradients, soil properties, biological community composition, and biogeochemical processes over relatively small land areas. The specific slope angle facilitates transient lateral subsurface flow while largely avoiding overland flow and rapid gully erosion. The minimized initial physical complexity, weatherability, and nutrient-content of these landscapes' soil allow emergent formation of flow pathways, soil spatial heterogeneity, surface morphology, and vegetation patterns to be observed over time.

The experiments performed on these landscapes utilize the control capabilities of the Biosphere 2 facility to generate a variable climate forcing over a period of 10 years. Rainfall sequences are selected to inform specific scientific hypotheses and induce significant spatial and temporal soil moisture variability in a climate that contains wet-dry transitions in both warm and cool seasons. As the initially abiotic landscapes evolve to increasingly complex ecosystems, scientists have the opportunity to address fundamental topics that integrate physical, chemical, and biological processes. These include (i) characterizing the patterns of interdependency between the evolving physicochemical properties of landscapes, the biological communities that inhabit them, and their water, biogeochemical, and energy cycles, (ii) determining whether simple or complex biological communities, weathering patterns, and flow networks arise as a consequence of differing climate regimes, (iii) understanding how point- to plot-scale processes impact integrated fluxes of mass and energy at the whole-landscape level, (iv) assessing whether knowledge of how and why multiscale heterogeneity in landscape structure forms can enable improved prediction of landscape function under change, and whether existing or new predictive tools are required, and (v) determining whether our present knowledge and observational capacity allow closing the balances of mass and energy at the landscape scale.

The controlled experimentation and dense observations at LEO are combined with mathematical modeling to effectively advance our understanding of landscape evolution and its effects on mass and energy cycling between the landscapes and their overlying atmospheres. Modeling is used to help formalize newly gained process knowledge, infer aspects of the system that are difficult to measure, refine the experimental and instrumental design, and generate cross-disciplinary hypotheses that can be tested in the experiments. Specifically, we are using a "learning cycle" approach in which models are used to predict system response before an experiment is performed, and subsequent targeted experiments are used to improve the models' accuracy. Modeling efforts at LEO focus specifically on (i) integrating existing community model representations of hydro-biogeochemical and ecological processes into a coupled modeling framework and (ii) representing landscape behavior over a wide spectrum of elementary scales, ranging from highly resolved small-scale parameterizations to whole-hillslope integrations. Ultimately, model development provides us with the opportunity to transpose the knowledge gained through the LEO experiment into natural environments and therefore forms a critical basis to improving the accuracy of forecasts of landscape change in the real world.

LEO was fundamentally designed as a community resource to effectively meet targets of scientific merit and broader impact. Acknowledging the complexity of the project and maximizing its use for the Earth sciences and general public, the LEO project seeks to (i) facilitate open and real-time availability of sensor network data, (ii) provide a framework for community collaboration and facility access that includes integration of new or comparative measurement capabilities into existing facility cyber-infrastructure, (iii) foster a community-guided approach to science planning, and (iv) develop novel education and outreach programs. This strategy has already proven successful in informing the landscapes' design and first climate experiments, as well as in informing the public through numerous general media publications, the more than 1000 site visitors per year and hundreds of students trained in Earth sciences.

This book chapter provides a detailed overview of the Landscape Evolution Observatory project. We first present the LEO large-scale controllable research infrastructure, its instrumentation and support facilities, and the integrated modeling framework being developed in feedback with experiments. We then describe LEO's combined capabilities to track important mass and energy balances as well as changes in physicochemical landscape structure and biological communities. This is followed by a discussion of LEO's potential to serve as an experimental platform to study interactions between hydrological, biogeochemical, microbiological, plant-ecological, and geomorphological processes associated with incipient hillslope coevolution. Characterizing those interactions is critical for the advancement of hydrologic and coupled-process models at LEO, which are also discussed. We conclude by summarizing the unique opportunities that this long-term institutional experiment provides for the interdisciplinary Earth science community to make significant progress toward understanding and predicting integrated landscape response to environmental change.

2. The Landscape Evolution Observatory large-scale controllable infrastructure to study coupled Earth-surface processes

2.1. The model landscapes, their atmosphere, and associated infrastructure

The Landscape Evolution Observatory (LEO) is comprised of three model landscapes that were designed and constructed to be exact replicas of each other and to contain landscape features emblematic of upland zero-order basins (hereafter ZOBs). The model ZOBs are contained within steel structures oriented parallel to one another within three adjacent, enclosed bays along the western extent of the University of Arizona—Biosphere 2 facility (**Figure 1**). Individually, the bays include more than 596 m^2 of floor space. The total volume of air that interacts with the landscapes is 12,956 m^3, including 10,712 m^3 within the bays where the model ZOBs are located and 2244 m^3 within the underlying basement where air passes before recirculating to the bay above. The steel structures are generally shaped like rectangular trays, with an average slope of 10° and a southerly aspect. Their interior length and width are 30 m and 11 m, respectively (**Figure 2**). The base of the interior volume of the trays is formed by concrete board, secured to steel slats, and mounted atop structural steel beams. That base is not planar but contains an 18-m long depression that is deepest at the downslope extent of the steel trays and diminishes in depth as it expands upslope. The sidewalls of the trays are all built vertically. The whole interior surface area of the trays was coated with an epoxy primer, which was covered with an elastomeric membrane, then an aggregate-filled urethane coating. At 144 locations along the base of the tray, holes were drilled to allow passage of sensor cables. A length of acrylic tubing

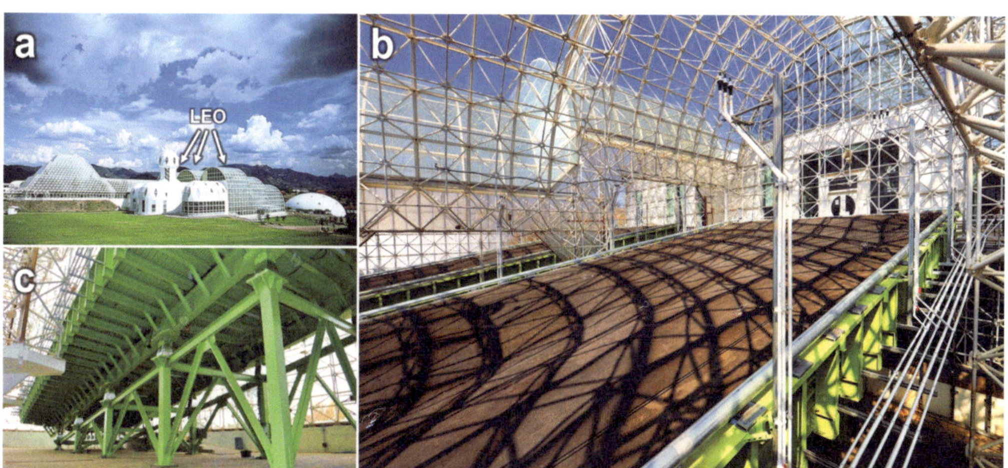

Figure 1. The Landscape Evolution Observatory is housed in adjacent bays within the Biosphere 2 facility (a) and comprises three replicate model landscapes (b) embedded into elaborate steel structures (c).

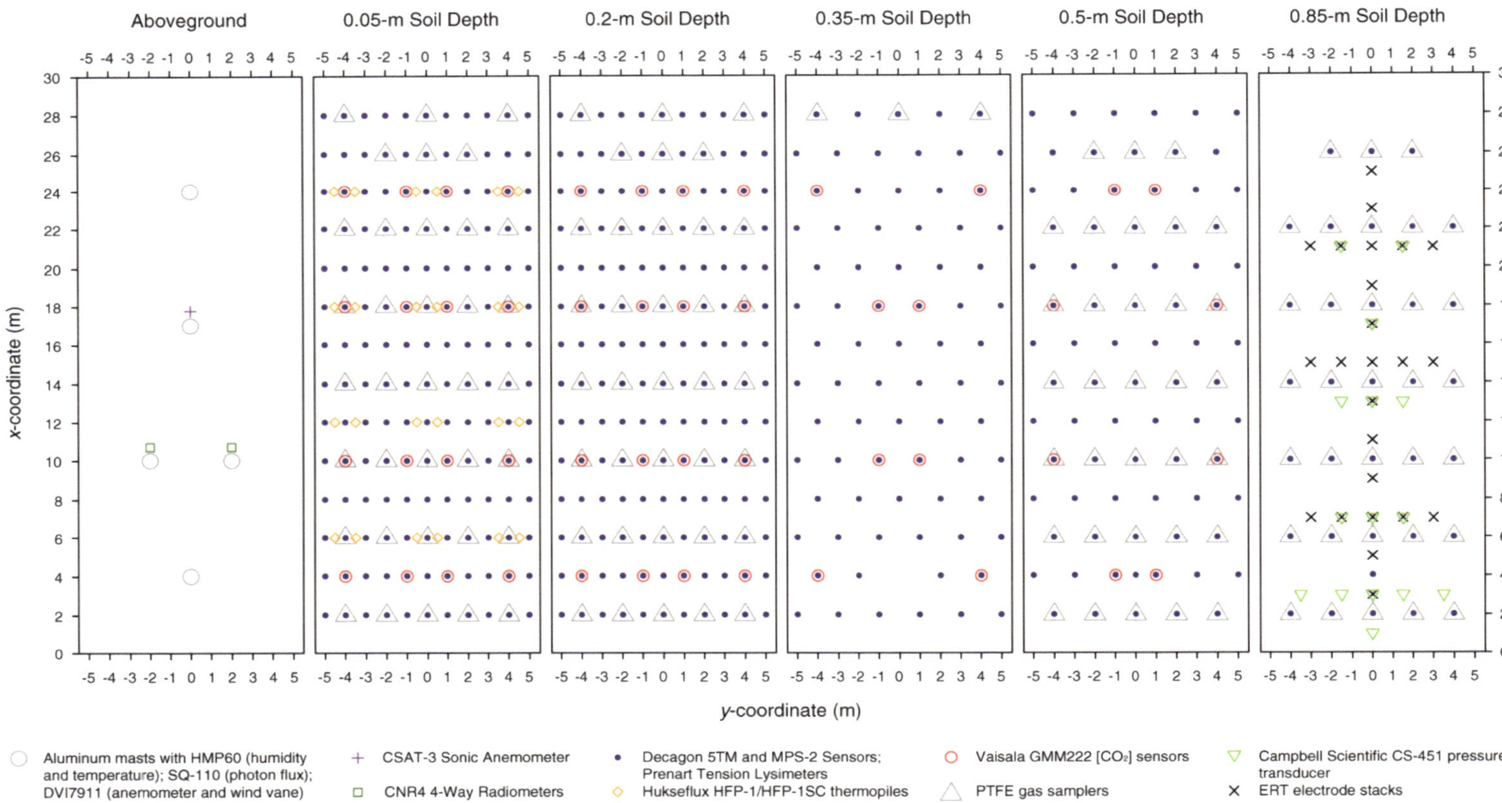

Figure 2. Diagram showing the orientation of the sensor and sampler network in and above the LEO landscapes. The left-most panel shows the approximate lateral (*xy*-coordinate plane) locations of instrumentation aboveground, where sensors are installed at one specific (CSAT-3, CNR4) or five different (all other sensors) heights along vertical aluminum masts. Each of the five remaining panels shows the lateral locations of belowground instrumentation at one specific depth (*z*-coordinate) below the soil surface. Note that CS-451 pressure transducers are actually installed within the structural base of the landscapes at 1-m soil depth, and ERT electrode stacks extend over multiple depths. Dashed gray lines indicate the axes of slope convergence zones.

was fitted inside a bulkhead fitting that was sealed into the hole. After final passage of sensor cables through these tubing lengths, they were sealed at their bottom end using expanding foam insulation and epoxy. The base and three of the interior walls of the steel trays are thus impermeable to water. The wall at the downslope extent of the trays was designed to allow water seepage out of the model landscapes. Steel supports and a lattice of steel slats form the primary retention structure of the downslope wall. Porous plastic sheeting was secured to that lattice, which allows water to readily seep out of the landscapes and into six partitioned drainage basins immediately bordering the downslope extent of the landscapes.

The interior volume of the steel structures was filled with crushed basaltic tephra, which comprises the original "parent material" of the model ZOBs. The material was mined from a subterranean deposit of basaltic tephra associated with Merriam Crater in northern Arizona. The mining company was contracted to crush the original rock material down to a loamy-sand texture. The average particle size distribution and elemental composition of the material were reported in [10]. The average sand (50–2000 μm), silt (2–50 μm), and clay (<2 μm) particle-size fractions are 84.6, 12.2, and 3.2%, respectively. The material contains exceptionally low amounts of organic carbon and nitrogen, which was desirable because the target initial condition was a landscape that was effectively abiotic and as spatially homogenous as could be achieved. The loamy sand was hauled from the mining site to LEO and eventually installed within the steel trays by conveyor belt and through manual dispersal and packing by workers. The packing procedure involved incremental installation of four discrete soil layers, each 0.32-m thick when filled into the tray and subsequently compacted to a thickness of 0.25 m. The final bulk density of the packed soil is 1.59 g cm^{-3}, and the porosity is 39% on average. Ground-based laser scans were performed after the installation of each layer, to ensure the greatest possible homogeneity of packing density among individual layers, and across the three replicate model landscapes. The final soil depth on each landscape is, on average, approximately 1 m throughout, though with spatial variability (see Figure 3 in [10]). Because of the designed shape of the underlying steel structure, the ZOBs' surfaces have convergent topography. A primary zone of convergence, or trough, lies near the center of the ZOBs and extends upslope approximately 18 m (**Figure 2**). At that point, the primary zone of convergence terminates at the base of a planar hillslope section that spans to the upslope extend of the ZOBs. Two smaller troughs emanate from this termination point and span toward the upslope corners of the model landscapes. This topographic signature is similar to that observed across many ZOBs in nature [11]. Though the model ZOBs have an average slope of 10°, maximum slopes of approximately 17° exist at that transition from hillslope segments to the primary zone of convergence. The soil should not ever be disturbed by foot traffic or any other source other than natural erosive processes. To accomplish that goal, a personnel transport system was designed and constructed on each landscape (**Figure 3a**). The system allows workers to travel within a railed cart to any point on the soil surface without ever stepping on the surface. This capability is imperative for point-scale measurements of soil and vegetation properties integral to the long-term research agenda.

Irrigation rates of less than 3 mm h^{-1} to approximately 40 mm h^{-1} may be applied individually to each landscape via the engineered irrigation system. In the basement below the landscapes, there are seven storage tanks, each with 8037-liter storage capacity (**Figure 3e**). The tanks are filled with water that is pumped from a local well and passes through a reverse-osmosis

Figure 3. Selected elements of the extensive instrumentation array at LEO: Atmospheric sensors along vertical masts, and personnel transport system in the background (a); high-resolution imaging system mounted to track system (b); multivalves for subsurface gas sampling (c); vacuum box with 55 sample vials for automated collection of soil solution from pore water samplers (d); reverse osmosis system and tanks storing irrigation water (e); sprinkler system for applying irrigation water (f); tipping buckets and flow meters for measuring discharge from six discrete lateral sections of the seepage face (g); 1 of 10 load cells per LEO hillslope measuring changes in total landscape mass.

purification system before storage. The tanks are plumbed such that water can be drawn from any individual tank, or from any combination of tanks, controlled through simple valve selections. Three water pumps deliver water to the landscape level. Here again, the plumbing was installed so that any pump may deliver water to any particular landscape at any time, which provides maximum operational flexibility. At the landscape level, there are five independent plumbing circuits that the water may be delivered to. Each circuit is opened and closed by a programmable valve. Each plumbing circuit dispenses through a unique combination of sprinkler heads (MP Rotator, Hunter Industries, San Marcos, CA, USA) that are located on risers resting approximately 3.3 m above the soil surface (**Figures 1b** and **3f**). There are seven such risers mounted on both the west and east edges of the steel structures containing the model ZOBs. Irrigation water may be dispensed onto the landscapes through any circuit individually or through any combination of the circuits. The spatial homogeneity of applied irrigation varies among the circuits. In general, spatial variability is greatest at low irrigation rates (coefficient of variation, CV > 0.5) and becomes more homogenous at high irrigation rates (CV ≈ 0.2). Disdrometer measurements show that the distribution of water-drop velocities created by the irrigation system approach realistic values of terminal velocity for naturally occurring precipitation. The drop sizes are somewhat smaller than real precipitation [10].

Each bay is enclosed by space frame construction. The exterior of the space frame is covered with 0.011-m thick duo-laminated glass with an interior Mylar sheet. Total solar radiation transmission through the glass is 50–60%, with less than 1% of ultra-violet radiation transmission. The glass is cleaned intermittently. Information about specific transmissions of different wavelengths of radiation was published previously [12, 13]. The space frame directly above the model ZOBs is arranged in three tiers. At the downslope extent, the space frame is approximately 9 m above the soil surface. Across the upper portions of the landscape, the distance to the overlying space frame is greater than 10 m above the surface. At five locations over the landscapes, aluminum masts hang vertically from attachment points on the space frame

(**Figures 2** and **3a**) to an approximate distance of 0.15 m from the soil surface. The masts serve as attachment structures for meteorological instruments that monitor the LEO atmosphere (described in Section 2.1.1). The bottom segment of each mast is telescopic and may be raised to variable heights above the soil surface to accommodate future vegetation growth. A hinged mount connects the masts to the space frame, allowing the masts to be rotated upward to the point of being nearly parallel with the space frame. That action is controlled by motorized winches and braided metal cable that passes through a series of pulleys mounted to the space frame and attaches to the masts at two locations along their length. In this way, the masts can be raised high enough to allow passage of the personnel transport system and to avoid any accumulation and dripping of water during irrigation events, without ever having to remove or rewire any of the meteorological instruments. Air circulation over the landscapes is driven by three air-handler units located in the basement of each bay. These units pull air from a vertical duct that connects the basement and landscape level along the southern extent of the bay. The units push air from south to north through the basement, then vertically through a similar duct on the northern extent of the bay. The air then circulates predominantly from upslope to downslope over the surface of the landscapes. These air handlers are capable of producing maximum air velocities exceeding 1 m s^{-1} over the landscape surfaces—with air velocity being greater near the soil surface and lesser toward the space frame. Additional portable fans can be installed at any time, and at varying locations within the bay, to create greater velocities or turbulence as needed. Coiled radiators within the air-handling units allow circulation of heated or cooled water for temperature and humidity control. These systems are operated through proportional-integral-derivative (PID) controllers, enabling real-time manipulation of air velocities, temperature, and humidity, within some operational constraints. The air space immediately surrounding each landscape is isolated from that in adjacent bays by air partitioning structures consisting of aluminum-pipe framing and twin-wall polycarbonate plastic sheeting mounted to the frame. The air space of each bay is also isolated within the underlying basement by wood-framed walls composed of either plastic sheeting or regular sheetrock.

In addition to the three full-scale model landscapes, a much smaller version of a LEO landscape was built within the central bay of LEO. This scaled-down model, referred to as the miniLEO, was designed as a rectangular cuboid with length of 2 m, width of 0.5 m, uniform soil depth of 1 m, and a 10° slope [14, 15]. Its construction otherwise resembles that of the full-scale models, and it is filled with the same basaltic tephra and equipped with equivalent basic instrumentation, including an irrigation system. The miniLEO is typically used for shorter term experimentation, and destructive soil sampling and complete soil excavations are possible. The primary purpose of the miniLEO is to evaluate measurement and interpretation techniques, optimize experiments (e.g., irrigation sequences), and test specific hypotheses with comparatively low cost and risk before extensive large-scale experiments are launched.

2.1.1. *The network of automated, electronic sensors operating at LEO*

Here, we outline the array of automated, electronic sensors that are installed in, on, or around the model ZOBs and the measurement capabilities they provide (**Figures 2** and **3**). Other manual sensing and sampling devices utilized within LEO are described in subsequent sections (e.g., an electrical-resistivity tomography system, soil solution, and soil gas sampling). All instruments described here are connected to a network of Compact Reconfigurable Input-Output devices (CRIOs; National Instruments Corp, Austin, TX, USA), ultimately with data transmission and storage to two onsite servers. Unless otherwise stated, data from each instrument are recorded at 15-min intervals, although this recording interval can be readily adjusted as needed. The instruments and quantities listed below are replicated on each of the three model ZOBs.

Aboveground instrumentation includes:

- *2 Kipp and Zonen CNR4 4-Way Radiometers*—These instruments are mounted at 2-m height above the soil surface on the masts overlying the east- and west-facing hillslope facets adjacent to the central convergence area. Separate radiometers record incoming and outgoing short- and long-wave radiation. The spectral range of measurements provided by the set of radiometers is 300–2800 μm and 4500–42,000 μm. The paired instruments allow comparison of all radiation components on portions of the model ZOBs with different slope aspect.
- *1 Campbell Scientific CSAT-3 Sonic Anemometer*—The instrument is mounted at 2-m height on the mast overhanging the centermost location of the ZOB and oriented upslope. The instrument measures the three-dimensional air-velocity field in high temporal resolution. The instrument is co-located with gas-intake tubing that is routed to an infra-red gas analyzer housed underneath the centermost model ZOB. The sonic anemometer records data at a frequency of 60 Hz.
- *24 Davis Cup Anemometers*—These instruments are located at up to 5 heights along each mast: 0.25, 1, 3, 6, and 10 m above soil surface. No instruments are available at the maximum height at the most downslope mast position, due to the closer proximity of the space frame. They measure wind speed and direction, with an initiation speed of approximately 1.3 m s^{-1}.
- *24 Vaisala HMP60 Temperature and Humidity Sensors*—These instruments are located at the same heights on all masts and measure air temperature and relative humidity within the ranges −40 to 60°C and 0 to 100%, respectively.
- *24 Apogee Instruments Quantum Sensors*—These sensors are also co-located with others on the masts. They measure photon-flux density within wavelengths spanning 410–655 nm.

Instruments installed within the subsurface of the model ZOBs include:

- *24 Huskeflux HPF-1 and HPF-1SC Soil-Heat-Flux Plates*—These instruments are located as 12 pairs, individual sensors within pairs spaced at 1 m, and broader spacing between pairs. They are thermopiles that measure conductive heat transport into the soil profile. They are buried at 0.05-m depth, with an accompanying thermocouple buried at 0.025-m depth, and co-located with soil water content sensors. Actual conductive heat transport into the soil is estimated based on an algorithm provided by the sensor manufacturer, which considers the water content-dependent heat capacity of the soil-water-air mixture.
- *496 Decagon 5TM Soil Water Content and Temperature Sensors*—These sensors measure the dielectric permittivity of wetted soil and convert to volumetric water content using a calibration equation. They also measure temperature with an installed thermistor. A soil-specific

calibration of the sensors was performed by the manufacturer, using the LEO soil, which enables estimates of volumetric water content with accuracy of ±0.025 or better.

- *496 Decagon MPS-2 Soil Water Potential Sensors*—These sensors also measure dielectric permittivity of an attached, wetted ceramic disc that is in equilibrium with soil water. Based on a calibrated relationship between permittivity and water content of the disc, they report soil water pressure across a range from −6 to −500 kPa. The manufacturer-stated accuracy is ±25% of reading.
- *48 Vaisala GMM222 [CO_2] Sensors*—Based on Vaisala's CARBOCAP technology, these sensors are able to measure [CO_2] within air in the soil pore space. They are buried at 48 locations within each model ZOB.

Other instruments installed external to the landscapes include:

- *15 Campbell Scientific CS-451 Vented Pressure Transducers*—These vented pressure transducers record gauge pressure. They are installed within sealed bulkhead fittings with compression caps at 15 locations along the base of the steel structure. The portion of the bulkhead fitting exposed to soil on the interior of the steel structure is screened to allow water intrusion, but not soil. These transducers measure pressure heads of 0 to 2 m and allow monitoring of the spatial dynamics of water table height.
- *10 Honeywell Model 3130 Load Cells*—These loads cells were installed at a nexus point between the primary vertical supports and the lateral beams forming the perimeter of the steel trays. Collectively, the 10 load cells enable accurate and precise measurements of changes in total-landscape mass (due to additions or losses of water). The manufacturer-reported repeatability is 0.05% of full scale.
- *6 SeaMetrics PE102 Flow Meters*—Groundwater seepage from the landscapes flows into one of six partitioned troughs at the downslope extent and then through tubing that leads to these meters. The manufacturer-reported accuracy is 1% relative error across flow rates of 0.11–11.4 liters per minute.
- *6 NovaLynx 26-2501-A Tipping Bucket Gauges*—After passing through the flow meters, the seepage water then flows into one of these tipping bucket gauges. They are calibrated onsite multiple times per year. These provide superior accuracy at very low flow rates compared to the flow meters.

2.1.2. *Collection and analysis of soil solution, rainfall, and discharge samples*

Sampling of soil solution is facilitated through 496 suction cup pore water samplers (Super Quartz, Prenart Equipment, Frederiksberg, Denmark; pore size <2 μm) that are embedded in each of the LEO hillslopes. The pore water samplers are co-located with the moisture and matric potential sensors (**Figure 2**). Each sampler is connected via a PTFE sampling line to one of 11 custom vacuum sampling modules distributed across the base structure of each hillslope (**Figure 3d**). Each sampling line is equipped with an individual shut-off valve. The modules consist of a Plexiglas vacuum box equipped with a manual pressure regulator, a vacuum gauge, and a tray capable of holding 55 centrifuge tubes of 50 mL volume for sample collection. Each module is connected to a vacuum manifold. Vacuum is supplied by a total of two single-stage rotary vane vacuum pumps (Model RC0100, Busch LLC, Virginia Beach, VA, USA) located outside of the LEO domain. The vacuum system allows simultaneous sample collection from all 496 samplers on each slope. However, when needed, vacuum application and sampling can be limited to any subset of modules or even any subset of individual samplers. This allows adapting the timing of sampling to in-slope processes, e.g., to the progression of a wetting front. Furthermore, the suction to be used in each module can be adjusted to matric potential recorded by the sensors co-located with solution samplers to ensure that adequate suction is applied to obtain a sufficient sample volume while avoiding excessive draw beyond the immediate vicinity of the samplers. In addition to soil solution sampling, water samples from the seven common irrigation tanks and from two to four custom precipitation collectors per hillslope are collected manually during irrigation events. Finally, custom-built, Arduino-based autosamplers for discrete sample collection at set intervals are in place to collect water samples from the outflow at the base of each hillslope (seepage flow and potentially overland flow) when it is generated.

All collected water samples are archived in freezers at a temperature of −9°C. Biogeochemical sample analysis is performed in an onsite analytical laboratory and can be complemented by isotopic analysis in the isotope and trace gas facility (Section 2.1.3). Prior to analysis samples are centrifuged at 4816 relative centrifugal force for 20 min to remove particulates (Sorvall Legend XTR, Thermo Fisher Scientific Inc., Waltham, MA, USA). The analytical lab is equipped with a Dionex ICS 5000 ion chromatography system (Thermo Fisher Scientific Inc.) with two conductivity detectors for high throughput and concurrent sample analysis for major anions and cations. Further capabilities include solution analysis for dissolved organic and inorganic carbon and total dissolved nitrogen (TOC-L Series total organic carbon and nitrogen analyzer, equipped with TOC-LCSH autosampler; all Shimadzu, Kyoto, Japan), as well as for pH and electrical conductivity (sympHony multimeter, VWR International, Radnor, PA, USA). In addition, selected samples are analyzed offsite for major, trace, and rare earth elements using inductively coupled plasma mass spectrometry (Elan DRC-II, Perkin Elmer, Waltham, MA, USA). Characterization of organic compounds in collected solutions is performed at the Environmental Molecular Sciences Laboratory (Richland, WA, USA) using Fourier transform ion cyclotron resonance mass spectrometry (FTICR-MS).

2.1.3. Stable isotope and trace gas monitoring

Analysis of the concentration and/or stable isotope composition of key molecular species across the hydrologically and biogeochemically relevant landscape compartments (soil, atmosphere, and outflow) of LEO is facilitated by extensive onsite equipment arrays, encompassing state-of-the-art probing interfaces and analyzing instruments. All central equipment is housed in a custom laboratory facility constructed below the central LEO landscape structure. To facilitate stable and gap-free operation of sensitive instrumentation onsite, the laboratory was equipped with uninterrupted power supply, air-conditioning, and zero air (i.e., air that is free of water, CO_2, and contaminants) supply from a lab generator (Aadco Instruments Inc., Cleves, OH, USA), utilizing compressed facility air via a 60-gal in-line back-up reservoir. The facility currently houses four laser-based gas analyzers. The first instrument is an off-axis integrated cavity output spectrometer (OA-ICOS; IWA-35EP, Los Gatos Research Inc., Mountain View, CA. USA) for continuous measurement of the hydrogen ($\delta^2H–H_2O$; $1\sigma < 0.2\%$)

and oxygen ($\delta^{18}O–H_2O$; $1\sigma < 0.05\%$) isotope composition of water vapor [16]. The second instrument is a dual quantum cascade laser absorption spectrometer (QCLAS; TILDAS-D, Aerodyne Research Inc., Billerica, MA, USA) with two tunable lasers to measure continuously and simultaneously the $\delta^2H–H_2O$ ($1\sigma < 0.1\%$) and $\delta^{18}O–H_2O$ ($1\sigma < 0.03\%$) composition of water vapor as well as the carbon ($\delta^{13}C–CO_2$; $1\sigma < 0.03\%$) and oxygen ($\delta^{18}O–CO_2$; $1\sigma < 0.03\%$) isotope composition of CO_2 [17]. The third instrument is a carbonyl sulfide (COS) monitor (mini QCLAS, Aerodyne Research Inc.) for continuous high-sensitivity ($1\sigma < 2$ pptv) trace gas analysis. Finally, a bench-top differential infrared gas analyzer (LI-7000; LI-COR Biosciences, Lincoln, NE, USA) is available for simultaneous high-speed measurements of water vapor and CO_2 concentrations. Details on the respective technologies can be found elsewhere [18–21]. The gas analyzers are interfaced with sophisticated sample conveyance and control systems utilizing custom LabView (National Instruments Corp.) software for automated in-situ monitoring of liquid and gas compositions across the LEO domains.

A multisource liquid sampling system was devised for high-frequency sampling and real-time analysis of seepage water outflow from the three LEO ZOBs utilizing the OA-ICOS instrument. The sampling system uses a four-channel peristaltic pump (Minipuls 3, Gilson Inc., Middleton, WI, USA) to continuously deliver liquid water from a given source to one of four ports of a stainless-steel sampling manifold that is mounted on the tray-holder of an autosampler (LC PAL, CTC Analytics AG, Zwingen, Switzerland) for liquid injection into a vaporization unit and subsequent vapor delivery into the isotope analyzer. This setup delivers robust performance [22] and facilitates water isotopologue analysis of outflow from each LEO landscape at approximately 30-min intervals. To obtain an even higher temporal resolution of discharge isotope composition, as well as for complementary analysis of irrigation, discharge, and soil solution samples, the OA-ICOS instrument and autosampler setup can be used to analyze collected water samples (Section 2.1.2) offline.

Gas sampling is performed in the atmospheres and subsurface soil of each LEO landscape utilizing extensive valving and probing arrays that can be simultaneously linked to the dual-QCLAS, COS, and benchtop gas analyzers. Twenty-four air intake lines, with sheltered inlets at four to five different heights along each of the five masts distributed over each slope surface (Section 2.1; **Figures 2** and **3a**), are available for atmospheric gas sampling. Subsequent sample intake from each of the lines is facilitated by three stream selector valves (VICI Valco Instruments Inc., Houston, TX, USA) situated at the central onsite laboratory, with flow driven by a downstream vacuum pump. To eliminate temporal delay associated with sample gas transport from air inlet to analyzer, the intake line sections upstream of the valves can be constantly purged with fresh atmospheric air using branch-off lines connected to a purge pump via custom vacuum manifolds. To extract air samples from the subsurface, 141 custom soil gas samplers installed in a uniform grid at multiple soil depths are available within each of the model landscapes (**Figure 2**). The samplers were constructed from microporous, hydrophobic PTFE tubing sealed at both ends to gas transport lines with epoxy and heat shrink tubing. A multilevel conveyance system, comprising 27 sub-level stream selector valves mounted at varied locations across the landscapes' structural bases (**Figure 3c**), three main-level selector valves (VICI Valco Instruments Inc.), and several digital mass flow and pressure controllers (Alicat Scientific, Tucson, AZ, USA), allows for automated sampling from the dense probe network according to programmed sequences. The best achievable sample interval is approximately 1 h for sampling all atmospheric inlets and 36 h for sampling all subsurface probes. Actual sampling frequencies for atmospheric and soil gases vary depending on experimental priorities. All measured trace gas concentrations and isotope abundances in water vapor and CO_2 are calibrated to reference scales using self-built delivery systems for calibration standards (e.g., NOAA and IAEA standards). In addition, and because the isotopic liquid-vapor equilibrium in soils is mainly temperature dependent [23], the measured vapor-phase isotope composition in soil air can be utilized along with measured soil temperatures to infer the liquid soil water isotope composition [24–26] throughout the LEO subsurface.

2.1.4. *Remote sensing instrumentation*

Each hillslope is equipped with a custom-designed camera imaging system consisting of a hyperspectral imager and a thermal infrared camera. The hyperspectral imager (SOC710VP, Surface Optics Corp, San Diego, CA, USA) produces images within the wavelength range of 400–1000 nm with 4.7-nm bands (128 total bands). The thermal infrared camera (ICI 9640 P-Series, Infrared Cameras Inc., Beaumont, TX, USA) produces images with a 7–14 μm spectral response with a ±1°C accuracy. The images will be used to estimate spatiotemporal patterns of bare soil evaporation, plant transpiration, and photosynthesis (see, e.g., [27]).

In order to image the entire surface of the LEO hillslopes in a precise and repeatable way, the cameras are installed on a 35-m long belt drive linear actuator (MSA-14S, Misumi, Schaumburg, IL, USA), which is suspended from the space frame of Biosphere 2 and hangs 7 m above the surface of the LEO hillslopes at a 10° pitch (**Figure 3b**). This linear actuator allows movement of the camera boxes in the *x*-direction of the hillslopes and provides sub-millimeter linear resolution. A custom robotic stage allows the cameras to pan right and left and provides a 0.2° angular resolution or 1.5 cm at the soil surface. The hyperspectral camera uses a f = 8 mm lens producing an image area of 4.6 × 4.6 m with 0.9-cm resolution at the slope surface, while the infrared camera uses a f = 16 mm lens producing an image of 4.6 × 3.4 m with 0.7-cm resolution. The cameras are housed in an environmentally controlled enclosure to minimize temperature fluctuations using chilled compressed air. The linear actuator and robotic stage allow recording the 27 images required (9 along the *x*-axis and 3 along the *y*-axis) to cover the entire LEO soil surface. The camera system and data offload are controlled using custom LabView software, and software is being developed to stitch the images together to create complete images of the LEO surfaces.

In addition to imaging of the LEO surfaces, a LiDAR scanner (ScanStation C10, Leica Geosystems, Heerbrugg, Switzerland) is used to map surface topography changes in the LEO hillslopes and will be used in the future to measure vegetation growth as plants are introduced. The scanner provides a 6-mm spatial accuracy and 2-mm modeled surface precision. A 6-month scan interval of all hillslopes will provide us with a time series of geomorphologic change.

2.1.5. *Electrical resistivity tomography system*

Electrical resistivity tomography (ERT) is used for minimal-invasive monitoring of the subsurface at LEO. An ERT survey is performed by injecting an electrical current through a pair of electrodes and measuring the potential at several other electrode pairs. Several current injections are performed at various locations in order to create a set of redundant potential measurements, which are used to derive a set of apparent electrical resistivity values. The apparent electrical resistivity values are then converted to a "true" electrical resistivity set through an inversion procedure [28], which can ultimately be related to soil physical properties (e.g., porosity), water content, and solute concentration [29].

Each LEO landscape is equipped with two current-injecting electrodes and 24 custom potential-measuring electrode stacks (**Figure 4a**). Each electrode stack is installed vertically into the LEO soil and comprises five stainless steel electrodes, separated by insulating acrylic cylinders, for potential measurements at varied soil depths (**Figure 4b**). Each of the 120 total electrodes can be connected to a Supersting R8 (Advanced Geosciences Inc., Austin, TX, USA) electrical resistivity meter and induced polarization and self-potential system. The fully automated eight-channel system allows rapid three-dimensional surveying and sequential imaging of dynamic processes with high accuracy and low noise levels. Specialized software (RES3DINVx64, Geotomo Software, Kardinya, Australia) is used for data processing. The system can provide a spatial resolution of a few centimeters in proximity to the electrodes and is coarser at increased distance from the electrodes.

The Supersting electrical resistivity meter is also used to perform geophysical surveys of the miniLEO small-scale replicate system, where intensive soil sampling is possible to validate ERT measurement and interpretation methods for use on the full-scale LEO landscapes. Specifically, we seek to establish relationships between measured resistivity fields and features of subsurface heterogeneity associated with, e.g., changes in porosity, clay formation, moving wetting fronts,

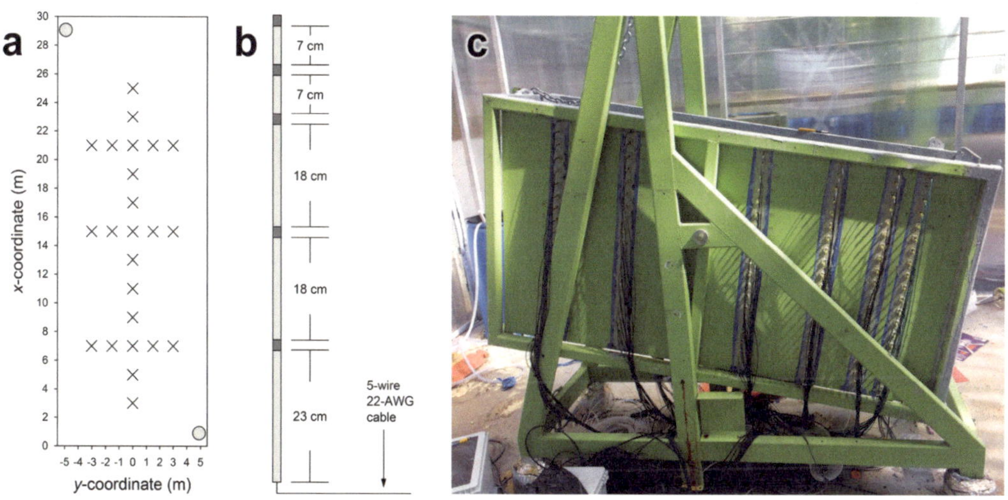

Figure 4. Overview of electrical resistivity tomography (ERT) instrumentation at LEO. (a) Horizontal (xy-plane) arrangement of the 2 current-injecting electrodes (gray dots) and the 24 potential-measuring electrode stacks (black x's) across a LEO hillslope. (b) Schematic of an electrode stack, where dark gray sections indicate stainless steel electrodes and light gray sections indicate insulating acrylic rods. (c) The miniLEO small-scale replicate of the LEO hillslopes, equipped with 228 electrodes installed through its walls along 12 vertical transects. This miniLEO setup is used to test and develop ERT surveying and interpretation methods that can ultimately be applied to nondestructively monitor incipient subsurface heterogeneity on the large-scale LEO landscapes.

and plant root distributions. The miniLEO was therefore equipped with 228 electrodes distributed along its walls (**Figure 4c**), allowing for subsurface mapping with particularly high resolution.

2.1.6. Sampling of soil material for physicochemical and biological analysis

Regular soil sampling is required to assess the spatiotemporal microbiological variations and geochemical transformations occurring in the hillslopes. However, soil coring in the landscapes has to be performed conservatively to preserve the topography of the slopes and avoid preferential flow paths following soil removal. Therefore, coring has thus far been limited temporally to a twice-yearly basis (including times before and after irrigation events) and spatially to 4–6 locations corresponding to the zones that are expected to experience divergent evolution (including head slope, convergence zone, toe slope, and side slopes). We use the personnel transport system to access coring locations without disturbing the soil surface. A 1-m long steel corer with 1-inch internal diameter powered by drill and fitted with 1 × 37 3/4-inch plastic liner (AMS Inc., American Falls, ID, USA) is used to collect the samples. For each location, a new clean liner is fitted into the corer to prevent cross-contamination between samples. The plastic sleeve is extracted post-coring and sealed at both ends to prevent soil loss. After core extraction, the resulting hole in the soil is backfilled with the same amount of original tephra material, which is stored in barrels. To ensure that the backfill material is as similar in composition and extent of weathering to the extracted slope material as possible, the barrels receive irrigation water at similar rates as the ZOB soil.

The cores are brought to the lab where their lengths are measured, and they are sub-sectioned according to the depth profile recovered. Each subsection is partitioned into two halves to obtain samples for microbiological and geochemical analyses. Soil sampling areas are co-located with solution samplers and sensors to obtain complementary physicochemical measurements, and modeled variables needed to perform coupled hydro-geochemical and mineralogical analyses.

After extraction, samples for microbiological analyses are either stored on ice or flash-frozen for DNA and RNA extractions, respectively. High-throughput analyses of DNA provide total community composition (amplicon sequencing) and predictions of functional potential (metagenome sequencing), while RNA sequencing (metatranscriptomics) provides expressed function of the community. Soil samples are also used to analyze copy numbers of important functional genes using quantitative real-time polymerase chain reactions (qPCRs).

Samples for geochemical and mineralogical analyses are air-dried, with weight recorded before and after drying to determine moisture content of the sample. Air-dried samples are analyzed to quantify and characterize accumulation of organic compounds and to quantify inorganic carbon accumulation through weathering processes. Content of total nitrogen and total and organic carbon (following treatment with phosphoric acid to remove inorganic carbon) is determined using a total carbon and nitrogen analyzer (TOC-L Series, Shimadzu; see also Section 2.1.2) coupled with a solid sample combustion unit (SSM-5000A, Shimadzu). Molecular characterization of soil organic matter (SOM) to observe carbon stabilization and fractionation during weathering processes is performed using Fourier transform ion cyclotron resonance mass spectrometry (FTICR-MS) at the William R. Wiley Environmental Molecular

Sciences Laboratory, a national scientific user facility at the Pacific Northwest National Laboratory (PNNL). Selected soil samples from various depths and along specific flow paths are subjected to solvent exactions (H_2O, MeOH, and $CHCl_3$), and soil extracts are analyzed together with co-located solution samples by FTICR-MS. Air-dried samples are also used for sequential extraction followed by inductively coupled plasma mass spectrometry (ICP-MS) and X-ray diffraction analysis (following size fractionation to concentrate newly formed minerals) to quantify changes in mineralogical composition of the hillslopes. Mossbauer spectroscopy is employed to characterize the oxidation state and bonding environment of iron in weathering basalt in order to trace weathering of Fe-containing phases and precipitation of new minerals.

2.1.7. Integrated mathematical modeling framework

Experiments at LEO are iterated with coupled Earth system modeling. The Terrestrial Integrated Modeling System (TIMS) takes advantage of existing state-of-the-art community models to study interactions and feedbacks between various physical, geochemical, and biological processes by communicating fluxes and states of energy, water, and mass between various component models (**Figure 5**). A model developed in a specific discipline is typically limited by the scope of the developers' expertise and limited knowledge of other disciplines. To compensate

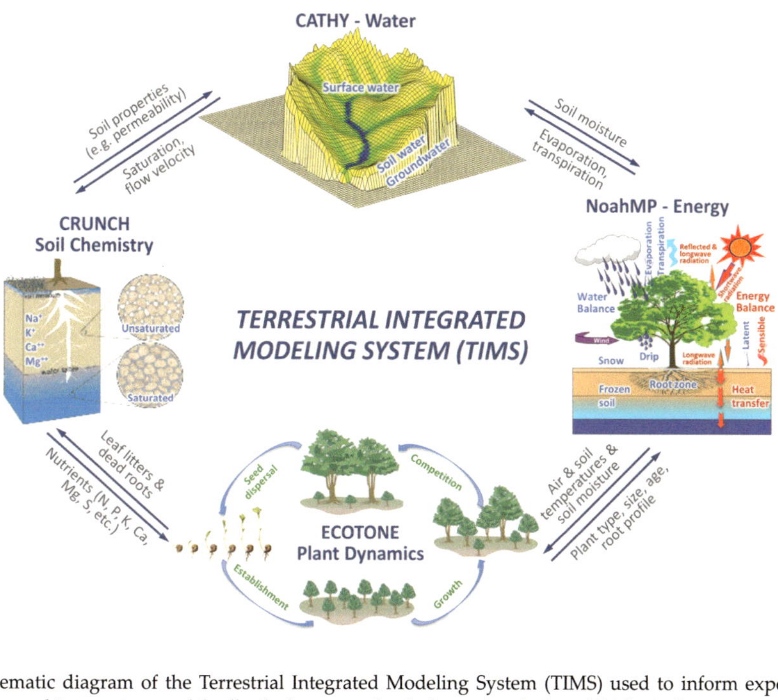

Figure 5. Schematic diagram of the Terrestrial Integrated Modeling System (TIMS) used to inform experiments and formalize observed interactions and feedbacks between physical, geochemical, and biological processes at LEO. TIMS couples existing state-of-the-art community models and aims to simulate the dynamics of (1) surface water in rivers, lakes, and wetlands, subsurface water in the vadose zone and aquifers, (2) organic and inorganic solute transport driven by surface and subsurface flow, volumes of various minerals, and porosity, (3) plant species and biomass distribution over a landscape, and (4) land surface energy, water, and carbon exchanges with the overlying air through atmospheric turbulent transfer and radiation transfer.

for these limitations in the models of individual disciplines, TIMS integrates advanced knowledge and expertise from various disciplines and may thereby exceed the sum of its parts. TIMS has integrated a physically based hydrological model (CATchment Hydrology model, CATHY) and a land-atmosphere energy, water, and carbon exchange scheme (NoahMP) [30, 31]. In addition, it has integrated newly developed modules such as a radiation correction model, which accounts for the effects of topographic shading and scattering [32], and a six-carbon pool microbial enzyme model, which provides the capability to model the responses of soil microbial respiration to soil moisture dynamics [33]. TIMS aims to further integrate an individual-based ecological model (e.g., ECOTONE) and a geochemical model (e.g., CrunchFlow).

CATHY [34] is a 3D, physically based, surface-subsurface coupled flow model. The subsurface flow module in CATHY solves the pressure-based 3D Richards equation describing flow in variably saturated porous media [35], while the quasi-2D surface flow module solves the diffusive wave equation describing surface flow propagation over hillslopes and in stream channels and lakes identified using terrain topography and the hydraulic geometry concept [36]. This model has undergone long-term development and represents one of the most thoroughly developed, physically based flow models. NoahMP [37] represents the land surface energy (e.g., radiation, sensible, and latent heat fluxes), water (e.g., transpiration and evaporation), and carbon fluxes exchanging with the atmosphere and provides multiple physical options for hypothesis testing. It is a community land surface model developed through collaborations among scientists in national centers (e.g., NCAR, NCEP, and NASA) and universities for use in water, weather, and short-term climate predictions (e.g., the National Water Model and the Weather Research and Forecasting Model). It also represents assimilation of carbon through photosynthesis, carbon allocation to various parts of the plant, autotrophic and heterotrophic respiration, leaf litter, root exudates, and dead roots, as well as leaf and root dynamics. ECOTONE [38] is an individual-based ecological model simulating changes in species and associated biomass of individual plants within a patch of land (1–10 m^2 area). Seed germination, establishment of seedlings, and mortality are described by stochastic elements (e.g., seed dispersal, local disturbance), but growth is deterministic based on root distribution and access to water and nitrogen within a competitive context. Resources are distributed to each plant based on the proportion of active roots at each depth relative to total root biomass of all plants. The yearly time step computation is changed to a daily time step to update the biomass in response to daily soil moisture dynamics that are aggregated from CATHY operating at sub-hourly time step. CrunchFlow [39] is a multicomponent reactive transport model describing advective, dispersive, and diffusive transport of solutes, resulting from various chemical reactions such as aqueous complexation, mineral precipitation and dissolution, ion exchange, surface complexation, radioactive decay, and biologically mediated reactions. It also deals with burial, erosion, and compaction of porous media, with an explicit treatment of spatially variable advection of solids as well as reaction-induced porosity and permeability feedbacks to both diffusion and flow.

2.2. Monitoring of hydrologic cycling and flow pathways

Each landscape and its surrounding atmosphere are extensively equipped to close the terrestrial water balance, written as follows:

$$\frac{\partial S}{\partial t} = I(t) - Q(t) - ET(t) \quad (1)$$

where S represents the volume of water stored within the landscape (L^3), I represents the irrigation inflow ($L^3\ T^{-1}$), Q represents the discharge outflow at the downslope end of the landscape ($L^3\ T^{-1}$), and ET represents the combined vapor-phase flux associated with bare soil evaporation and plant transpiration from land to atmosphere ($L^3\ T^{-1}$). All variables are functions of time t.

All water balance terms in Eq. 1 can be measured as integrated, landscape-scale states and fluxes at LEO—a capability critical for characterizing hydrologic partitioning under variable environmental conditions yet not achieved in any other experiment at the hillslope scale (**Figure 6a–c**). Temporal changes in water storage within the entire landscape are monitored via 10 load cells installed under the only load-bearing points connecting the main hillslope with the supporting structure. This makes the LEO landscapes the world's largest weighing lysimeters. Volumetric irrigation flow rates are monitored with electromagnetic flow meters. The total specific flux and the spatial distribution associated with each of the five irrigation circuits were characterized through a series of manual calibration trials. The discharge term can comprise both subsurface and overland outflows of water from the landscape, depending on intensity and duration of irrigation forcing. Subsurface flow can exit through one of six lateral subsections of the seepage face at the downslope extent of each landscape and is then routed through

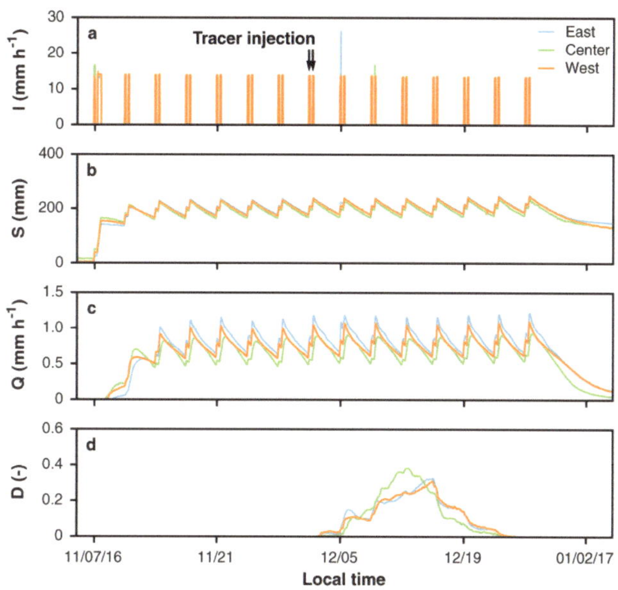

Figure 6. Measured hydrological and tracer dynamics for the three LEO landscapes (East, Center, and West) over the course of a two-month long experiment with periodic forcing and deuterium tracer application during two subsequent irrigation pulses (indicated by black arrows; preliminary data). Data shown are whole-landscape means of irrigation influx (I), water storage (S), discharge (Q), and deuterium tracer abundance in discharge (D; values are relative to the injected deuterium abundance in irrigation water).

a plumbing system with inline electromagnetic flow meters and tipping bucket gauges. Each subsection is measured separately to capture spatial variability of flow, particularly during high flow conditions, and two different types of instruments are used to achieve optimal precision over the full range of possible flow rates. If present, overland flow will be routed over a flume structure and through a plumbing system into an open reservoir, where a pressure transducer continuously monitors changes in water depth. The single remaining water balance component, the combined evapotranspiration flux, can be estimated as the residual term of Eq. 1 using the directly measured rates of all other terms as described above.

The landscape-scale hydrologic cycling is the product of inherently variable surface and subsurface hydrological fluxes and dominantly controlled by landscape heterogeneity [1, 40]. Even in a simplified model system such as the initial LEO landscapes, water movement is not homogeneous [41], and continued coevolution and variable forcing are anticipated to induce increasingly complex flow patterns. The landscape-scale measurements of water storage and fluxes are therefore complemented by spatially resolved measurements utilizing conventional hydrometric as well as innovative, minimal-invasive geophysical and optical techniques. A laterally (154 locations in the xy-plane) and vertically (five different depths) dense grid comprising 496 co-located soil water content and matric potential sensors (**Figure 2**) provide meter-scale lateral and sub-meter-scale vertical resolution of water storage, availability, and retention characteristics in continuous time. An even higher spatial resolution of subsurface water dynamics can be achieved using electrical resistivity tomography (ERT) measurements (Section 2.1.5). Three-dimensional time-lapse ERT scans from 24 potential-measuring electrode stacks installed into each hillslope can be geophysically inverted [42] and coupled with hydrological models [43] to resolve decimeter-scale variations in water content and flow processes.

Direct measurements of spatially distributed soil evaporation and plant transpiration fluxes can, in principal, be obtained based on flux-gradient and eddy covariance techniques commonly used in field studies (e.g., [44, 45]). The vapor-phase surface fluxes are mainly determined by wind speeds and a vertical gradient of atmospheric vapor pressure deficit (VPD, a function of air temperature and humidity), and the atmospheric instrumentation array at LEO delivers all required data. Profiles of air temperature, absolute and relative humidity, and wind speed are measured along five vertical masts (at heights of 0.25, 1, 3, 6, and 10 m above the land surface) distributed over each landscape's surface, and high-frequency measurements of the three-dimensional wind vector are available for a central location over each landscape (**Figure 2**). However, application of conventional flux-estimation methods is challenging under the indoor climate conditions at Biosphere 2 [46], as stable atmospheric stratification and associated turbulent intermittency, waves, and other processes make specific methodological adaptations necessary (e.g., [47]). The closed nature of the LEO atmospheres makes it possible, in turn, to use mass balance calculations to approximate whole-landscape evapotranspiration fluxes and their spatial heterogeneity from the spatially stratified measurements. An additional opportunity for measuring spatially resolved evaporation fluxes is through high-resolution thermal imaging. An infrared camera system moving precisely along a track system mounted to the space frame of each LEO bay maps whole-slope soil surface temperature at centimeter-scale resolution (Section 2.1.4). When coupled with atmospheric measurements and known soil properties, the thermal imagery facilitates calculation of surface evaporation using methods similar to those applied by [27]. Given the high density of these spatially resolved aboveground and belowground measurements, good approximations of landscape integrated hydrological states, and fluxes can be attained to corroborate the direct landscape-scale measurements outlined above.

Characterizing the origin, flow pathways, and transit times of water through landscapes is a particular challenge that none of the above instrumentation can meet. LEO was therefore additionally equipped with a state-of-the-art stable isotope facility that operates the first hillslope-scale real-time isotope monitoring network (Section 2.1.3) and performs isotope analysis of collected water samples (Section 2.1.2). Irrigation water can be analyzed and labeled to create a known and time-variable isotope tracer input to the landscapes. Using equilibrium calculations, online measurements from the dense soil gas probing system (141 samplers per hillslope; **Figure 2**) can then be used to track the labeled liquid water through the subsurface soil in continuous time [24]. Isotopic analysis of pore water samples provides additional, spatially detailed (496 samplers per hillslope) snapshots of tracer plumes. Tracer finally leaving the landscapes through evaporation and transpiration can be identified through online gas monitoring along the masts throughout the LEO atmospheres (24 gas inlets per slope), and tracer leaving the landscapes as discharge is recorded using online or offline high-frequency liquid water sampling and isotope analysis (**Figure 6d**). These landscape-integrating and spatially resolved isotope measurements can thus be integrated with mixing models [25], transfer functions [14], and coupled-process models [48, 49] to characterize the pathways and fate of water molecules entering the hillslopes at a given time throughout the LEO domains.

2.3. Monitoring of land-surface energy exchange

The exchange of energy is a key component of the coupling between the landscape surface and the overlying atmospheric boundary layer. The surface energy balance of a LEO landscape can be described as follows:

$$R_{si}(t) + R_{li}(t) + R_{so}(t) + R_{lo}(t) = H(t) + \lambda ET(t) + G(t) \tag{2}$$

where R represents radiative fluxes associated with shortwave and longwave (subscripts s and l) radiation that is incoming or outgoing (subscripts i and o) to or from the landscape ($P\ L^{-2}$), H represents the sensible heat flux between land and air ($P\ L^{-2}$), λET is the product of the latent heat of vaporization ($E\ M^{-1}$) and the magnitude of evapotranspiration ($M\ T^{-1}\ L^{-2}$), and G represents the conductive heat transport and storage into the landscape ($P\ L^{-2}$).

The latent heat flux is the only term in Eq. 2 that is measured at the landscape scale. This is accomplished using the known value of heat of vaporization and the whole-landscape evapotranspiration flux estimates, which are based on load cell measurements and mass balance calculations (see Section 2.2). All other terms of the energy balance are measured at several discrete locations across each LEO landscape, and landscape-scale fluxes can be estimated based on the point measurements (**Figure 7c**).

All radiative flux terms on the left-hand side of Eq. 2, and thus the net radiative flux, are measured directly by a pair of four-way net radiometers installed at 1-m height above the soil surface on the two masts located off-axis over the east- and west-facing hillslope segments adjacent to the convergence zone (**Figure 2**). Photosynthetically active radiation (i.e. visible light) is additionally measured at 4–5 heights along all five vertical atmospheric masts. Added uncertainties exist in utilizing these point measurements to represent the average landscape-scale fluxes due to the effects of the windows and structural frames of the climate-controlled bay on solar radiation and mismatched source areas. The conductive heat flux into and out of the ground is measured by 12 pairs of heat flux plates (buried at 0.08 m depth, with associated averaging soil thermocouple probes installed at 0.02 m depth). Those devices are arranged in an approximately uniform grid spanning most of the landscape surface, including monitoring locations below the atmospheric masts (**Figure 2**). Finally, the sensible heat flux from land to atmosphere can be estimated as the residual term of Eq. (2). An alternative possibility for monitoring H, as well as λET, is through application of modified flux-gradient or eddy covariance methods (e.g., [47]). As described in Section 2.2, those methods would utilize the array of aboveground meteorological instruments, but their application under often stable adiabatic conditions above the hillslopes' surfaces poses a challenge and requires methodological developments. Finally, the aboveground and belowground instrumentation allows tracking the spatial and temporal variations in kinetic energy (in terms of temperature, measured at 496 soil locations and 24 atmospheric location; **Figure 7a**, **b**) and latent energy (in terms of humidity, measured at 24 atmospheric location) contained across the LEO domain that result from the local balance of energy fluxes and importantly control hydro-biogeochemical flux and reaction processes.

2.4. Monitoring of biogeochemical cycling

The LEO landscapes are uniquely suited to examine biogeochemical cycling (with the potential to close elemental mass balances) due to the comprehensive array of sensors and samplers installed in and above the hillslopes. In general, a cycle of any element on the landscape can be described in the following common terms:

$$\frac{\partial E}{\partial t} = E_p(t) + E_a(t) - E_r(t) - E_q(t) \tag{3}$$

where E is the element storage in different forms on or within the landscape (M), E_p represents the transfer rate to the landscape with precipitation (M T^{-1}), E_a represents the transfer from the atmosphere through different mechanisms (M T^{-1}), E_r represents the release back into (or new emissions to) the atmosphere (M T^{-1}), and E_q represents the loss with water discharge from the landscape (M T^{-1}). All of these fluxes can be measured, integrated at the landscape scale, and spatially resolved across the landscape using existing instrumentation for major and trace elements. A particular focus lies on carbon, elements essential for plant nutrition, and those indicative of soil formation processes, such as weathering.

Given known rainfall duration and intensity (as well as changes in the mass of the hillslopes), the E_p term can be quantified for the whole landscape by measuring the total element content in rainfall. Measuring element concentrations in the seepage and overland flow (if any) as

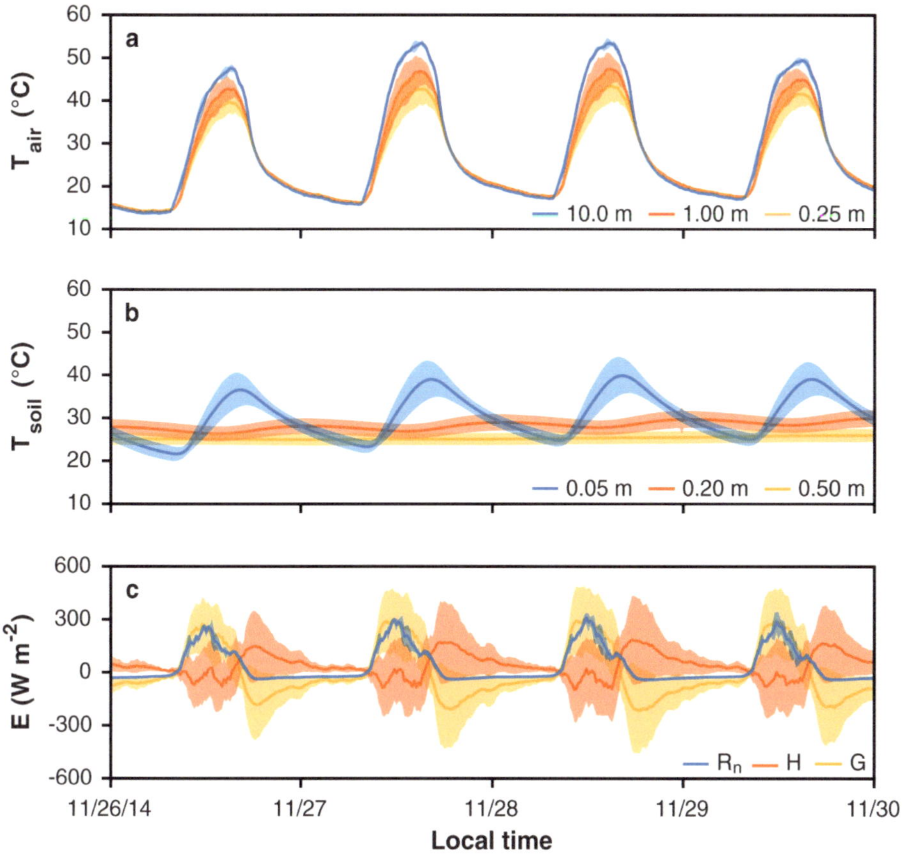

Figure 7. Diel cycles of temperatures and land-surface energy fluxes measured across the domain of one LEO landscape. Air temperatures (T_{air}) are based on all available sensors installed at three selected heights along masts above the land surface, and soil temperatures (T_{soil}) are based on all available sensors installed at three selected depths below the surface. Energy flux (E) observations include the net downward radiative flux (R_n), the upward surface sensible heat flux (H), and the downward ground heat flux into the subsurface medium (G). The latent heat flux was negligible due to extremely dry soil conditions. Solid lines represent whole-landscape means and shaded bounds indicate standard deviations associated with spatial variability across the landscape's domain.

well as seepage volumes per time enables determination of E_q. Autosamplers are installed to capture these water samples during and following all irrigation events on the landscapes. Collected rainfall and seepage samples are analyzed for major cations and anions, as well as total, organic, and inorganic carbon, in addition to pH and electrical conductivity, and a subset is analyzed for major and trace elements (Section 2.1.2).

Exchange with the atmosphere, E_a and E_r, is measured using profiles of gas sampling ports installed above and within the LEO slopes. Atmospheric air can be drawn into the benchtop infrared gas analyzer to measure CO_2 concentrations at approximately hourly intervals (24 samples per slope; see also Section 2.1.3). The gas analysis system is flexible and can also be used for analysis of CO_2 isotope ratios and mole fractions of other gases (such as methane, carbon monoxide, nitrous oxide, hydrogen, carbonyl sulfide, etc.) when connected to alternative analytical instrumentation such as continuous and discrete gas samplers (e.g., laser spectrometers and gas chromatographs, respectively). The anticipated transfer processes from the atmosphere, E_a, that affect CO_2 dynamics at LEO include ecosystem uptake by photosynthesis, as well as via silicate weathering. Relevant emission processes to the atmosphere, E_r, include a sum of autotrophic and heterotrophic respiration. In systems with well-developed atmospheric turbulence, E_a and E_r are routinely determined from changes in atmospheric concentrations using eddy covariance [50–52] and flux-gradient [45] techniques. However, application of these flux-estimation methods is challenging given frequent temperature and wind speed inversions in the LEO atmospheres, and specific methodological adaptations are required to monitor ecosystem trace gas cycling (see Section 2.2). Currently, with limited expected contributions of biological activity, estimation of the weathering flux of CO_2 can be achieved through in-slope sensors and samplers, as well as through carbonate mass balance of seepage face discharge (see the following paragraph). Microbial respiration is anticipated to increase over time, and once plants are introduced on the landscapes, root respiration, and enhanced microbial activity associated with plant organic matter inputs will increase the biological influence on soil gas concentrations. However, the organic and inorganic carbon export and conserved element export can be used as an estimate of total weathering. The increased complexity of CO_2 net exchange between soil and atmosphere following the introduction of plants will require application and further development of sophisticated flux partitioning approaches to quantify abiotic and biotic components. Currently available approaches focus on partitioning net exchange of CO_2 into the biological gross primary productivity and ecosystem respiration components based on nighttime versus daytime assumptions (e.g., [51]). Nighttime approaches assume that only respiration occurs at night; however, they do not account for CO_2 uptake through weathering reactions. Daytime approaches, in turn, poorly represent daytime respiration and neglect carbonate precipitation. Combining these approaches with use of relationships between CO_2 flux and measured photosynthetically active radiation (PAR) as well as with available measurements of trace gases homologous to CO_2, such as COS [53] and CO_2 isotopologues [50, 52], can help constrain the spatial and temporal mechanics of CO_2 flux processes at LEO and improve empirical and process-based models.

Changes in landscape storage of the elements obtained using estimates of influxes and outfluxes of the hillslopes can be verified by direct measurements of the storage in the landscapes through time. This is made possible by analyzing element concentrations in (i) the solution phase collected from 496 suction lysimeters distributed across each hillslope, (ii) the gas phase obtained using 141 soil gas samplers and 48 Vaisala [CO_2] sensors, and (iii) the solid phase extracted by coring of the soil for subsequent analysis. Solution and gas sampling are nondestructive, and the main limitations on frequency and density of sampling are cost and time of the analyses. Soil coring, in turn, is destructive and has to be used conservatively in order to avoid impacting hillslope behavior (see Section 2.1.6). The unique ability to close elemental mass balances was demonstrated for carbon by [54]. The study was conducted early in the LEO project, when no vegetation was present and most of the carbon fluxes were assumed to be controlled by abiotic processes, driven by weathering of basalt substrate. We quantified atmospheric CO_2 consumption by basalt weathering using pore gas concentration data (from Vaisala sensors) and carbon input with rainfall and export with seepage by analysis of rainfall and seepage solutions for inorganic carbon. Storage of total inorganic carbon (TIC) in

http://dx.doi.org/10.5772/intechopen.72325

the solution phase was quantified as the product of TIC concentrations obtained via the pore water samplers over time and the moisture contents measured by co-located sensors. The study demonstrated that the estimated change in storage was consistent with the difference between incoming and outgoing carbon fluxes (**Figure 8**), validating the capacity of the LEO system to close the carbon mass balance. Employing various densities of pore water sampler data for these calculations further showed that a decrease in data density by one order of magnitude (from about 350 to 35 samples per hillslope) does not significantly affect solution storage estimates for carbon at LEO.

For lithogenic elements, such as Si, Na, Ca, K, Al, and Fe, inputs and outputs to and from the atmosphere are negligible (dust deposition is minimized by the superstructure), but their mass balance is strongly impacted by weathering processes that release a fraction of these elements from rock into solutions that are exported as seepage water (i.e., "from land to rivers"). A fraction of the element mass released from rock by weathering is retained within the landscape in secondary mineral form, promoting the formation of reactive soil interfaces and transforming the pore size distribution with important feedbacks to hydrologic flows. Measuring concentrations of these elements in seepage waters enables a quantification of terrestrial-to-aquatic effluxes [54], while in-slope measurements from pore water samplers

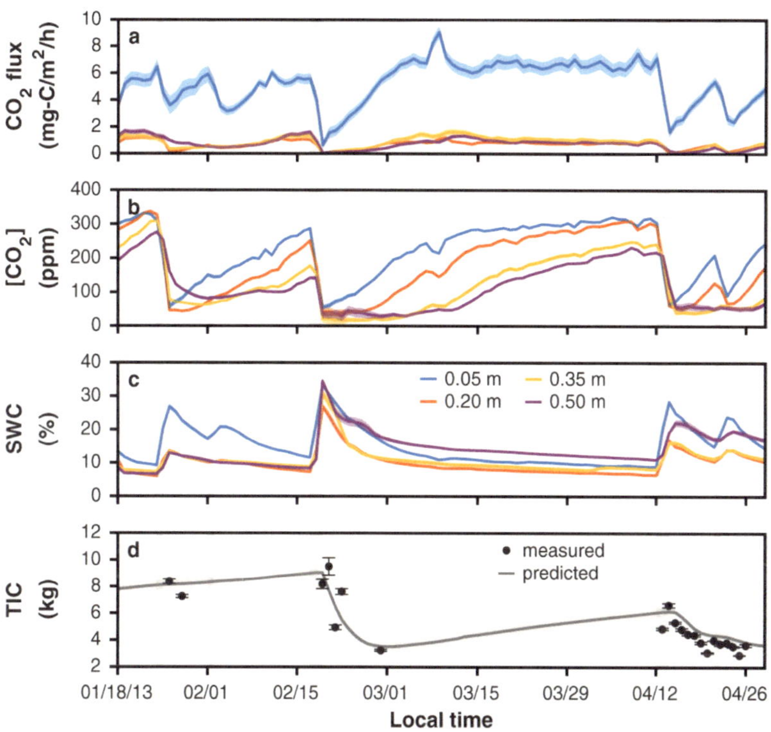

Figure 8. Time series of calculated CO_2 fluxes (a; positive values indicate fluxes from the atmosphere into the soil), gas-phase CO_2 concentrations (b), soil water content (SWC; c), and total inorganic carbon storage (TIC; d) of a LEO landscape over several rainfall events (indicated by gray vertical bars). TIC storage was predicted using incoming and outgoing fluxes (solid line) as well as integrated from measured concentrations in pore solutions (dot markers). Shaded bounds and error bars indicate 95% confidence intervals associated with each variable.

enable characterization of the spatial and temporal variability in the trajectory of mineral transformations and soil formation [55].

2.5. Monitoring of biological community composition and function

Microbial community dynamics on LEO are monitored via soil core collection (Section 2.1.6) to identify pioneering microbial species and metabolisms and to describe the response of microbial communities to environmental forcing (e.g., rainfall) and their longer term successional patterns. This approach provides information on the long-term reciprocal impact of landscape evolution on microbial community composition and function. High-throughput amplicon sequencing of 16S rRNA genes to target the bacterial and archaeal communities revealed significant differences in relative abundances for samples collected from the LEO landscapes before and after rainfall (**Figure 9**) with distinct depth-dependent community distribution profiles of the soils (**Figure 10**). We expected microbes to be distributed nonrandomly along environmental gradients in accordance to their metabolic strategies as observed in the scaled-down mini-LEO model (**Figure 11**; [56]). The phylogenetic distribution observed in the cores extracted from LEO soils was significantly different from that of the parent material [56] and revealed heterogeneous microbial community establishment and development in an otherwise nearly homogeneous soil system. Furthermore, significant differences in pre- and post-rainfall community structure suggest a dynamic system that responds to rainfall events, which may have implications for accelerating bio-weathering rates. Additional genomic and transcriptomic sequencing efforts will reveal functional diversity (gene abundance) and gene-expression profiles, respectively, in the hillslopes. An integrated understanding of microbial community diversity, gene abundance, and functional potential alongside geochemical changes, CO_2 fluxes, and hydrological flow paths can potentially reveal predictive patterns of landscape evolution.

Plant community function, composition, and organization will be monitored through a blend of direct and remote-sensing approaches. Using the personnel transport system that operates over the LEO structure, we will measure leaf-level carbon and water exchange to inform our mass balance equations and to examine interspecific variation in plant function, as it is extensively done outside model systems in critical zone research [57, 58]. Coupling these point-specific measures with multispectral and thermal imaging (Section 2.1.4) will provide spatial patterns of function across the artificial landscape, and the repeated image acquisition through time will provide insights into temporal dynamics. Hyperspectral analysis of ecosystems can yield remarkable insights into vegetation function, and "signatures" of spectra specific to plant species can also be used for mapping distribution across the landscape. The use of narrow (<5 nm) band spectrometers allows for the observation of many ecological features such as pigment composition and content, canopy water content, dry plant litter and wood, and foliar chemistry (e.g., [59] and references therein). Photosynthesis, or gross primary production (GPP), is the largest global land surface carbon flux [60, 61], but the spatially explicit approaches to quantify GPP based on meteorology-driven land surface carbon cycle models, MODIS-based remote sensing, and typical eddy flux tower measurement driven models carry large uncertainties [62, 63]. GPP is directly correlated with solar-induced chlorophyll fluorescence (ChF), because both are driven by absorbed radiation [64–67]. This correlation

http://dx.doi.org/10.5772/intechopen.72325

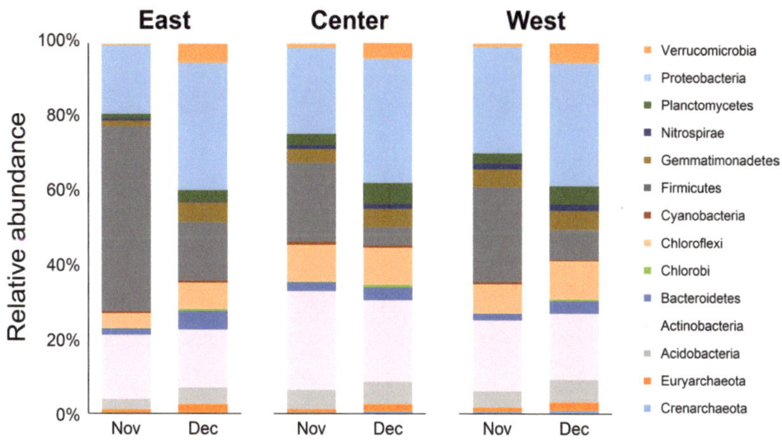

Figure 9. Relative abundance of the 14 most abundant bacterial and archaeal phyla in the three LEO landscapes (East, Center, and West) before (November) and after (December) a series of irrigation events (preliminary data).

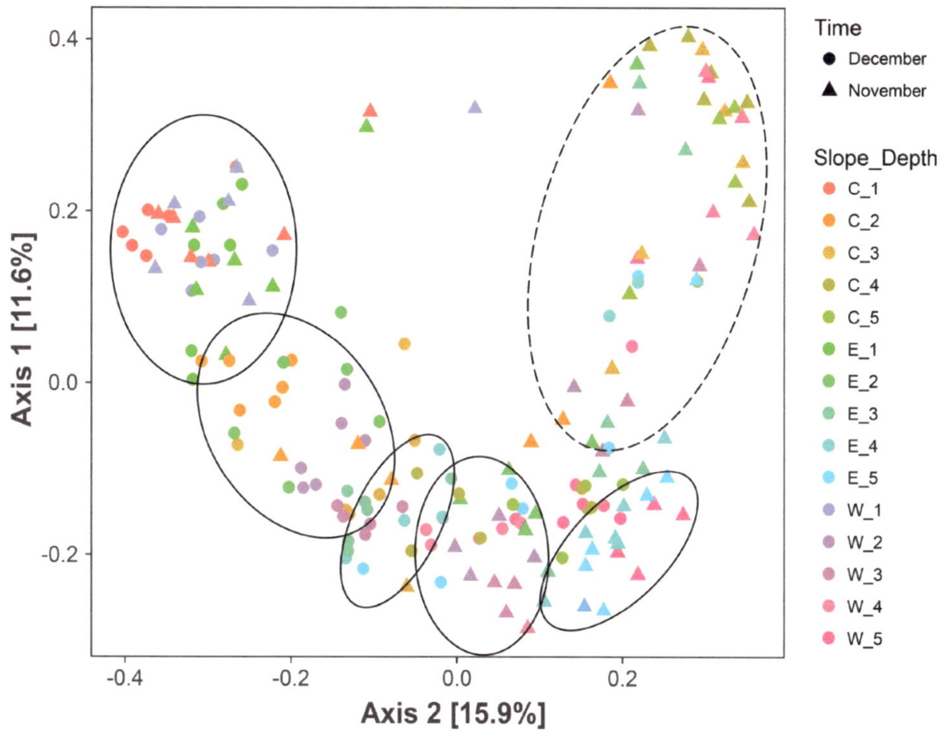

Figure 10. Bray-Curtis ordination of microbial community composition in the East (E), Center (C), and West (W) landscapes before (November) and after (December) a series of irrigation events (preliminary data). Soil depths are indicated by integers representing increments of 15 cm and starting at the surface (number 1). Depth-dependent clustering is observed for December samples and most November samples (solid-line ellipses), while November samples from intermediate depths reveal a more widely distributed pattern (dashed-line ellipse).

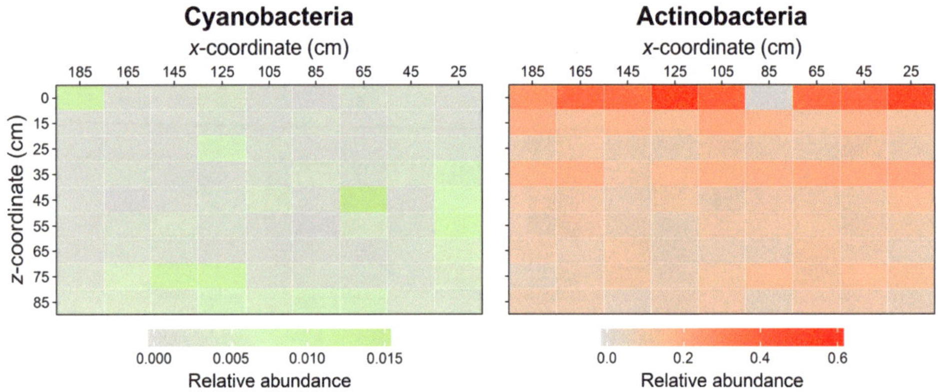

Figure 11. Relative abundance of two major phyla (*Cyanobacteria* and *Actinobacteria*) across the lateral (*x*-coordinate; long axis) and vertical (*z*-coordinate) dimensions of the miniLEO small-scale replicate model. Rectangles indicate unique voxels of volume 200 cm^3 (increments of 20 cm in the lateral and 10 cm in the vertical direction). Samples were collected from each voxel, followed by DNA extraction and high-throughput sequence analysis.

is the result of a reduction in ChF and photosynthesis yield following increases in heat dissipation under high light conditions. Chlorophyll fluorescence is the re-emission of absorbed photosynthetically active radiation (400–700 nm) by the plants at higher wavelengths in the visible red and near infrared (660–800 nm). We will use spectral features around the higher wavelength because the lower band is affected by the re-absorption of chlorophyll pigments, while the higher one is minimally affected by chlorophyll re-absorption effects [68].

2.6. Monitoring of subsurface structural development and pedogenesis

The dense sensor and sampler arrays at LEO offer the capabilities (i) to characterize in great detail the spatial and temporal variability of solution chemistry for a small ZOB (via solution sampling and analysis; see Sections 2.1.2 and 2.4) and (ii) to monitor in high resolution the subsurface structural development (via electrical resistivity tomography surveys and soil coring; see Sections 2.1.5 and 2.1.6) critical during incipient stages of landscape evolution. Together, these capabilities enable the establishment of cause-and-effect relations in soil formation. Known inputs of rainwater, which act as both solvent and transport vector, drive the dissolution of primary mineral surfaces and increase in pore water saturation with respect to secondary mineral phases, including carbonates. Precipitating solids can passivate the surfaces of primary minerals against further dissolution [69]. They also contribute to stabilization of organic carbon against leaching, to mineralization by interacting with newly formed minerals [70–72], and to soil retention of plant-available water and surface-exchangeable nutrients in plant-available form.

Dissolution of primary minerals and precipitation of clay-sized secondary minerals lead to shifts in particle size distribution of the soils toward finer materials. This is predicted to be correlated to flow patterns across the landscapes [8]. In addition, precipitation of poorly crystalline silicates, phyllosilicate clays, and sesquioxides as well as additions of carbon, initially from microbial activity and later from plant root exudation, promote aggregation of the primary particles resulting in changes of the hillslopes' physical structure. The application

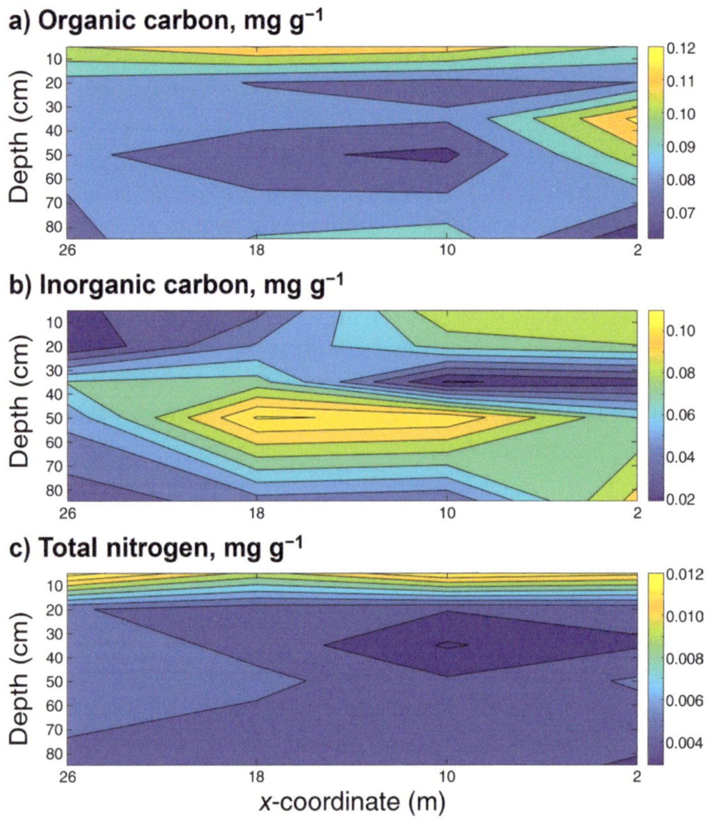

Figure 12. Distribution of organic carbon (a), inorganic carbon (b), and total nitrogen (c) over a cross section of a LEO hillslope 3 years after inception (preliminary data). The cross section spans the depth (z-coordinate) and length (x-coordinate) of the hillslope through its central plane (i.e., y-coordinate equal to zero).

3. Research on integrated hillslope coevolution to improve predictions of landscape-scale change

The LEO landscapes and their extensive instrumentation and control capabilities allow us to track every step along the evolutionary trajectory of a ZOB-scale system, from purely abiotic substrate to living, breathing ecosystems. Emulating their real-world archetypes, the convergent topography of these landscapes promotes spatially variable substrate and resource availability, which is anticipated to eventually facilitate biological diversity and influence how the landscapes filter precipitation and sequester carbon from the atmosphere. Refining our understanding of and our ability to predict how these and other significant ecosystem services are affected by landscape evolution, climatic variability, and long-term environmental change is the central goal of Earth scientists working in the LEO project. This section discusses current foci of research at LEO that target this goal by advancing understanding of how hydrological and geochemical (Section 3.2), microbiological (Section 3.3), and plant-ecological (Section 3.4) processes interact to drive the coevolution of incipient hillslopes and their mass and energy cycling. Concepts underlying these research foci and early results from initial experiments are presented. The process research at LEO is complemented by the development of hillslope-integrated parametrizations and distributed coupled-process modeling approaches that are hoped to ultimately allow improved prediction of real-world systems' behavior in a changing environment. These modeling approaches are described first (Section 3.1) and with emphasis on hydrologic predictions.

3.1. Changing paradigms for hydrologic prediction at the hillslope scale

In recent years, there has been a paradigm shift in our understanding of flow and transport at watershed scales and in our approaches to prediction. The complexity and heterogeneity of water movement within individual landscape units have been recognized in hillslopes [73, 74], riparian areas [75, 76], and within streams and their hyporheic zone [77]. This has led to calls for new predictive approaches that go beyond the traditional continuum models (i.e., Richards and Saint-Venant equations for flow and convection-dispersion/diffusion equations for transport) [40, 78–80], as these generally rely on calibrated "effective" property values to replace the spatially distributed properties of the landscape—those are essentially unknowable at catchment scales using current technology.

New approaches have sought ways to represent flow and transport directly at the scales of interest, with the expectation that the new equations may differ in form, not just in the parameters [79], from the continuum-scale equations. The concept of a representative elementary watershed, or REW [78, 81–83], provides a framework for representing flow through individual landscape elements and in a river network based on a rigorous time-space averaging of the conservation laws for mass, energy, momentum, and entropy. However, these equations are not complete. They require specification of "closure relations" that specify the boundary fluxes exchanged between these landscape elements in terms of their states and are parameterized by measurable properties of the landscape. These closure relations must represent the aggregate effect of the unresolved sub-REW heterogeneities and flow complexity without resolving them explicitly, and they have been termed the "Holy Grail" of scientific hydrology [78].

An overarching objective of the LEO project is to develop closure relations for hillslope-scale hydrologic flux and transport [7]. These closure relations are parameterizations of the fluxes that cross boundaries between hydrologically relevant units of the landscape [78, 81]—an elementary example is a storage-discharge relation [84, 85]. Taken broadly, such parameterizations are a component of all hydrologic models, but LEO is a useful experimental tool for developing closure relations at the scale of hillslopes [7]. At LEO, it is possible to observe boundary fluxes—and how they emerge from the distribution of internal state variables—with a precision not possible in real landscapes and at a scale not possible in bench-top experiments. Through experimentation and iterative modeling using both lumped and highly resolved models, efforts at LEO are driving toward a suite of hillslope closure relationships that (ideally) can be parameterized on the basis of the observable physical structure of the landscape.

Work at LEO has focused on developing hillslope-scale closure relationships for discharge and for transport both by building on and testing existing theory, as well as by developing new approaches. Hillslope closure relations predicting discharge have been developed from the principles of hydraulic groundwater theory [86–88] and are being compared to both the results of physical experiments and numerical models. Three-dimensional Richards equation-based models have been implemented to simulate flow and transport dynamics in the LEO hillslopes ([41, 49]; see also Section 2.1.7) and calibrated to reproduce the flow data with the physically justifiable parameter sets. **Figure 13** illustrates the observed storage-discharge relationship from one of the LEO hillslopes and a modeled relationship. The observed relationship shows a large degree of hysteresis, and the simulated relationship captures most of the features of this relationship. This type of hysteresis can be captured by the theoretical frameworks of Troch [88] and others, but not by the typical one-to-one storage-discharge relationships used in hydrologic models to simulate baseflow.

Projects at LEO have also driven the development of new parameterizations of hillslope-based transport that build on the concept of rank-storage selection functions (rSAS; [89]). This is a highly promising approach for a new generation of transport models [90]. rSAS theory depends on parameterization of probability distributions that capture the emergent effect of finer-scale processes determining transport through the hillslope. These functions extend the idea of a storage-discharge relationship: the function specifies not the overall discharge as a function of storage but rather the way the age distribution of discharge is selected from the age distribution of storage.

The PERTH (PERiodic Tracer Hierarchy) was developed to allow rSAS functions to be directly determined from the results of physical tracer experiments [14]. The key requirement is that the flow varies in a periodic way, so that the progressive breakthrough of a single tracer injection reveals information about the contribution to discharge of multiple ages at each time in the cycle. The LEO hillslope hydrodynamics can be controlled to allow precise observations of the flow and transport dynamics, and a large-scale PERTH-type experiment was conducted between 1 December 2016 and 28 December 2016. All three slopes were irrigated in

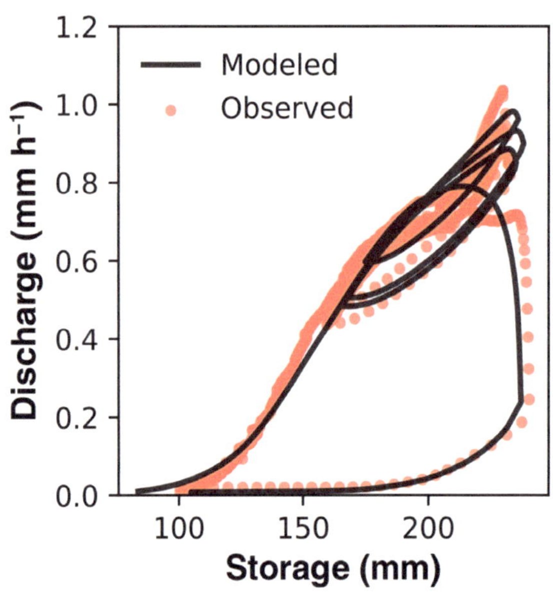

Figure 13. Observed and model storage-discharge relationship of a LEO landscape (for the period of 1–30 June 2015).

an identical fashion every 3.5 days over the course of 4 weeks—a total of about 580 mm irrigations. The results are revealing much about the nature of hillslope rSAS functions.

These LEO experimental results can be extended to a wider class of idealized hillslopes through fine resolution modeling of system-scale flow and transport dynamics in "virtual hillslopes". The effect of variations in hillslope morphology, soil properties, and climate forcing on flow and transport closures can be examined using a three-dimensional Richards equation-based model and particle tracking algorithm validated against the LEO dataset. Moreover, the sensitivity of the parameterizations to observable physical properties (and their associated uncertainties) can be deduced.

However, LEO was inspired in part by efforts to go beyond the typical approach in hydrology of treating hydrologic properties as fixed features of the landscape, and instead ask deeper questions about why a landscape has the properties it does and how it came to be that way [91]. Physical, chemical, and biological data are also collected at LEO to connect the hydrologic behavior to interacting critical zone processes and ultimately to the coevolution of the system [3]. We aim to understand how the landscape internal structure evolves over time, feeding back on the flow and transport processes and modifying the emergent behavior that is the basis of flow and transport closure relations.

To develop improved predictive ability of the hydrological as well as biogeochemical and ecological responses in evolving landscapes, a second focus of the modeling at LEO is therefore the development of a coupled-process modeling framework. This modeling framework, referred to as TIMS (Section 2.1.7), aims to not only parameterize water flow and transport as functions of static landscape properties. Instead, it focuses specifically on coupling hydrological, microbial, geochemical, geomorphological, and ecological processes and considers the landscape properties themselves as dynamic. By attempting to predict dynamic landscape parameters based on a fundamental understanding of how they are created in the coevolution of a landscape, we may stand a chance to overcome the need for assumptions on unmeasurable "effective" model parameters and the inherent inability of most current modeling tools to represent integral adaptation to change (e.g., in climate).

At present, however, TIMS and other state-of-the-art Earth system models are still inadequate to represent many key Earth system processes that control long-term land-atmosphere exchanges of energy, water, and carbon. This is due to a lack of predictive understanding of the interactions between the relevant physical, chemical, and biological processes. Therefore, the modeling system will be continuously extended, synthesizing results from the physical experimentation at LEO to develop representations of critical process couplings that are currently not adequately represented. These include, for example, schemes to represent the evolution of landscape heterogeneity associated with transport and deposition of particles (colloids and sediments) and biogeochemical weathering, as well as associated feedbacks with the hydraulic properties of the LEO soils. Other areas of active model development include parameterizing leaf and root dynamics adaptive to environmental changes (e.g., high temperature and drought) and soil carbon dynamics as affected by geochemical reactions, microbial enzymes, climate, and water flow [33].

The observational and modeling results of experiments conducted under the present conditions will eventually be compared to results of identical experiments performed after the hillslopes have been altered either through endogenous changes or through the introduction of new factors, such as the establishment of plants. As the LEO hillslopes evolve over time, we will iterate the physical and the numerical experiments to test hypotheses about the feedbacks between flow and transport dynamics and hillslope evolution. Lessons learned from failure and success in reproducing observations in the physical landscapes (see, e.g., [41]) will then be used to refine the mathematical representations and reduce uncertainties in model structures and parameters. The LEO landscapes and diverse modeling approaches are thereby anticipated to help fill the gap between plot-scale studies and larger scale (hillslope to global) model developments by constructing relationships between varying fluxes and states at the ZOB-scale, both through direct inference of closure relations and scaling of coupled-process modeling schemes. These developments may finally serve to project impacts of climate change on water resources as well as ecological processes and landscape evolution in varied environmental contexts.

3.2. Linking hydrological and geochemical processes in evolving landscapes

The nature of hydrological and geochemical interactions is a primary determinant of hillslope structure formation, available water quantity, and water quality along the entire evolutionary trajectory of a landscape—from pristine abiotic substrate subjected to a first rain pulse, when microbial (Section 3.3) and vascular plant (Section 3.4) life only begin to establish, to a matured landscape adjusting to variable climatic forcing. Examining and predicting the time evolution of subsurface structure through biogeochemical processes and its effect on hydrological partitioning and water residence time are therefore a key focus of the research at LEO. The LEO experiment provides unprecedented capability to examine the complex interactions that are integral to the hillslope evolution and soil formation processes because of the high density of hydrological and geochemical measurements in space and time, control of some of the drivers of weathering such as rainfall or temperature, and the ability to perform coupled hydrological-biogeochemical modeling.

In any developing landscape, including the LEO landscapes, the spatial structure of flow pathways along hillslopes determines the rate, extent, and distribution of geochemical reactions (and biological colonization) that drive weathering, the transport and precipitation of solutes and sediments, and the evolution of soil structure. The resulting structure and process evolution, in turn, induces spatiotemporal variability of hydrological states and flow pathways. Weathering reactions are strongly influenced by the time water spends along the subsurface flow paths [92, 93] at any stage of the landscape development. Dissolution of primary silicates is kinetically limited, and increasing residence time, and thus the duration of contact time of water with rocks, therefore potentially increases the concentration of lithogenic elements in the soil solution. The resulting relative saturation of soil solution (as measured by the saturation index) with respect to soil minerals will affect both primary mineral dissolution and secondary mineral precipitation [94]. The farther a solution is from equilibrium, the higher the rates of both dissolution and precipitation processes. In addition to water residence times, rock dissolution rates are also influenced by formation of secondary minerals [95]. Plants and microorganisms can also strongly influence weathering rates through production of organic acids and other complexing agents [72, 96], through respiration and uptake of water and dissolving nutrients [97], and through enzymatic promotion of bicarbonate formation from CO_2 [98, 99].

Since the distribution of microorganisms and vegetation on the slopes in space and time is driven by water and nutrient availability (Sections 3.3 and 3.4), biota will further complicate the already complex relationships between water flow and weathering in evolving hillslopes.

Precipitation of secondary minerals during incongruent weathering decreases the particle size of the soil, which affects its pore size distribution and hydraulic conductivity, and consequently the water transit times [100, 101]. Mineral precipitation can also feedback to further weathering [102, 103], nutrient and carbon retention, and microbial and plant distribution. Coupled geochemical-hydrological modeling (combined with pedotransfer functions) was performed for the LEO hillslopes to estimate mineral transformations and changes in soil hydraulic properties over a 10-year period with rainfall amount derived from several real-world locations in the southwestern United States [8]. The predictions suggested a significant increase in the fraction of secondary minerals and, as a result, a decrease in hydraulic conductivity over time. Predicted changes were highly variable in space, closely mirroring flow and saturation patterns on the hillslopes. It was further observed that solution-phase concentrations of lithogenic elements calculated by the models could provide early indication of the soil formation processes, even before changes in the solid phase would have become readily measurable. Since then, the physical experiments performed at LEO [54, 55] confirm both development of heterogeneity in the solution composition as a function of flow patterns and spatially resolved precipitation of secondary minerals as indicated by saturation indices for a number of minerals that would influence soil structure and flow patterns. For example, it can be seen for calcite (**Figure 14**) that there is pronounced spatial heterogeneity in the measurements linked to soil water content, and that there is a temporal progression in the saturation indices as the hillslope gets drier. Results from LEO soil cores further support spatially resolved precipitation of inorganic carbon through direct measurements (Section 2.5).

The geochemical evolution of a hillslope is also strongly linked to runoff generation. The fraction of incoming rainfall that eventually leaves the landscapes as seepage outflow (or overland flow, if any) affects the export of inorganic and organic compounds from the landscapes. This export process presents an integral part of soil formation. When soils age, their composition shifts toward elements such as iron and aluminum that form poorly soluble oxides and hydroxides, as these are retained. More soluble elements in turn are lost, such as silicon, which presents an originally predominant element on the landscape. Seepage generation additionally has significant implications for the carbon balance in the environment. The export of inorganic carbon that was previously captured in the landscape as a result of weathering (predominantly as bicarbonate and carbonate), and its transport and further storage in the oceans, serve as important mechanisms of carbon sequestration from the atmosphere. Weathering of basaltic rocks, such as those comprising the LEO landscapes, occurs rapidly [104], and our experiments showed that as much as 5 kg of carbon was lost from the hillslopes in a single rain event (see **Figure 8**; [54]). Finally, feedbacks between water flow, mineral weathering and soil formation, and biological activity can have important implications for water quality. These feedbacks control the off-site transport not only of lithogenic and biogenic compounds but also of potential anthropogenic and possibly hazardous compounds into streams and other downstream landscape elements.

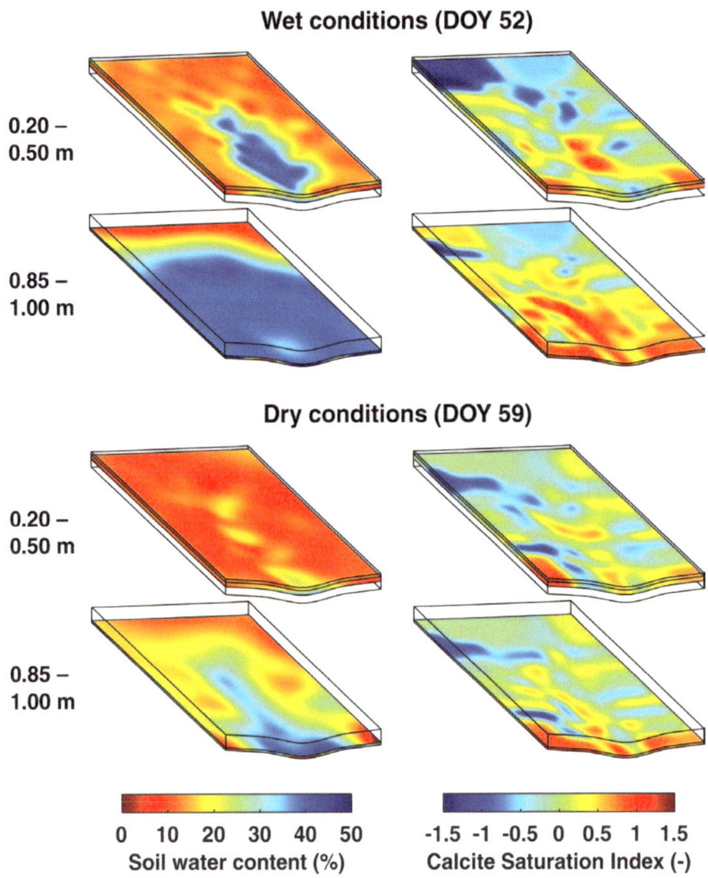

Figure 14. Three-dimensional interpolations of volumetric soil water content (SWC; left) and calcite saturation index (right) projected onto a LEO hillslope model at two depth intervals (0.20–0.50 m and 85–1.00 m) and two sampling dates (DOY 52 and DOY 59). For SWC: warm colors represent areas of relatively dry conditions while cool colors represent wet conditions. For calcite saturation index: warm colors represent areas of supersaturation while cool colors represent areas of reduced supersaturation or undersaturation. Note the unique scales for SWC (%) and saturation index (i.e., log Q/K).

3.3. Linking hydrological and microbial processes in evolving landscapes

Microorganisms typically represent the life pioneering initially abiotic, developing hillslopes. Microbial populations affect the trajectory and speed of evolution of physicochemical landscape properties and functions and may be critical in facilitating the establishment of vascular plant life (Section 3.4). Yet, we lack a clear picture of the manifold feedbacks between evolving patterns of microbial (and plant) ecology and physicochemical system dynamics across scales. In particular, our understanding of early-stage oligotrophic landscape elements is limited, as we rarely have the opportunity for their study in nature (but see related work by, e.g., [105, 106]) and typically lack the multifaceted observational density required. Microbial dynamics across incipient landscapes, such as the LEO hillslopes, are tightly linked to hydrological and geochemical signatures. Hydrological flux processes themselves (e.g., infiltration, lateral redistribution of vadose zone water and groundwater, and evapotranspiration) and resulting soil moisture spatial and temporal dynamics are primarily controlled by structural properties of the subsurface and the driving forces of surface exchange. Hydrological dynamics, in turn, impact the time spent by a parcel of water in contact with microbial cells, the dissolution and flow of nutrients, and the movement of microbes in the system [107]. Additionally, hydro-geochemical processes, such as weathering and dissolution of primary minerals, dissolution of reactive phases, and reprecipitation of weathered elements, influence nutrient availability and may facilitate microbial colonization [108]. The concurrent microbial signatures of growth and function impact pore structure, weathering rate, and labile carbon deposition and thereby establish feedbacks of coupled hydro-biogeochemical processes influencing soil formation. Microbial life is therefore both a follower and facilitator of water flow paths in incipient landscape systems.

The heavily instrumented LEO landscapes, with the dense network of sensors and samplers and diverse modeling approaches, provide the opportunity to develop a mechanistic understanding of integrated hydrological and microbial processes occurring in the hillslopes as a result of their mutual interactions and to identify signatures that can be used to quantitatively characterize hydrology-microbiology relationships. For instance, it is important to understand how microbial cells and the biomolecules produced by the microbes are transported with the water. The time spent by a microbial community in a unique microenvironment impacts species-species interactions and accessibility of nutrients. Water movement also impacts the turnover of microbial community assemblies and has been shown to influence microbial ecological assembly processes [109]. However, many open questions remain regarding the nature of these interactions between hydrology and microbiology. Using **Figure 11** as an example, an unexpected trend was observed in the relative microbial abundances in the miniLEO small-scale model. While *Actinobacteria* were predominantly abundant at the surface, *Cyanobacteria* were found at greater depths in the model system. *Cyanobacteria* are photoautotrophs, i.e., they need sunlight to survive, and yet they occurred preferentially in deep and dark layers of the soil (highest abundance at 40 cm below the surface). In contrast, *Actinobacteria* were consistent in their localization toward the surface despite the potential for transport with water from the surface to deeper layers. Hypothesizing that water flow caused *Cyanobacteria* to travel deeper fails to explain why *Actinobacteria* maintained a constant relative abundance at the surface. So, are microbial species transported passively with the water depending on specific parameters such as cell size, or do they differ in their active movement (i.e., motility) and surface attachment (e.g., in biofilms)? Or are the relative abundance patterns simply imposed by the local environment and maintained by constant cell turnover rates, despite water flow? Tackling these questions requires combining hydrological tools with microbiological methods. For example, it will be relevant to incorporate microbial processes into reactive transport models and to evaluate average water volumes, transit times, and upstream flow path lengths associated with the microbial abundance patterns (e.g., for each voxel in the example **Figure 11**). Such distributed indices may facilitate the interfacing of hydrology and microbiology to study subsurface ecosystems.

3.4. Linking hydrological and plant ecological processes in evolving landscapes

The spatiotemporal interplay between water and vegetation dynamics, and their feedbacks to physicochemical landscape (Section 3.2) and microbial community (Section 3.3) organization, profoundly impacts the evolution of a hillslope and its cycling of mass and energy. Hydrological processes exert primary controls on the establishment, distribution, structure, and function of ecosystems, while biotic processes directly (e.g., through transpiration) and indirectly (e.g., through alteration of soil properties) affect the cycling of water through the landscape [110, 111]. A substantial body of ecohydrological research has examined plant-water interactions with respect to one-dimensional (vertical) water and nutrient fluxes at the patch scale [112], however, without incorporating lateral redistribution of water and nutrients imposed by hillslope morphology. Similarly, biogeography has been mindful of spatial patterns of drivers of vegetation distribution but has paid little attention to the inherent coevolution of the physical and biological realms as a process that drives the form and function of the biological and physical landscape. Due to their large scale and imposed gradients in topography and environmental conditions, as well as their climate control and monitoring capabilities, the LEO hillslopes provide the unique opportunity to tackle the gaps in our understanding of how physical-biological interactions drive landscape evolution in space and time. Working at the hillslope scale forces integration between one- and two-dimensional conceptualizations of ecohydrological processes and provides the topological structures that connect patches on the landscape by gravitational fluxes organized by hillslope morphology. From the patch or pedon perspective, the hillslope provides nonlocal controls on water and nutrient fluxes, while local controls at the patch scale influence downslope patches in the ecohydrological system.

The basic elements that define hillslope morphology, such as shape, gradient, aspect, and slope complexity (i.e., nonuniformity), affect water and energy availability (e.g., [3]). Hillslope shape expresses the convergence or divergence of surface flow paths in the planform (across slope) and profile (normal to slope) directions and therefore relates directly to soil moisture redistribution. Hillslope shape and gradient both affect runoff processes and erosion rates. Hillslope aspect directly influences irradiance and hence energy availability for evapotranspiration. Complex interactions at the hillslope scale between topography, soil development, runoff processes, and vegetation create self-reinforcing positive feedbacks in ecohydrological processes that must be considered to develop a comprehensive understanding of ecohydrological patterns and processes.

The effects of vegetation on physical processes will depend on the structure of the community [25]. One might hypothesize that shallow-rooted plants would have a much less significant impact on the ecohydrology of a hillslope than a deep-rooted shrub because they lack a physical integration with as much of the soil profile. But are ecohydrological process more influenced by percent cover of the soil (presumably higher in a lower-stature forb or grassland system), higher photosynthetic function that derives organic acids that can drive soil processes, or simply the water use efficiency of the vegetation, regardless of type? Aboveground-belowground linkages are so inherently complex that when we add discussion of connections among soil pedons in space or consider the various members of a vegetative community, how little we know about the ecohydrology at the hillslope scale becomes glaringly apparent.

http://dx.doi.org/10.5772/intechopen.72325

LEO is ideally suited to investigate the coevolution of water flow paths and plant ecological processes, soil properties, and microbial dynamics. After an initial period of bare soil conditions, seeds will be dispersed on the hillslope surfaces and vascular plant growth will strongly modify the surface and subsurface properties of the LEO landscapes. Differences in available water and nutrients in the subsurface, resulting from the coevolution of hydrological and biogeochemical processes prior to vegetation, will create niches that different plant species can occupy. These adaptation and selection mechanisms will result in whole ecosystem dynamics that are different from a uniform distribution of plant species across the slopes. The aboveground and belowground instrumentation at LEO will allow the detailed monitoring of water, carbon, and energy fluxes at the land-atmosphere interface and throughout the hillslopes. Understanding these connections between the physical environment and germination and survival rates of various species will yield important information on characteristics of plant establishment. Following these patterns through time will allow for insights into those key processes of coevolution of plant-soil dynamics in the profile and planform.

4. Conclusions and outlook

This chapter has presented the research infrastructure, facilities, and initial experimental results of the Landscape Evolution Observatory (LEO) project at the Biosphere 2. LEO is a carefully designed and massively outfitted macrocosm experiment of an unprecedented scale and ambition of scope. Each of the three model landscapes emulates a pristine, sloping zero-order basin consisting of more than 500 metric tons of homogenous basaltic tephra housed within climate-controlled bays. The infrastructure operates dense arrays of more than 1900 sensors and samplers per landscape that are complemented by state-of-the-art research support systems, including an isotope and trace gas analysis network, electrical resistivity tomography instrumentation, high-resolution remote imaging systems, and advanced analytical capabilities for analyzing liquids and solids. The combined capacity of these structures allows tracking states, fluxes, and pathways of water, energy, and critical elements such as carbon at sub-meter to landscape scales, arguably representing the most successful attempt to closing hydro-biogeochemical budgets for a hillslope-size system to date. The system importantly allows key developmental processes to be rigorously tracked, including changes in subsurface and the development of microbial and ultimately vascular plant communities.

LEO is fully operational and has recently (October 2016) entered its 10-year long institutional experiment. During this experiment, variable climate forcing (mainly rainfall treatments) will drive the initially simple, abiotic model landscapes into life-sustaining ecosystems. After a first phase of bare soil surface conditions that is scheduled to last approximately 2 to 3 years, the landscapes will be colonized by vascular plants able to germinate and grow on the poorly developed LEO soil. While the landscapes evolve to increasingly complex states, Earth scientists will have the opportunity to iteratively build knowledge on the interactions and feedbacks between hydrological, geochemical, geomorphological, microbial, and ecological processes that control landscape form and function, and to formalize this knowledge into distributed coupled-process models and closure relations at the hillslope-scale. The threefold replication of initial landscape conditions and climate treatment will thereby allow to develop and rigorously test (i.e., to accept or reject) laws of fundamental natural processes (e.g., flow and transport) at space-time scales relevant for prediction. In later phases, varying experimental treatments across the replicate slopes (e.g., different rainfall distribution, temperature or CO_2 levels) may drive divergent landscape evolutionary trajectories, which allow further evaluation and refinement of the knowledge and predictive capabilities gained.

The LEO infrastructure is designed as a community resource with open data availability and seeks to foster broad interdisciplinary collaboration and science planning. During the next 10 years, scientists from across the world will have the opportunity to propose smaller research projects that can be implemented without loss of objectives of the institutional experiment. For instance, researchers who would like to study certain rainfall-runoff dynamics can propose a sequence of rain events, or those commanding specific measurement or analysis capabilities are welcome to integrate those into existing efforts. Similarly, the reader is encouraged to contact the authors to share their ideas about research opportunities with respect to the planned evolutionary forcing of the landscapes. For example, the composition of the seed pool for the upcoming vascular plant colonization is still under debate. By rapidly iterating dense experimental measurement with community-based planning, data analysis, and model development, we envision that our understanding and ability to predict the coevolution of hydrological, biogeochemical, and ecological processes and their interactions under variable climate can be significantly improved. This vision will be tested when we ultimately extrapolate our understanding of abiotic-biotic system coevolution into the complex reality of natural environments to meet the challenge of predicting landscape-scale response to global change.

Acknowledgements

The authors gratefully acknowledge support from the Philecology Foundation of Fort Worth Texas, the National Science Foundation (NSF; NSF-funded project 1344552, NSF-funded Hydrologic Synthesis Project: Water cycle dynamics in a changing environment: advancing hydrologic science through synthesis, NSF grant EAR-0636043, NSF grant EAR-1340912, NSF grant EAR-1417097), and the Department of Energy (Joint Genome Institute small scale Community Science Program grant 502880). Additional funding support was provided by the Office of the Vice President of Research at the University of Arizona and by the Technology and Research Initiative Fund (TRIF) Water, Environmental, and Energy Solutions (WEES) initiative at the University of Arizona (Shared Equipment Enhancement Funds). PT is grateful for financial support from the Agnese Nelms Haury Program.

http://dx.doi.org/10.5772/intechopen.72325

Author details

Till H. M. Volkmann[1*], Aditi Sengupta[1], Luke A. Pangle[2], Katerina Dontsova[1], Greg A. Barron-Gafford[3], Ciaran J. Harman[4], Guo-Yue Niu[5], Laura K. Meredith[6], Nate Abramson[1], Antonio A. Meira Neto[5], Yadi Wang[7], John R. Adams[1], David D. Breshears[6,8], Aaron Bugaj[1], Jon Chorover[6,7], Alejandro Cueva[1], Stephen B. DeLong[9], Matej Durcik[5], Ty P. A. Ferre[5], Edward A. Hunt[1], Travis E. Huxman[10], Minseok Kim[4], Raina M. Maier[7], Russell K. Monson[6], Jon D. Pelletier[11], Michael Pohlmann[7], Craig Rasmussen[7], Joaquin Ruiz[1,11], Scott R. Saleska[8], Marcel G. Schaap[7], Michael Sibayan[1], Markus Tuller[7], Joost L. M. van Haren[1], Xubin Zeng[5] and Peter A. Troch[1,5]

*Address all correspondence to: tillv@email.arizona.edu

1 University of Arizona, Biosphere 2, Tucson, AZ, USA

2 Georgia State University, Department of Geosciences, Atlanta, GA, USA

3 University of Arizona, School of Geography and Development, Tucson, AZ, USA

4 Johns Hopkins University, Department of Geography and Environmental Engineering, Baltimore, MD, USA

5 University of Arizona, Department of Hydrology and Atmospheric Sciences, Tucson, AZ, USA

6 University of Arizona, School of Natural Resources and the Environment, Tucson, AZ, USA

7 University of Arizona, Department of Soil, Water and Environmental Science, Tucson, AZ, USA

8 University of Arizona, Department of Ecology and Evolutionary Biology, Tucson, AZ, USA

9 United States Geological Survey, Menlo Park, CA, USA

10 University of California at Irvine, Center for Environmental Biology, Irvine, CA, USA

11 University of Arizona, Department of Geosciences, Tucson, AZ, USA

References

[1] Troch PA, Carrillo GA, Heidbüchel I, Rajagopal S, Switanek M, Volkmann THM, et al. Dealing with landscape heterogeneity in watershed hydrology: A review of recent progress toward new hydrological theory. Geography Compass. 2009;**3**(1):375-392

[2] Troch PA, Lahmers T, Meira A, Mukherjee R, Pedersen JW, Roy T, et al. Catchment coevolution: A useful framework for improving predictions of hydrological change? Water Resources Research. 2015;**51**(7):4903-4922

[3] Pelletier JD, Barron-Gafford GA, Breshears DD, Brooks PD, Chorover J, Durcik M, et al. Coevolution of nonlinear trends in vegetation, soils, and topography with elevation and slope aspect: A case study in the sky islands of southern Arizona. Journal of Geophysical Research - Earth Surface. 2013;**118**(2):741-758

[4] Brantley S, White TS, White AF, Sparks D, Richter D, Pregitzer K, et al. Frontiers in Exploration of the Critical Zone: Report of a NSF-Sponsored Workshop. VA, USA: Arlington; 2006:**30**

[5] Chorover J, Troch PA, Rasmussen C, Brooks PD, Pelletier JD, Breshears DD, et al. How water, carbon, and energy drive critical zone evolution: The Jemez–Santa Catalina critical zone observatory. Vadose Zone Journal. 2011;**10**(3):884-899

[6] Ivanov VY, Fatichi S, Jenerette GD, Espeleta JF, Troch PA, Huxman TE. Hysteresis of soil moisture spatial heterogeneity and the "homogenizing" effect of vegetation. Water Resources Research. 2010;**46**(9):W09521

[7] Hopp L, Harman C, Desilets SLE, Graham CB, McDonnell JJ, Troch PA. Hillslope hydrology under glass: Confronting fundamental questions of soil-water-biota co-evolution at Biosphere 2. Hydrology and Earth System Sciences. 2009;**13**(11):2105-2118

[8] Dontsova K, Steefel CI, Desilets S, Thompson A, Chorover J. Solid phase evolution in the Biosphere 2 hillslope experiment as predicted by modeling of hydrologic and geochemical fluxes. Hydrology and Earth System Sciences. 2009;**13**(12):2273-2286

[9] White T, Brantley S, Banwart S, Chorover J, Dietrich W, Derry L, et al. The role of critical zone observatories in critical zone science. In: Giardino R, Hauser C, editors. Principles and Dynamics of the Critical Zone. Amsterdam, Netherlands: Elsevier; 2015. pp. 15-78

[10] Pangle LA, DeLong SB, Abramson N, Adams J, Barron-Gafford GA, Breshears DD, et al. The Landscape Evolution Observatory: A large-scale controllable infrastructure to study coupled earth-surface processes. Geomorphology. 2015;**244**:190-203

[11] Dietrich WE, Reneau SL, Wilson CJ. Overview: Zero-order basins and problems of drainage density, sediment transport and hillslope morphology. International Association of Scientific Hydrology. 1987;**165**:27-37

[12] Finn M. The mangrove mesocosm of Biosphere 2: Design, establishment and preliminary results. Ecological Engineering. 1996;**6**(1):21-56

[13] Marino BDV, Mahato TR, Druitt JW, Leigh L, Lin G, Russell RM, et al. The agricultural biome of Biosphere 2: Structure, composition and function. Ecological Engineering. 1999;**13**(1):199-234

[14] Kim M, Pangle LA, Cardoso C, Lora M, Volkmann THM, Wang Y, et al. Transit time distributions and StorAge selection functions in a sloping soil lysimeter with time-varying flow paths: Direct observation of internal and external transport variability. Water Resources Research. 2016;**52**(9):7105-7129

[15] Pangle LA, Kim M, Cardoso C, Lora M, Meira Neto AA, Volkmann THM, et al. The mechanistic basis for storage-dependent age distributions of water discharged from an experimental hillslope. Water Resources Research. 2017;**53**(4):2733-2754

[16] Los Gatos Research Inc. Isotopic Water Analyzer (δ^2H, $\delta^{17}O$, $\delta^{18}O$) - Enhanced Performance. 2015. Available from: www.lgrinc.com/documents/LGR_IWA-45EP.pdf

[17] McManus JB, Nelson DD, Zahniser MS. Design and performance of a dual-laser instrument for multiple isotopologues of carbon dioxide and water. Optics Express. 2015;**23**(5):6569-6586

[18] Nelson DD, McManus JB, Herndon SC, Zahniser MS, Tuzson B, Emmenegger L. New method for isotopic ratio measurements of atmospheric carbon dioxide using a 4.3 μm pulsed quantum cascade laser. Applied Physics B. 2008;**90**(2):301-309

[19] Tuzson B, Mohn J, Zeeman MJ, Werner RA, Eugster W, Zahniser MS, et al. High precision and continuous field measurements of $\delta^{13}C$ and $\delta^{18}O$ in carbon dioxide with a cryogen-free QCLAS. Applied Physics B. 2008;**92**(3):451-458

[20] Baer DS, Paul JB, Gupta JB, O'Keefe A. Sensitive absorption measurements in the near-infrared region using off-axis integrated-cavity-output spectroscopy. Applied Physics B: Lasers and Optics. 2002;**75**(2-3):261-265

[21] McManus JB, Zahniser MS, Nelson DD, Shorter JH, Herndon SC, Jervis D, et al. Recent progress in laser-based trace gas instruments: Performance and noise analysis. Applied Physics B. 2015;**119**(1):203-218

[22] Pangle LA, Klaus J, Berman ESF, Gupta M, McDonnell JJ. A new multisource and high-frequency approach to measuring δ^2H and $\delta^{18}O$ in hydrological field studies. Water Resources Research. 2013;**49**(11):7797-7803

[23] Mathieu R, Bariac T. A numerical model for the simulation of stable isotope profiles in drying soils. Journal of Geophysical Research. 1996;**101**(D7):12685-12696

[24] Volkmann THM, Weiler M. Continual in situ monitoring of pore water stable isotopes in the subsurface. Hydrology and Earth System Sciences. 2014;**18**(5):1819-1833

[25] Volkmann THM, Haberer K, Gessler A, Weiler M. High-resolution isotope measurements resolve rapid ecohydrological dynamics at the soil–plant interface. The New Phytologist. 2016;**210**(3):839-849

[26] Volkmann THM, Kühnhammer K, Herbstritt B, Gessler A, Weiler M. A method for in situ monitoring of the isotope composition of tree xylem water using laser spectroscopy. Plant, Cell and Environment. 2016;**39**(9):2055-2063

[27] Shahraeeni E, Or D. Thermo-evaporative fluxes from heterogeneous porous surfaces resolved by infrared thermography. Water Resources Research. 2010;**46**(9):W09511

[28] Vereecken H, Huisman JA, Pachepsky Y, Montzka C, van der Kruk J, Bogena H, et al. On the spatio-temporal dynamics of soil moisture at the field scale. Journal of Hydrology. 2014;**516**:76-96

[29] Samouëlian A, Cousin I, Tabbagh A, Bruand A, Richard G. Electrical resistivity survey in soil science: A review. Soil and Tillage Research. 2005;**83**(2):173-193

[30] Niu G-Y, Paniconi C, Troch PA, Scott RL, Durcik M, Zeng X, et al. An integrated modelling framework of catchment-scale ecohydrological processes: 1. Model description and tests over an energy-limited watershed. Ecohydrology. 2014;**7**(2):427-439

[31] Niu G-Y, Troch PA, Paniconi C, Scott RL, Durcik M, Zeng X, et al. An integrated modelling framework of catchment-scale ecohydrological processes: 2. The role of water subsidy by overland flow on vegetation dynamics in a semi-arid catchment. Ecohydrology. 2014;**7**(2):815-827

[32] Fang Y-H, Zhang X, Niu G-Y, Zeng W, Zhu J, Zhang T. Study of the spatiotemporal characteristics of meltwater contribution to the total runoff in the upper Changjiang River basin. Water. 2017;**9**(3):165

[33] Zhang X, Niu G-Y, Elshall AS, Ye M, Barron-Gafford GA, Pavao-Zuckerman M. Assessing five evolving microbial enzyme models against field measurements from a semiarid savannah—What are the mechanisms of soil respiration pulses? Geophysical Research Letters. 2014;**41**(18):6428-6434

[34] Camporese M, Paniconi C, Putti M, Orlandini S. Surface-subsurface flow modeling with path-based runoff routing, boundary condition-based coupling, and assimilation of multisource observation data. Water Resources Research. 2010;**46**(2):W02512

[35] Paniconi C, Putti M. A comparison of Picard and Newton iteration in the numerical solution of multidimensional variably saturated flow problems. Water Resources Research. 1994;**30**(12):3357-3374

[36] Orlandini S, Rosso R. Parameterization of stream channel geometry in the distributed modeling of catchment dynamics. Water Resources Research. 1998;**34**(8):1971-1985

[37] Niu G-Y, Yang Z-L, Mitchell KE, Chen F, Ek MB, Barlage M, et al. The community Noah land surface model with multiparameterization options (Noah-MP): 1. Model description and evaluation with local-scale measurements. Journal of Geophysical Research, [Atmospheres]. 2011;**116**(D12):D12109

[38] Peters DPC. Plant species dominance at a grassland–shrubland ecotone: An individual-based gap dynamics model of herbaceous and woody species. Ecological Modelling. 2002;**152**(1):5-32

[39] Steefel C. GIMRT, Version 1.2: Software for Modeling Multicomponent, Multidimensional Reactive Transport. User's Guide, UCRL-MA-143182. Lawrence Livermore National Laboratory: Livermore, CA, USA; 2001

[40] Sivapalan M. Pattern, process and function: Elements of a unified theory of hydrology at the catchment scale. In: Anderson MG, editor. Encyclopedia of Hydrological Sciences. Chichester, England: Wiley; 2005. pp. 193-219

[41] Niu GY, Pasetto D, Scudeler C, Paniconi C, Putti M, Troch PA, et al. Incipient subsurface heterogeneity and its effect on overland flow generation – Insight from a modeling study of the first experiment at the Biosphere 2 Landscape Evolution Observatory. Hydrology and Earth System Sciences. 2014;**18**(5):1873-1883

[42] Nijland W, van der Meijde M, Addink EA, de Jong SM. Detection of soil moisture and vegetation water abstraction in a Mediterranean natural area using electrical resistivity tomography. Catena. 2010;**81**(3):209-216

of electrical resistivity tomography (ERT) provides an opportunity to observe nondestructively the subsurface structure changes of the initially homogenous crushed basalt tephra in response to hydrological (e.g., surface-subsurface water interactions) and biogeochemical processes. Through geophysical inversion of the three-dimensional resistivity field recorded by the ERT system, changes in soil physicochemical properties (e.g., pore volume) can be mapped at high spatial resolution. The combined above observations, especially within the saturated zone, can resolve the spatial distribution of biogeochemical transformations due to abiotic and biotic processes in a pedogenic environment.

Pore water geochemistry can be used to predict the initial stages of basalt weathering, while soil coring samples can validate predictions based on pore water geochemistry. The coupling of pore water data and soil coring data enables measurements of elemental partitioning during the basalt weathering. Therefore, the cycle of a given element on the landscape can also be described by rewriting Eq. 3 as follows:

$$\frac{\partial E}{\partial t} = \frac{\partial E_{aq}}{\partial t} + \frac{\partial E_{s}}{\partial t} + \frac{\partial E_{g}}{\partial t} \tag{4}$$

where subscript aq represents the aqueous phase (where concentrations E can be measured from pore water samplers), subscript s represents the solid phase (where concentrations E are derived from soil coring), and subscript g represents the gas phase (where concentrations E can be obtained from gas sensors and samplers). The gas-phase term only applies when quantifying the partitioning of elements that form gaseous compounds, such as most importantly carbon.

Figure 12 demonstrates the onset of carbon and nitrogen accumulation on the LEO hillslopes as measured from soil cores. While organic carbon and total nitrogen accumulations are mostly limited to the soil surface, inorganic carbon tends to precipitate in a lens in the center of the hillslope. This supports observations of developing heterogeneity in calcite saturation indices on the slopes [55]. Collected soil samples are also analyzed by synchrotron-based (for higher sensitivity) X-ray diffraction to examine changes in mineral composition of the soil, particularly formation of secondary crystalline minerals. By further examining the change of element availability in the solid phase by sequential extraction, it is possible to characterize operationally the formation of X-ray amorphous phases and to better understand geochemical transformations in the soil. In addition to bulk measurements, stabilization of organic carbon in the soils—an integral part of soil formation and important mechanism of carbon sequestration from the atmosphere—is being examined for selected soil samples by Mossbauer spectroscopy, high-resolution transmission electron microscopy (HRTEM), scanning electron microscopy (SEM), high spatial resolution secondary ion mass spectrometry (NanoSIMS) analysis, and X-ray photoelectron spectroscopy (XPS). Release and fractionation of dissolved organic matter are also being examined using Fourier transform ion cyclotron resonance mass spectrometry (FTICR-MS) analysis of pore waters and soil sample extracts. The microbiological analysis of the soil cores (Sections 2.1.6 and 2.5) can address variations in carbon cycling and nutrient availability in addition to linking biological and abiotic processes during the initial stages of basalt weathering and pedogenesis.

[43] Ferré T, Bentley L, Binley A, Linde N, Kemna A, Singha K, et al. Critical steps for the continuing advancement of hydrogeophysics. Eos, Transactions American Geophysical Union. 2009;**90**(23):200-200

[44] Baldocchi D, Falge E, Gu LH, Olson R, Hollinger D, Running S, et al. FLUXNET: A new tool to study the temporal and spatial variability of ecosystem-scale carbon dioxide, water vapor, and energy flux densities. Bulletin of the American Meteorological Society. 2001;**82**(11):2415-2434

[45] Meredith LK, Commane R, Munger JW, Dunn A, Tang J, Wofsy SC, et al. Ecosystem fluxes of hydrogen: A comparison of flux-gradient methods. Atmospheric Measurement Techniques. 2014;**7**(9):2787-2805

[46] Altaf Arain M, James Shuttleworth W, Farnsworth B, Adams J, Lutfi SO. Comparing micrometeorology of rain forests in Biosphere-2 and Amazon basin. Agricultural and Forest Meteorology. 2000;**100**(4):273-289

[47] Zeng X, Wang Z, Wang A. Surface skin temperature and the interplay between sensible and ground heat fluxes over arid regions. Journal of Hydrometeorology. 2012;**13**(4):1359-1370

[48] Sprenger M, Volkmann THM, Blume T, Weiler M. Estimating flow and transport parameters in the unsaturated zone with pore water stable isotopes. Hydrology and Earth System Sciences. 2015;**19**(6):2617-2635

[49] Scudeler C, Pangle LA, Pasetto D, Niu GY, Volkmann THM, Paniconi C, et al. Multiresponse modeling of variably saturated flow and isotope tracer transport for a hillslope experiment at the Landscape Evolution Observatory. Hydrology and Earth System Sciences. 2016;**20**(10):4061-4078

[50] Reichstein M, Falge E, Baldocchi D, Papale D, Aubinet M, Berbigier P, et al. On the separation of net ecosystem exchange into assimilation and ecosystem respiration: Review and improved algorithm. Global Change Biology. 2005;**11**(9):1424-1439

[51] Phillips CL, Gregg JW, Wilson JK. Reduced diurnal temperature range does not change warming impacts on ecosystem carbon balance of Mediterranean grassland mesocosms. Global Change Biology. 2011;**17**(11):3263-3273

[52] Bowling DR, Pataki DE, Randerson JT. Carbon isotopes in terrestrial ecosystem pools and CO_2 fluxes. The New Phytologist. 2008;**178**(1):24-40

[53] Berry J, Wolf A, Campbell JE, Baker I, Blake N, Blake D, et al. A coupled model of the global cycles of carbonyl sulfide and CO_2: A possible new window on the carbon cycle. Journal of Geophysical Research – Biogeosciences. 2013;**118**(2):842-852

[54] van Haren J, Dontsova K, Barron-Gafford GA, Troch PA, Chorover J, Delong SB, et al. CO_2 diffusion into pore spaces limits weathering rate of an experimental basalt landscape. Geology. 2017;**45**(3):203-206

[55] Pohlmann M, Dontsova K, Root R, Ruiz J, Troch P, Chorover J. Pore water chemistry reveals gradients in mineral transformation across a model basaltic hillslope. Geochemistry, Geophysics, Geosystems. 2016;**17**(6):2054-2069

[56] Sengupta A, Stegen JC, Nielson JW, Meira-Neto A, Wang Y, Troch PA, et al. Assessment of microbial community patterns under incipient conditions in a basalt soil system. Manuscript in preparation

[57] Barron-Gafford GA, Sanchez-Cañete EP, Minor RL, Hendryx SM, Lee E, Sutter LF, et al. Impacts of hydraulic redistribution on grass–tree competition vs facilitation in a semi-arid savanna. The New Phytologist. 2017;**215**(4):1451-1461

[58] Potts DL, Minor RL, Braun Z, Barron-Gafford GA. Photosynthetic phenological variation may promote coexistence among co-dominant tree species in a Madrean sky island mixed conifer forest. Tree Physiology. 2017;**37**(9):1229-1238

[59] Ustin SL, Roberts DA, Gamon JA, Asner GP, Green RO. Using imaging spectroscopy to study ecosystem processes and properties. Bioscience. 2004;**54**(6):523-534

[60] Beer C, Reichstein M, Tomelleri E, Ciais P, Jung M, Carvalhais N, et al. Terrestrial gross carbon dioxide uptake: Global distribution and covariation with climate. Science. 2010;**329**(5993):834-838

[61] Zhao M, Running SW. Drought-induced reduction in global terrestrial net primary production from 2000 through 2009. Science. 2010;**329**(5994):940-943

[62] Zhao M, Heinsch FA, Nemani RR, Running SW. Improvements of the MODIS terrestrial gross and net primary production global data set. Remote Sensing of Environment. 2005;**95**(2):164-176

[63] Friedlingstein P, Cox P, Betts R, Bopp L, Bloh WV, Brovkin V, et al. Climate–carbon cycle feedback analysis: Results from the C4MIP model intercomparison. Journal of Climate. 2006;**19**(14):3337-3353

[64] Damm A, Elbers JAN, Erler A, Gioli B, Hamdi K, Hutjes R, et al. Remote sensing of sun-induced fluorescence to improve modeling of diurnal courses of gross primary production (GPP). Global Change Biology. 2010;**16**(1):171-186

[65] Frankenberg C, Fisher JB, Worden J, Badgley G, Saatchi SS, Lee J-E, et al. New global observations of the terrestrial carbon cycle from GOSAT: Patterns of plant fluorescence with gross primary productivity. Geophysical Research Letters. 2011;**38**(17):L17706

[66] Rascher U, Agati G, Alonso L, Cecchi G, Champagne S, Colombo R, et al. CEFLES2: The remote sensing component to quantify photosynthetic efficiency from the leaf to the region by measuring sun-induced fluorescence in the oxygen absorption bands. Biogeosciences. 2009;**6**(7):1181-1198

[67] Zarco-Tejada PJ, González-Dugo V, Berni JAJ. Fluorescence, temperature and narrow-band indices acquired from a UAV platform for water stress detection using a micro-hyperspectral imager and a thermal camera. Remote Sensing of Environment. 2012;**117**:322-337

[68] Amoros-Lopez J, Gomez-Chova L, Vila-Frances J, Alonso L, Calpe J, Moreno J, et al. Evaluation of remote sensing of vegetation fluorescence by the analysis of diurnal cycles. International Journal of Remote Sensing. 2008;**29**(17-18):5423-5436

[69] Stumm W, Morgan JJ. Aquatic Chemistry: Chemical Equilibria and Rates in Natural Waters. 3rd ed. Chichester, England: Wiley; 1996

[70] Torn MS, Trumbore SE, Chadwick OA, Vitousek PM, Hendricks DM. Mineral control of soil organic carbon storage and turnover. Nature. 1997;**389**(6647):170-173

[71] Chorover J, Amistadi MK, Chadwick OA. Surface charge evolution of mineral-organic complexes during pedogenesis in Hawaiian basalt. Geochimica et Cosmochimica Acta. 2004;**68**(23):4859-4876

[72] Dontsova K, Zaharescu D, Henderson W, Verghese S, Perdrial N, Hunt E, et al. Impact of organic carbon on weathering and chemical denudation of granular basalt. Geochimica et Cosmochimica Acta. 2014;**139**:508-526

[73] Jencso KG, McGlynn BL. Hierarchical controls on runoff generation: Topographically driven hydrologic connectivity, geology, and vegetation. Water Resources Research. 2011; **47**(11):W11527

[74] Nippgen F, McGlynn BL, Marshall LA, Emanuel RE. Landscape structure and climate influences on hydrologic response. Water Resources Research. 2011;**47**(12):W12528

[75] Buttle JM, Dillon PJ, Eerkes GR. Hydrologic coupling of slopes, riparian zones and streams: An example from the Canadian shield. Journal of Hydrology. 2004;**287**(1):161-177

[76] Detty JM, McGuire KJ. Topographic controls on shallow groundwater dynamics: Implications of hydrologic connectivity between hillslopes and riparian zones in a till mantled catchment. Hydrological Processes. 2010;**24**(16):2222-2236

[77] Bencala KE, Gooseff MN, Kimball BA. Rethinking hyporheic flow and transient storage to advance understanding of stream-catchment connections. Water Resources Research. 2011;**47**(3):W00H3

[78] Beven K. Searching for the holy grail of scientific hydrology: $Q_t=(S, R, \Delta t)A$ as closure. Hydrology and Earth System Sciences. 2006;**10**(5):609-618

[79] Kirchner JW. Getting the right answers for the right reasons: Linking measurements, analyses, and models to advance the science of hydrology. Water Resources Research. 2006;**42**(3):W03S4

[80] McDonnell JJ, Sivapalan M, Vaché K, Dunn S, Grant G, Haggerty R, et al. Moving beyond heterogeneity and process complexity: A new vision for watershed hydrology. Water Resources Research. 2007;**43**(7):W07301

[81] Reggiani P, Sivapalan M, Hassanizadeh SM. A unifying framework of watershed thermodynamics: Balance equations for mass, momentum, energy and entropy and the second law of thermodynamics. Advances in Water Resources. 1998;**22**(4):367-398

[82] Reggiani P, Sivapalan M, Hassanizadeh SM. Conservation equations governing hillslope responses: Exploring the physical basis of water balance. Water Resources Research. 2000;**36**(7):1845-1863

[83] Reggiani P, Sivapalan M, Hassanizadeh SM, Gray WG. Coupled equations for mass and momentum balance in a stream network: Theoretical derivation and computational experiments. Proceedings of the Royal Society London, A. 2001;**457**(2005):157-189

[84] Kirchner JW. Catchments as simple dynamical systems: Catchment characterization, rainfall-runoff modeling, and doing hydrology backward. Water Resources Research. 2009;**45**(2):W02429

[85] Wittenberg H. Baseflow recession and recharge as nonlinear storage processes. Hydrological Processes. 1999;**13**(5):715-726

[86] Brutsaert W. The unit response of groundwater outflow from a hillslope. Water Resources Research. 1994;**30**(10):2759-2763

[87] Harman CJ, Reeves DM, Baeumer B, Sivapalan M. A subordinated kinematic wave equation for heavy-tailed flow responses from heterogeneous hillslopes. Journal of Geophysical Research - Earth Surface. 2010;**115**(F1):F00A8

[88] Troch PA, Paniconi C, Emiel van Loon E. Hillslope-storage Boussinesq model for subsurface flow and variable source areas along complex hillslopes: 1. Formulation and characteristic response. Water Resources Research. 2003;**39**(11):1316

[89] Harman CJ. Time-variable transit time distributions and transport: Theory and application to storage-dependent transport of chloride in a watershed. Water Resources Research. 2015;**51**(1):1-30

[90] Rinaldo A, Benettin P, Harman CJ, Hrachowitz M, McGuire KJ, van der Velde Y, et al. Storage selection functions: A coherent framework for quantifying how catchments store and release water and solutes. Water Resources Research. 2015;**51**(6):4840-4847

[91] Harman C, Troch PA. What makes Darwinian hydrology "Darwinian"? Asking a different kind of question about landscapes. Hydrology and Earth System Sciences. 2014;**18**(2):417-433

[92] Steefel CI. Geochemical kinetics and transport. In: Brantley SL, Kubicki JD, White AF, editors. Kinetics of Water-Rock Interaction. NY, USA: Springer; 2008. pp. 545-589

[93] Maher K. The role of fluid residence time and topographic scales in determining chemical fluxes from landscapes. Earth and Planetary Science Letters. 2011;**312**(1):48-58

[94] Maher K, Steefel CI, White AF, Stonestrom DA. The role of reaction affinity and secondary minerals in regulating chemical weathering rates at the Santa Cruz soil Chronosequence, California. Geochim Cosmochim Acta. 2009;**73**(10):2804-2831

[95] Yokoyama T, Banfield JF. Direct determinations of the rates of rhyolite dissolution and clay formation over 52,000 years and comparison with laboratory measurements. Geochimica et Cosmochimica Acta. 2002;**66**(15):2665-2681

[96] Neaman A, Chorover J, Brantley SL. Implications of the evolution of organic acid moieties for basalt weathering over geological time. American Journal of Science. 2005;**305**(2):147-185

[97] Zaharescu DG, Burghelea CI, Dontsova K, Presler JK, Maier RM, Huxman T, et al. Ecosystem composition controls the early fate of rare earth elements during incipient soil genesis. BioRxiv. 2016;061846

[98] Lian B, Xiao L, Sun Q. Ecological effects of the microbial weathering of silicate minerals. Acta Geologica Sinica - English Edition. 2017;**91**(s1):150-152

[99] Xiao L, Lian B, Hao J, Liu C, Wang S. Effect of carbonic anhydrase on silicate weathering and carbonate formation at present day CO_2 concentrations compared to primordial values. Scientific Reports. 2015;**5**:7733

[100] Li L, Steefel CI, Yang L. Scale dependence of mineral dissolution rates within single pores and fractures. Geochimica et Cosmochimica Acta. 2008;**72**(2):360-377

[101] Pacheco FAL, Van der Weijden CH. Integrating topography, hydrology and rock structure in weathering rate models of spring watersheds. Journal of Hydrology. 2012; **428**:32-50

[102] Anbeek C. The effect of natural weathering on dissolution rates. Geochimica et Cosmochimica Acta. 1993;**57**(21):4963-4975

[103] Brantley SL, Mellott NP. Surface area and porosity of primary silicate minerals. American Mineralogist. 2000;**85**(11-12):1767-1783

[104] Gislason SR, Oelkers EH, Eiriksdottir ES, Kardjilov MI, Gisladottir G, Sigfusson B, et al. The feedback between climate and weathering. Mineralogical Magazine. 2008; **72**(1):317-320

[105] Valentín-Vargas A, Root RA, Neilson JW, Chorover J, Maier RM. Environmental factors influencing the structural dynamics of soil microbial communities during assisted phytostabilization of acid-generating mine tailings: A mesocosm experiment. Science of the Total Environment. 2014;**500**:314-324

[106] Neilson JW, Quade J, Ortiz M, Nelson WM, Legatzki A, Tian F, et al. Life at the hyperarid margin: Novel bacterial diversity in arid soils of the Atacama Desert, Chile. Extremophiles. 2012;**16**(3):553-566

[107] Wang Y, Bradford SA, Šimůnek J. Transport and fate of microorganisms in soils with preferential flow under different solution chemistry conditions. Water Resources Research. 2013;**49**(5):2424-2436

[108] Banfield JF, Barker WW, Welch SA, Taunton A. Biological impact on mineral dissolution: Application of the lichen model to understanding mineral weathering in the rhizosphere. Proceedings of the National Academy of Sciences. 1999;**96**(7):3404-3411

[109] Graham EB, Crump AR, Resch CT, Fansler S, Arntzen E, Kennedy DW, et al. Coupling spatiotemporal community assembly processes to changes in microbial metabolism. Frontiers in Microbiology. 2016;**7**:1949

[110] Angers DA, Caron J. Plant-induced changes in soil structure: Processes and feedbacks. Biogeochemistry. 1998;**42**(1-2):55-72

[111] Porporato A, Daly E, Rodriguez-Iturbe I. Soil water balance and ecosystem response to climate change. The American Naturalist. 2004;**164**(5):625-632

[112] Asbjornsen H, Goldsmith GR, Alvarado-Barrientos MS, Rebel K, Van Osch FP, Rietkerk M, et al. Ecohydrological advances and applications in plant–water relations research: A review. Journal of Plant Ecology. 2011;**4**(1-2):3-22

Chapter 3

Ecosystem Development in the Constructed Catchment "Chicken Creek"

Wolfgang Schaaf, Christoph Hinz, Werner Gerwin,
Markus K. Zaplata and Reinhard F. Huettl

Additional information is available at the end of the chapter

http://dx.doi.org/10.5772/intechopen.70546

Abstract

Landscapes and ecosystems are complex systems with many feedback mechanisms acting between the various abiotic and biotic components. The knowledge about these interacting processes is mainly derived from mature ecosystems. The initial development of ecosystem complexity may involve state transitions following catastrophic shifts, disturbances, or transgression of thresholds. We propose a conceptual framework of feedback processes in early states of ecosystem development affected by spatiotemporal environmental drivers. To test this concept, we used 10-year time series of hydrological, biological, geomorphological, and soil data from the constructed catchment Chicken Creek." The 6ha site was left to unrestricted development since 2005 and was intensively monitored. The data showed a very rapid development of the site with an increasing complexity and heterogeneity. In the first years, stochastic signals like the initial substrate conditions and external drivers like extreme weather events were the most important factors resulting in abiotic/abiotic feedback mechanisms shaping the morphology of the site and creating site diversity. Initial abiotic feedback mechanisms between water and substrate were soon followed by abiotic/biotic feedbacks between biological soil crusts, invading vegetation, geomorphology, and hydrology resulting in state transitions of catchment functioning.

Keywords: state transition, feedback mechanisms, succession, hydrology

1. Introduction

In 2001, the US National Academy of Sciences defined the understanding of the so-called Critical Zone as one of the most challenging central research topics. The Critical Zone is the near-surface part of the Earth's crust, which sustains all terrestrial life [1]. During the last decade, a large number of landscape observatories have been implemented in order to study the Critical Zone and its behavior under global change condition. First Critical Zone Observatories (CZOs) were established in the USA; later, additional sites have been launched all over the world [2]. This paper introduces one of these CZOs in Germany with very unique conditions. The site represents a constructed watershed designed to analyze the ecological feedback mechanisms during the initial development of an ecosystem.

Ecosystems have been shown in many studies to be utterly complex systems with many feedback mechanisms interacting between compartments [3, 4]. However, most of these studies were carried out in mature systems that had evolved over long time periods (e.g., [5–7]). Therefore, very little is known about how this complexity evolves over time and what role these feedbacks play during the development of ecosystems [8]. Most knowledge about early states of ecosystems was derived either from case studies after natural or anthropogenic disturbances (e.g., [9–14]) or from chronosequence studies (e.g., [15–18]).

In recent years, landscape development and ecosystem development have been viewed as complex interacting processes leading to self-organization and state transitions [19, 20]. Within this context the concept of multiple stable states depending on environmental settings and on the possibility for catastrophic shifts in ecosystem composition and functioning, especially after disturbances or after passing ecological thresholds, was created [21–26]. Unfortunately, most of the data supporting these concepts were recorded only post-event.

The role of feedback mechanisms for the development of ecosystems and their functioning may differ depending on internal dynamics and external drivers. To better understand the underlying mechanisms, we developed a simple conceptual framework of how feedbacks between the major system components substrate, water, and biota interact and are affected by stochastic spatiotemporal drivers. Studies from arid environments indicate that ecosystems under water limitation may display alternative stable states, in which the interactions between abiotic and biotic processes determine whether or not a degraded state will prevail after disturbance [27]. The interaction between substrate and water is clearly driven by rainfall and evaporative demand, in which soil particles are mobilized on the surface (erosion) and within the substrate (soffusion) and subsequently immobilized during dry periods in which, for example, seals are transformed into physical surface crusts irreversibly changing runoff properties. Such interactions represent abiotic feedbacks that can either yield into self-stabilizing surface structure or a continuously changing erosive surface. At the same time, the formation of biological crusts and the germination of seeds will generate feedback processes that will affect substrate properties. In more humid and temperate climates, eventually feedback mechanisms between abiotic components and the biota will dominate the system, which was shown in many ecosystem studies.

To study the role of feedbacks in state transitions during ecosystem development, we used 10-year time series of data from the constructed catchment Chicken Creek. This unique site offers the chance to observe state transitions in a relatively simple ecosystem from a very initial state with hardly any internal ecological memory [28] to more complex states.

2. Material and methods

The Chicken Creek ("Hühnerwasser") catchment is located within the lignite mining district of Lusatia in Northeastern Germany, about 150 km southeast from Berlin. It was constructed in 2004–2005 in the lignite mining area of Welzow-Süd, 30 km south of Cottbus, Germany (**Figure 1**) [29]. The region is characterized by temperate seasonal climate (563 mm mean annual precipitation, 8.9°C mean annual air temperature). The catchment was constructed by means of large mining techniques to establish a headwater for the Hühnerwasser, a small stream, which was destroyed by mining activities in the 1980s. This stream has to be restored during the reclamation process of the post-mining landscape. The site consists of two layers and has an inclination of 2.0–3.5% and a southeastern exposition. At the lowest part of the site, a basin was formed allowing the formation of small pond (originally about 60–70 m in diameter and 3 m depth in 2005/2006). Details of the construction process and of the substrates used are described [30].

The construction started with a 1–2-m-thick clay layer 8 (**Figure 2a**), which forms an aquiclude underlying the whole area of 60,000 m^2 (400 × 150 m). The material was separated from the tertiary overburden layers by bucket wheel excavators and dumped by a stacker. The clay surface was then leveled by bulldozers – but not compacted, as this material tends to self-seal when considerable swelling occurs. Initially, the freshly dumped clay consisted of large aggregates, but these vanished after wetting and subsequent swelling. In this state, the clay layer had an extremely low permeability (ksat. ~10^{-9} m s^{-1}). The clay layer was shaped into a shallow basin ascending from the center to the edges to form the subsurface boundaries of the catchment (**Figure 2b**). Belowground in the lower part of the catchment, additional clay dams were constructed on top of the clay layer

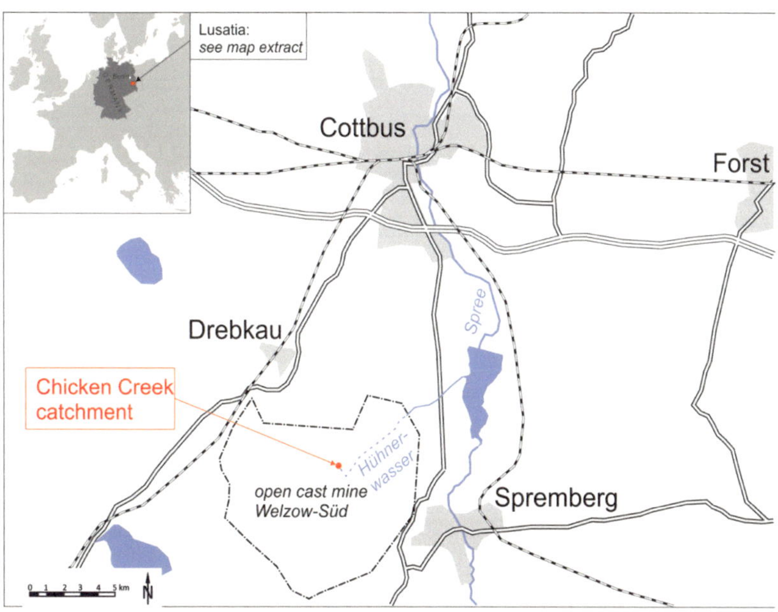

Figure 1. Location of the artificial catchment Chicken Creek.

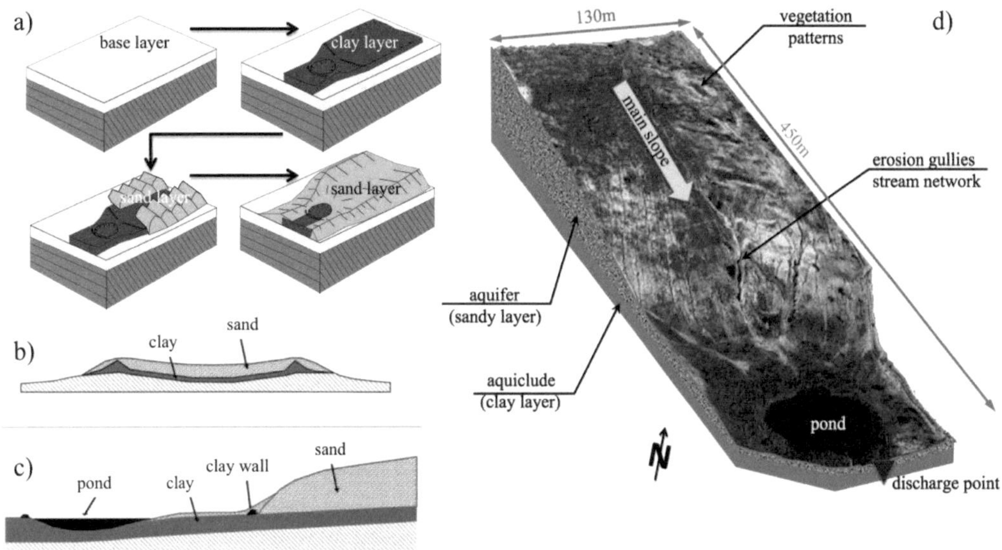

Figure 2. (a) Schematic sketch of the major construction steps of the Chicken Creek catchment, (b) schematic cross and (c) longitudinal profile of the catchment (d) aerial image of the catchment showing the main morphological characteristics.

perpendicular to the slope as a stabilization barrier to prevent the sandy substrate of the aquifer from sliding downhill on the clay layer and as a central groundwater discharge unit for the creation of an artificial spring for the reconstructed creek. A clay wall at the southern edge of the pond defines the lower boundary of the catchment and has a single defined outlet (**Figure 2c**).

On top of this clay layer, 117,500 m^3 of Pleistocene sandy material taken from the forefield of the open-cast mine (i.e., mainly C horizon substrates from the former landscape) was dumped to form the 2–3 m aquifer of the watershed (**Figure 2a**). The construction of this layer began in August 2004 with the eastern area of the hillside adjacent to the later hydrological catchment, followed by the central parts of the catchment during the next 2 months. The area of the catchment generally consists of three different sections which can be distinguished in terms of the overall construction procedure. The sandy material in the eastern part was dumped in August/September 2004. During the next construction phase in September/October 2004, the western part of the clay layer was completed and immediately covered with sandy substrate material. The central part of the site was left open as a "central trench" for a period of 7 months, before finally being filled in by bulldozers in May 2005 with substrate material from the eastern and western sections (**Figure 2b**). The surface level of the eastern part of the catchment was lowered in order to remove surface substrate that had been exposed to the atmosphere for more than 1 year and to restore the surface to an initial state. The surface layer was flattened and shaped into a shallow basin in order to define clear hydrological boundaries at the surface. As a final step, the surface of the sandy layer was homogenized, and the remaining surface structures from the construction were removed as good as possible.

The hillslope-shaped site with defined boundary conditions and well-documented inner structures allows for studying ecosystem development ab initio at the catchment scale (**Figure 2d**). No amelioration measures, fertilization, or planting was carried out. Since 2005, the unrestricted, unmanaged development of the catchment was intensively monitored. Sensors and monitoring plots were originally oriented along a regular 20 × 20 m grid and were successively complemented with more structure and pattern-oriented instrumentation adapted to the development of the catchment (**Figure 3**). In total, 3 flumes in the main erosion gullies and 2 weirs equipped with automated sampling devices, 3 weather stations, 42 groundwater wells, 9–18 deposition samplers, 16 suction plates, 88 FDR probes, 40 pF meter, and 1 multiparameter probe in the pond are installed. Vegetation is surveyed every summer at four 1 × 1 m plots around each of the 119 grid points. More details on installed sensors and analytical methods are described [31, 32].

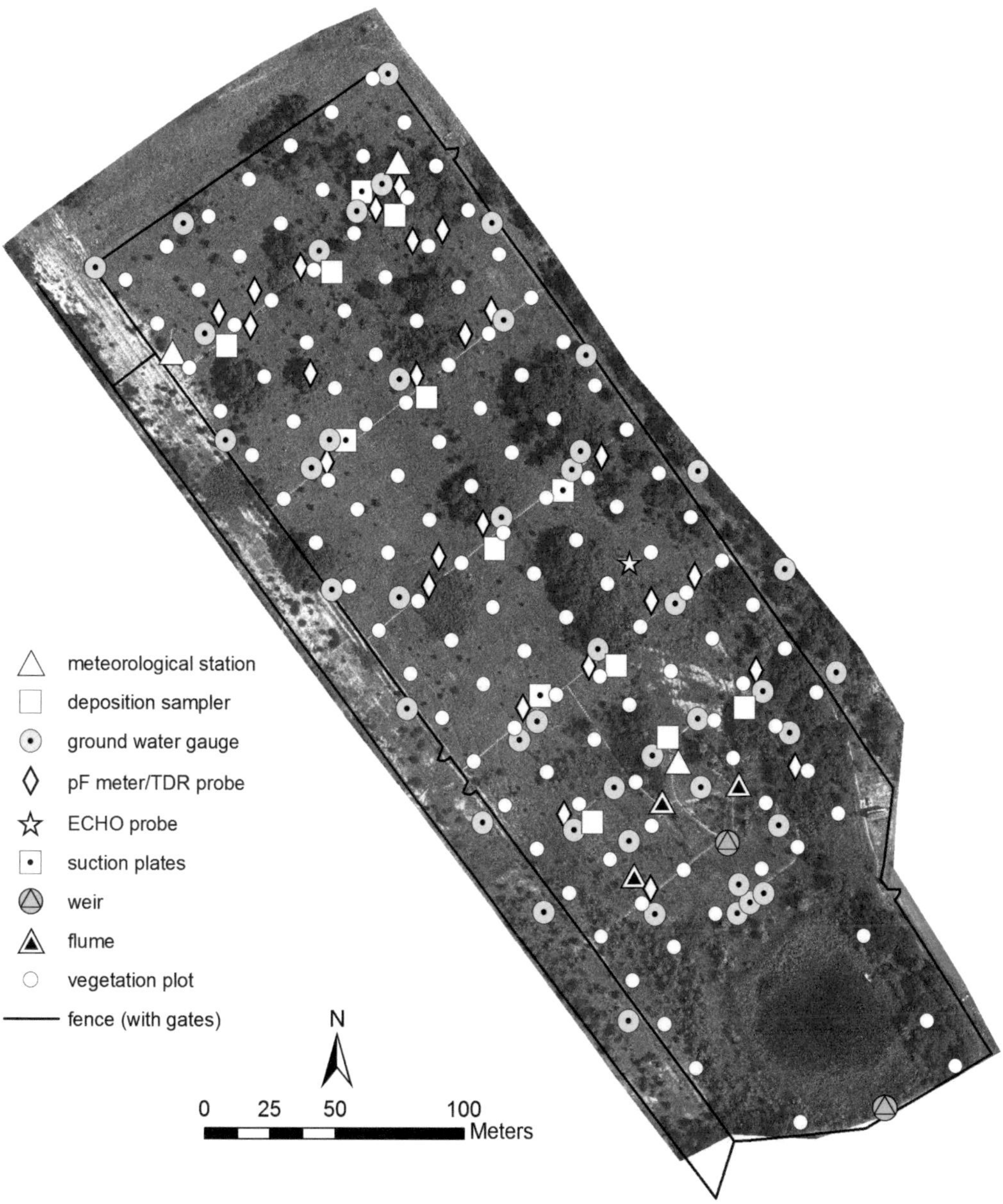

Figure 3. Overview of monitoring measurements at Chicken Creek catchment (aerial photo from 2013).

2.1. Results and discussion

Initial soil sampling from all 120 grid points revealed some slight spatial variation of the Pleistocene material with respect to texture and chemical parameters reflecting the natural variability of these postglacial deposits. The western part of the catchment had more loamy sands, whereas in the eastern part, pure sands dominated. The vertical distribution was very homogenous. Due to the carbonate contents of the material (0.6–1.1%), pH values were uniformly between 7 and 8. Organic carbon ($C_{org.}$) content was very low (1–2 mg g^{-1}).

Time series of meteorological, hydrological, biological, and soil data revealed a fast colonization of the catchment with invading vegetation that transformed the site from an initially abiotic system similar to arid systems to a state where abiotic/biotic feedback processes dominate [33, 34]. Aerial images of the catchment documented this rapid development and showed the increasing heterogeneity and the formation of surface structures and patterns (**Figure 4**). From 2010 woody plants increasingly formed clearly visible patches particularly in the eastern part of the catchment and in the surroundings of the pond. In 2015, more than 30 tree species were detected in the catchment.

Total vascular plant cover reached 58% in 2015, composed of over 170 plant species [35]. During the first 2–3 years, surface runoff, gully erosion, and sediment translocation were the dominating processes at the sparsely vegetated site that were triggered by single episodic events like

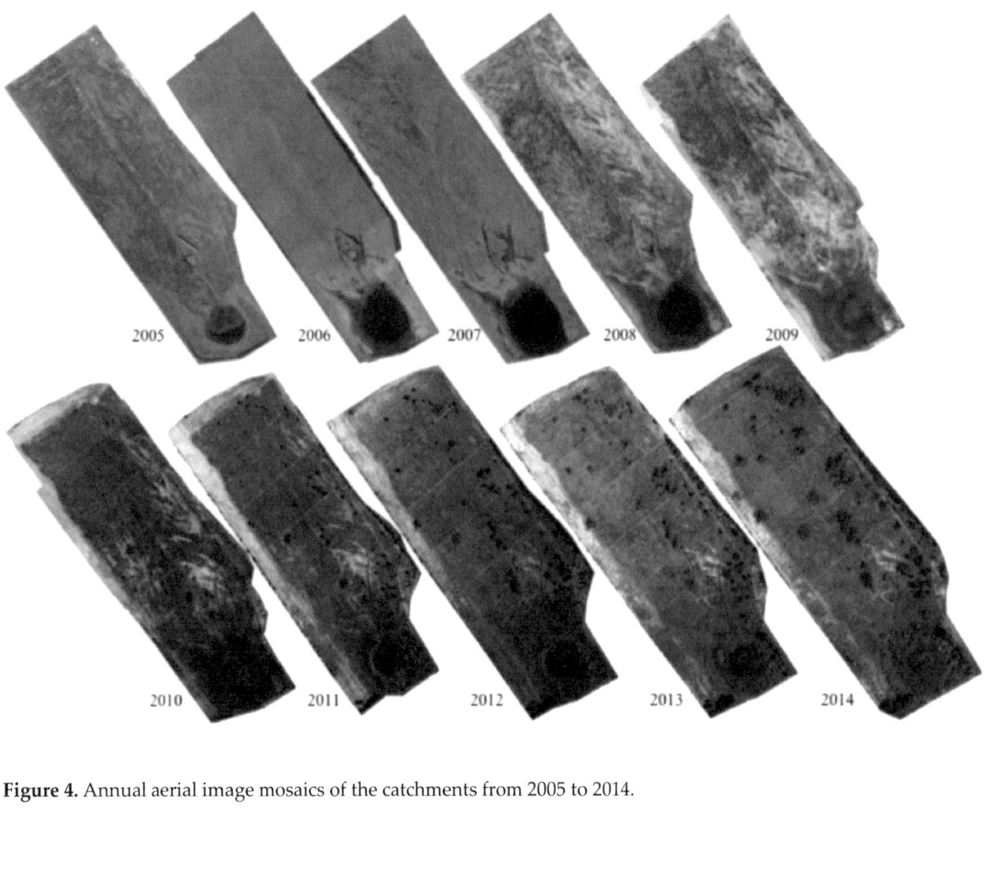

Figure 4. Annual aerial image mosaics of the catchments from 2005 to 2014.

heavy thunderstorms [32]. The slight textural variations in the initial substrates resulted in differences in morphological features of these surface structures, e.g., the frequency, depth, and width of erosion channels [36]. Surface runoff was promoted by the development of mechanical and biological surface crusts (BSC, [37]) sealing the soil surface. This sealing effect affected the catchment hydrology and runoff behavior far more than was predicted by hydrological models using mainly the textural composition of the substrate [38, 39]. Even though the establishment of BSC and its effects on infiltration and runoff is well described for arid and semiarid regions [40–44], the effect of BSC on hydraulic soil properties is discussed controversially, e.g., [45]. They conclude that the impact of BSC on water infiltration and conductivity is depending on surface properties of the crusts (roughness) as well as on specific crustal components. Due to regional climatic differences, BSC may either promote or inhibit infiltration and conductivity in soils. [46] showed the effects of different successional stages of soil crusts. They found that physical crusts, directly formed after a disturbance, lead to a homogenous and smooth soil surface and a promotion of runoff and soil erosion. The further development of BSC caused an increasing surface roughness (particularly well-developed BSC including mosses and lichens) and a decreasing potential for surface runoff. This is supported by soil moisture measurements in the topsoil along two gradients in the catchment showing highly significant increased soil moisture under BSC compared to bare soil (e.g., **Table 1**).

	Plot 1	**Plot 2**	**Plot 3**	**Plot 4**
(a) Grid point K5				
	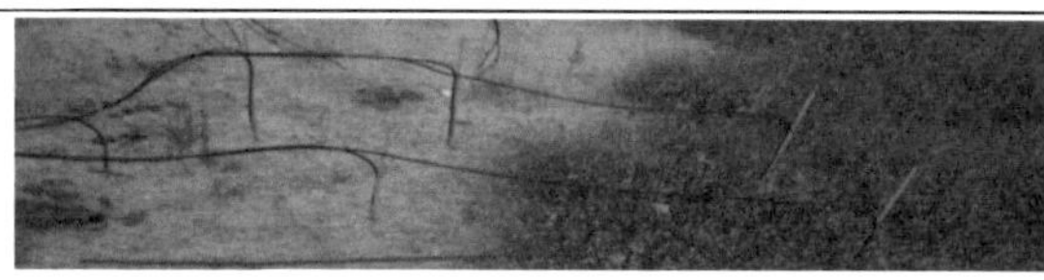			
Vascular plant cover 2012/2013 [%]	5/4		38/48	
Moss cover 2012/2013 [%]	1/1		80/60	
3 cm depth				
Sand [%]	91.4	94.1	82.0	81.6
Silt [%]	6.7	5.1	13.0	13.2
Clay [%]	1.9	0.8	5.0	5.2
$C_{org.}$ [mg g^{-1}]	0.8	0.7	2.0	0.7
Soil moisture [vol. %]				
Mean	1.6^{a}	7.5^{b}	11.5^{c}	14.3^{d}
Standard error	0.12	0.10	0.15	0.18
10 cm depth				
Sand [%]	93.7	95.0	79.7	80.4
Silt [%]	5.1	4.2	13.9	13.0
Clay [%]	1.2	0.8	6.4	6.6

	Plot 1	Plot 2	Plot 3	Plot 4
(a) Grid point K5				
$C_{org.}$ [mg g^{-1}]	<0.1	0.8	2.6	1.2
Soil moisture [vol. %]				
Mean	7.5^{a}	17.9^{b}	21.3^{c}	19.3^{d}
Standard error	0.06	0.07	0.11	0.10
(b) Grid point N5				
	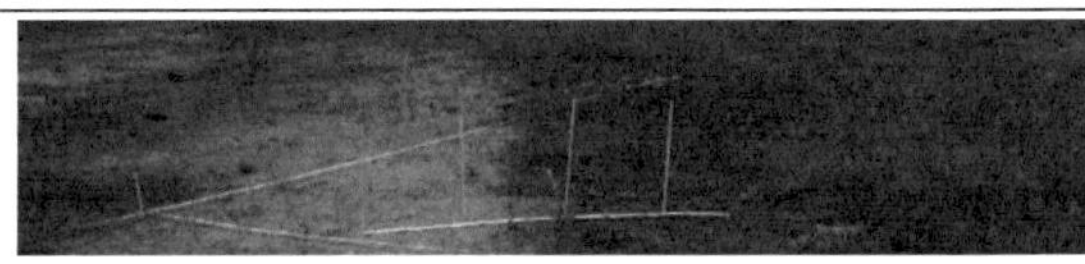			
Vascular plant cover 2012/2013 [%]	7/6	28/22		
Moss cover 2012/2013 [%]	3/5	92/80		
3 cm depth				
Sand [%]	95.5	83.3	92.1	83.5
Silt [%]	4.0	11.3	6.2	11.7
Clay [%]	0.5	5.4	1.7	4.8
$C_{org.}$ [mg g^{-1}]	<0.1	1.5	1.6	4.0
Mean soil moisture [vol. %]	8.2^{a}	5.9^{b}	11.9^{c}	21.8^{d}
Standard error	0.13	0.13	0.14	0.23
10 cm depth				
Sand [%]	95.2	93.9	73.8	91.5
Silt [%]	4.2	5.5	15.3	6.4
Clay [%]	0.6	0.6	10.9	2.1
$C_{org.}$ [mg g^{-1}]	<0.1	0.5	3.2	2.4
Mean soil moisture [vol. %]	11.4^{a}	10.7^{b}	23.4^{c}	21.6^{d}
Standard error	0.13	0.13	0.17	0.21

Measuring period: 06/2010–11/2013; different letter indicates significant differences at $p < 0.001$.

Table 1. Soil characteristics and soil moisture of a gradient from bare soil to vegetated site at grid points (a) K5 and (b) N5.

This was most prominent for the mean and maximum values in both 3 and 10 cm soil depth, but not for minimum soil moisture. At both sites, the moss cover of the BSC was very high. At the same time, the gradients showed increased silt, clay, and $C_{org.}$ contents below the BSC. These findings indicate that surface stability is probably one of the key factors for BSC establishment. Once established, they initially promoted surface runoff due to hydrophobicity and pore clogging [47]. During further development, the BSC were more and more covered by mosses, which provide a high potential for water storage and eventually for higher infiltration rates due to increasing surface roughness [48]. During recent years, most of the BSC areas were overgrown by higher vegetation. This development provides a good example for abiotic/biotic feedbacks and their spatiotemporal functioning.

Other examples for these kinds of feedback mechanisms are the spatial spreading of plant species within the catchment once they have established and the effect of surface structures like erosion gullies on plant species distribution. After the first years that were dominated by a disperse colonization of the site by prolific spreader species like *Conyza canadensis* [49], most of the newly arriving species first established in the western part of the catchment with the more loamy sands and then spread from there to other parts of the catchment [50, 51]. In some parts of the catchment, incision of erosion gullies increased to the point that the gully floor intersected with the groundwater surface leading to permanent groundwater seepage into these gullies. This transformed the ephemeral channels into more permanent stream networks [52]. *Phragmites australis*, a reed species that first established around the pond area in the lower part of the catchment used these stream networks to extend its distribution uphill preferentially along the channels (**Figure 5**) probably due to better water availability caused by more favorable physical properties of the stream bed sediments and by a closer proximity to the groundwater surface.

Despite the high surface runoff in the first years, an unconfined aquifer was formed above the clay layer across the hillslope (**Figure 6**). The groundwater table showed a seasonal and spatial variation but an overall increasing trend during the first 5 years.

In 2010, a very wet year with more than 900 mm precipitation compared to the long-term average of 563 mm, groundwater levels peaked and reached almost the soil surface in many parts of the catchment. Afterward, levels decreased again. The overall discharge from the catchment is

Figure 5. Spatial distribution and areal cover of *Phragmites australis* in the catchment from 2007 to 2011 (data from Jansone 2012, unpublished).

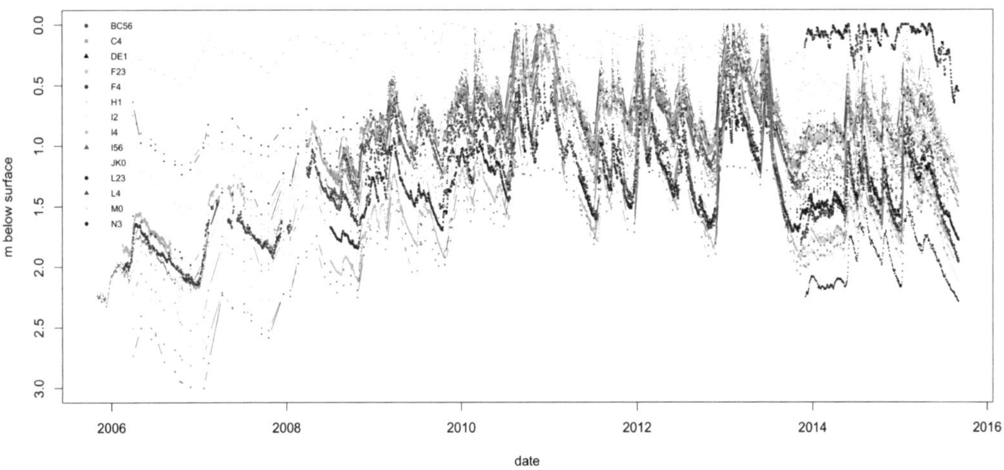

Figure 6. Groundwater levels below surface at different grid points in the Chicken Creek catchment from 2005 to 2015.

controlled by the water level in the pond above the outlet weir. Until spring 2007, no discharge was recorded, due to the filling processes at the hillslope and the pond. Then, spiky, episodic discharge events were induced by rain, surface runoff, and snowmelt events (**Figure 7**).

Discharge also peaked in 2010 (max. 21 L s^{-1}) and became more continuous afterward but still with long periods without discharge during summer. Since summer 2013 only very low discharge <1 L s^{-1} was recorded. Data from groundwater levels at the hillslope and water level in the pond that determines discharge showed no significant correlation during the first years (**Figure 8**).

	2006	2007	2008	2009	2010	2011	2012	2013	2014	2015
Σ precipitation[mm]:	432	635	572	600	829	726	681	707	664	554
Σ discharge [mm]:	0	89	78	122	567	289	243	202	40	27

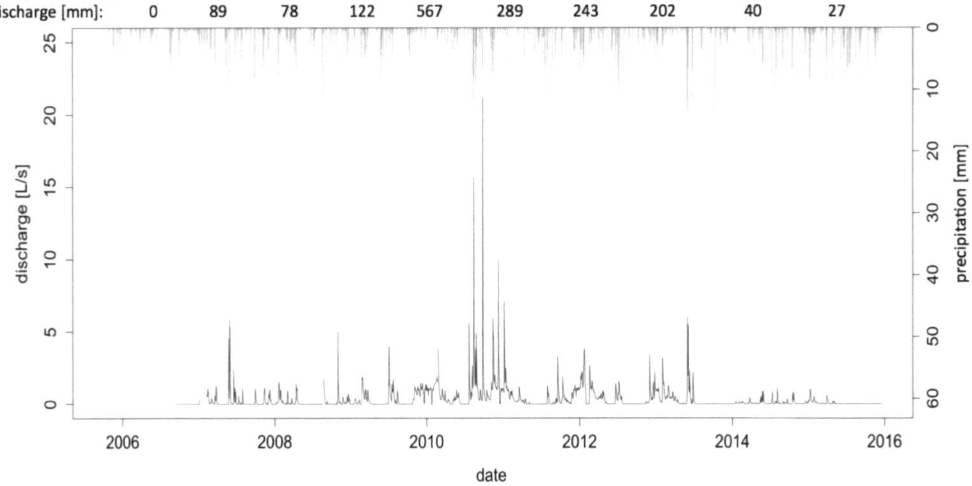

Figure 7. Discharge from the catchment (at the pond outlet) and precipitation from 2005 to 2015 (values on top are annual sums for hydrological year, i.e., November 1–October 31).

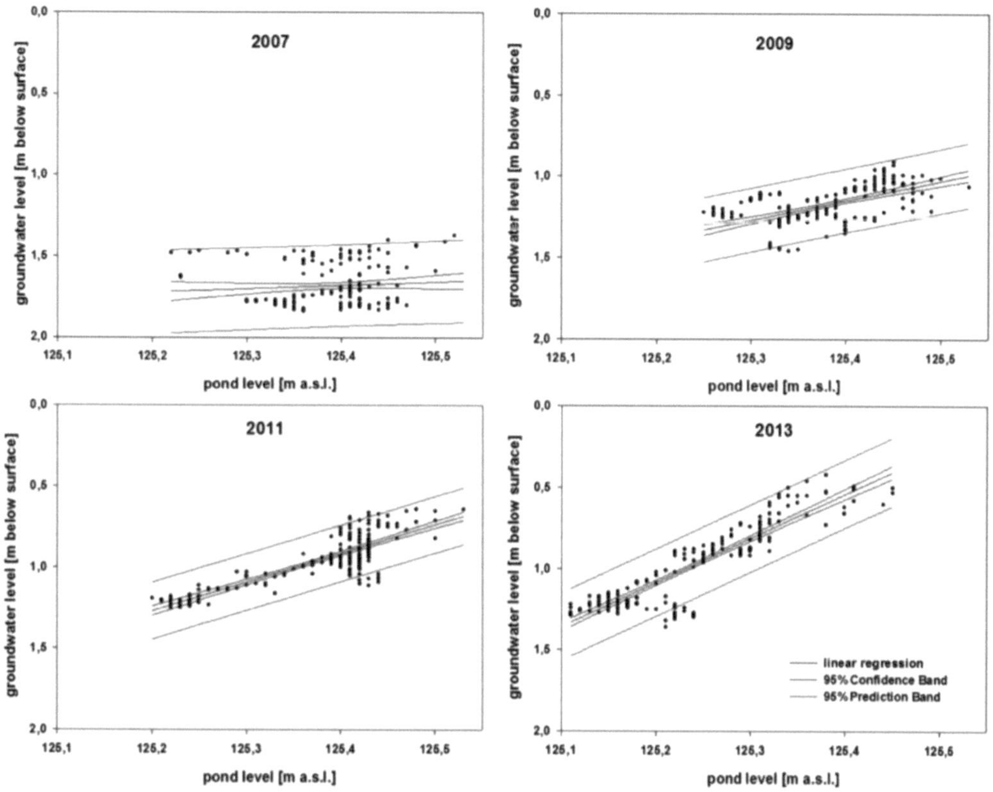

Figure 8. Correlation between groundwater levels and the water level in the pond for selected years.

With time, both the correlation between the coefficient and the slope of the linear regression increased. This indicated that discharge was not controlled by groundwater flow in the beginning but mainly by surface runoff from the hillslope. With development of the catchment, especially the buildup of groundwater and filling of the aquifer, base flow became likely the dominating source of discharge as is typical for many catchments [53]. Time series of R^2 values and slope of the regression between the groundwater and pond level showed a similar trend as the cover of vascular plants in the catchment (**Figure 9**).

These results indicate a transition of the hydrological regime from a surface runoff dominated discharge state to a state in which evapotranspiration and groundwater flow dominate the catchment response to rainfall.

Over the period of 10 years, we could define three phases of feedback controls on the catchment hydrology (**Figure 10**). In the very initial phase, these controls were mainly abiotic feedbacks between the initial substrate properties and rainfall events resulting in surface runoff, erosion, and gully formation (**Figure 10a**). These processes controlled discharge in the catchment and resulted in geomorphological changes of the hillslope surface. Surface runoff was promoted by the almost complete absence of vegetation and the sealing of surfaces due to the formation of various soil crusts. In this phase, substrate was transported from the hillslope downhill forming

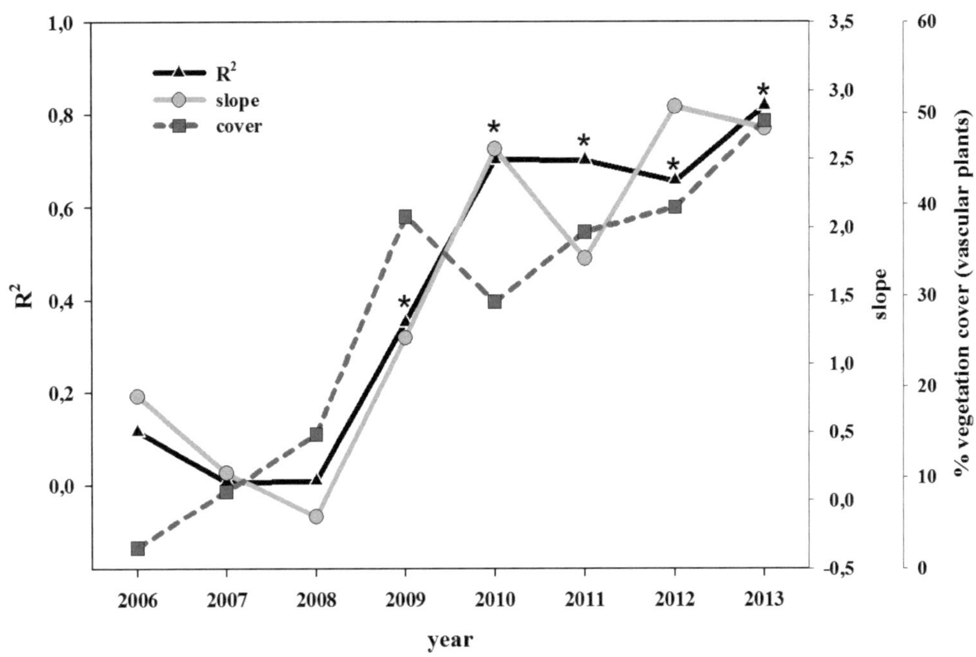

Figure 9. R^2 and slope parameters of the regression groundwater vs. pond water levels and vascular plant cover in the catchment for the period 2006–2013 (* indicates significant correlation at $p < 0.001$).

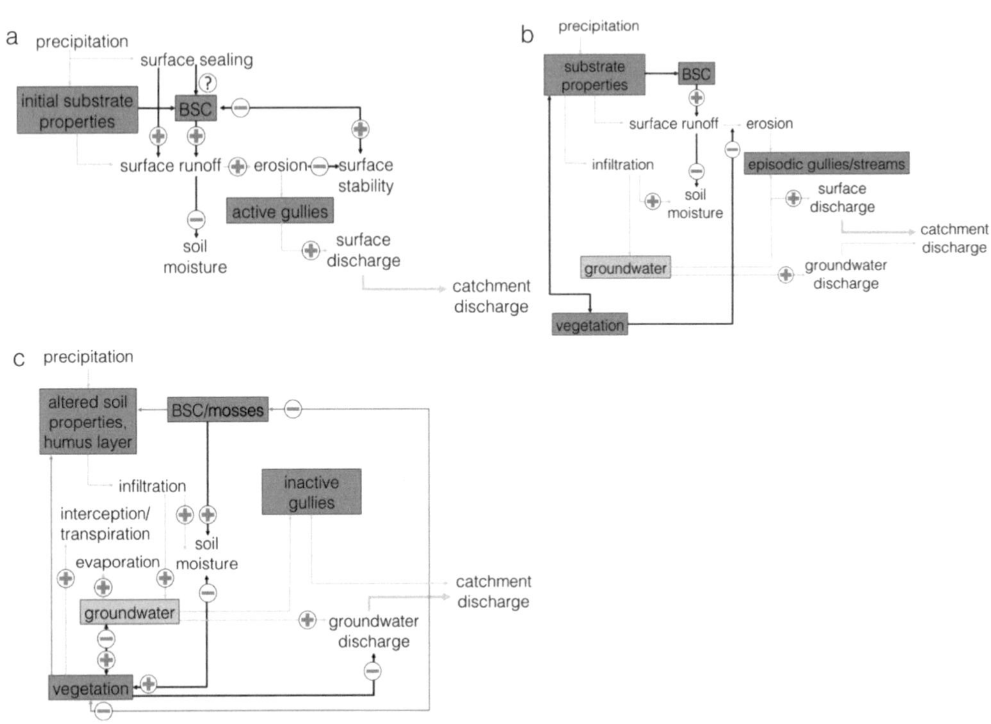

Figure 10. Feedback processes controlling the hydrology of the Chicken Creek catchment in different phases of development (a – c).

a sedimentation fan in the lower part above and in the pond. This resulted not only in a sorting of particle sizes within the fan but also in a residual accumulation of coarse-textured fragments at the soil surface uphill. Due to the various filling processes in the Pleistocene sediment body above the clay layer and in the pond, catchment discharge in this phase was very low and was dominated by single episodic events of surface runoff triggered by rainfall events or snowmelt.

After the rapid invasion and establishment of vascular plants, the second phase was mainly controlled by abiotic/biotic feedbacks (**Figure 10b**). As the vegetation cover increased, the soil surface was stabilized, and surface roughness increased. Groundwater levels increased and promoted plant growth especially within the gullies due to better water availability. Surface runoff and erosion were reduced and occurred only at episodic events after heavy rains or thunderstorms. After the pond was filled very quickly and the groundwater had increased to high levels, catchment discharge was high and more continuous in this phase.

In the third phase, catchment hydrology was still controlled by abiotic/biotic feedback mechanism, but biotic components had an increasing impact (**Figure 10c**). The vegetation and especially the woody species clearly profited from the high groundwater table and allowed a strong growth. The BSC developed to a moss dominated state, which changed their role in hydrological functioning from reducing infiltration and surface runoff promotion to increasing water storage, infiltration, and higher soil moisture below the BSC. At the same time, the spatial distribution of BSC was reduced due to overgrowing by higher vascular plants. Both vegetation and BSC altered the soil properties of the uppermost topsoil by increasing the organic matter contents and by accumulation of surface litter, which resulted in better infiltration. Evapotranspiration increased both due to evaporation in periods with groundwater tables close to the surface and the increase of vegetation cover throughout the catchment, which in turn lowered the groundwater. Especially, the higher woody vegetation increased canopy interception. Some of the mainly inactive erosion gullies that were incised deep enough to drain groundwater contributed to a higher groundwater discharge in this phase. Surface runoff was almost completely stopped. Due to the elevated evapotranspiration demand, catchment discharge decreased again and was mainly fed by groundwater.

These three phases of feedback controls on catchment hydrology have to be seen as an idealization of governing processes and the transition between them is gradual. The overall fast development within the catchment also increased spatial heterogeneity and diversity. Therefore, different parts and patches within the catchment may be in different phases at the same time or may remain in single phases over different periods. This behavior depends mainly on initial variations in substrate properties, on vegetation patterns formed and on spatial extent and intensity of the described feedback processes.

2.2. Conclusions

Ten years of data from the Chicken Creek catchment showed a very rapid development of the site with an increasing complexity and heterogeneity. In the first years, stochastic signals like the initial substrate conditions and external drivers like extreme weather were the most important factors resulting in abiotic/abiotic feedback mechanisms shaping the morphology of the site and creating site diversity. Invading vegetation from the regional species pool (representing external ecological memory) over time increased the role of abiotic/biotic feedbacks as could be shown in many examples.

Sites like Chicken Creek with known boundary conditions and structure information could help in disentangling feedback mechanisms between hydrological, pedogenic, biological, and geomorphological processes. Also, a more integrative view of succession and its drivers during the transition from initial, less complex systems to more mature ecosystems can be derived from such experimental sites developing from defined starting conditions [54–56]. Long-term time series of data are a key for a better understanding of these processes and the effects on ecosystem resilience [3, 57] and self-organization [19] as well as past and future effects of disturbance and global change [58].

Acknowledgements

This study was part of the Transregional Collaborative Research Centre 38 (CRC/TR 38), which was financially supported by the Deutsche Forschungsgemeinschaft (DFG, Bonn). The monitoring was funded by the Brandenburg University of Technology (BTU) and the Brandenburg Ministry of Science, Research, and Culture (MWFK, Potsdam). The authors thank Vattenfall Europe Mining AG for providing the research site and the many helping hands at the BTU contributing to the field and lab work.

Author details

Wolfgang Schaaf[1]*, Christoph Hinz[2], Werner Gerwin[3], Markus K. Zaplata[2] and Reinhard F. Huettl[1,4]

*Address all correspondence to: wolfgang.schaaf@b-tu.de

1 Soil Protection and Recultivation, Brandenburg University of Technology Cottbus – Senftenberg, Cottbus, Germany

2 Hydrology and Water Resources Management, Brandenburg University of Technology Cottbus – Senftenberg, Cottbus, Germany

3 Research Centre Landscape Development and Mining Landscapes (FZLB), Brandenburg University of Technology Cottbus – Senftenberg, Cottbus, Germany

4 Helmholtz Centre Potsdam – GFZ German Research Centre for Geosciences, Potsdam, Germany

References

[1] National Research Council (2001). Basic Research Opportunities in Earth Science. Washington D.C.: National Academies Press; 2001

[2] Brantley SL, White TS, White AF, Sparks D, Richter D, Pregitzer K, Derry L, Chorover J, Chadwick O, April R, Anderson S. Frontiers in Exploration of the Critical Zone: Report

of a Workshop Sponsored by the National Science Foundation (NSF), October 24-26. Newark DE; 2005

[3] Bennet EM, Cumming G, Peterson GD. A systems model approach to determining resilience surrogates for case studies. Ecosystems. 2005;**8**:945-957. DOI: 10.1007/s10021-005-0141-3

[4] Jørgensena SE, Patten BC, Straškraba M. Ecosystem emerging: 4. Growth. Ecological Modelling. 2000;**126**:249-284. DOI: 10.1016/S0304-3800(00)00268-4

[5] Campbell JL, Driscoll CT, Eagar C, Likens GE, Siccama TG, Johnson CE, Fahey TJ, Hamburg SP, Holmes RT, Bailey AS, Buso DC. Long-term trends from ecosystem research at the Hubbard brook experimental Forest. General Technical Report. NRS-17. 2007;41

[6] Ellenberg H, Robert M, Schauermann J. Ökosystemforschung: Ergebnisse des Sollingprojektes 1966-1986. Stuttgart: Ulmer; 1986. p. 507. DOI: 10.1002/jpln.19881510415

[7] Pennisi E. A groundbreaking observatory to monitor the environment. Science. 2010;**328**:418-210. DOI: 10.1126/science.328.5977.418

[8] Schaaf W, Bens O, Fischer A, Gerke HH, Gerwin W, Grünewald U, Holländer HM, Kögel-Knabner I, Mutz M, Schloter M, Schulin R, Veste M, Winter S, Hüttl RF. Patterns and processes of initial terrestrial-ecosystem development. Journal of Plant Nutrition and Soil Science. 2011;**174**(2):229-239. DOI: 10.1002/jpln.201000158

[9] Bishop JG. Early primary succession on mount St. Helens: Impact of insect herbivores on colonizing lupines. Ecology. 2002;**83**:191-202. DOI: 10.1890/0012-9658(2002)083[0191:EPSOMS]2.0.CO;2

[10] Bradshaw AD. The reconstruction of ecosystems. Journal of Applied Ecology. 1983;**20**:1-17

[11] Fridriksson S. Surtsey, Ecosystem Formed. Reykjavik: University of Iceland; 2005. p. 112

[12] Hüttl RF, Weber E. Forest ecosystem development in post-mining landscapes: A case study of the Lusatian lignite district. Die Naturwissenschaften. 2001;**88**:322-329. DOI: 10.1007/s001140100241

[13] Schaaf W. What can element budgets of false-time series tell us about ecosystem development on post-lignite mining sites? Ecological Engineering. 2001;**17**:241-252. DOI: 10.1016/S0925-8574(00)00142-7

[14] Walker LR. Ecosystems of Disturbed Ground. Ecosystem of the World. 16th ed. Amsterdam, The Netherland: Elsevier; 1999. p. 868. DOI: 10.1046/j.1365-2745.2000.00499-3.x

[15] Chadwick OA, Chorover J. The chemistry of pedogenic tresholds. Geoderma. 2001;**100**(3-4):321-353. DOI: 10.1016/S0016-7061(01)00027-1

[16] Lambers H, Mougel C, Jaillard B, Hinsinger P. Plant-microbe-soil interactions in the rhizosphere: An evolutionary perspective. Plant and Soil. 2009;**321**(1):83-115. DOI: 10.1007/s11104-009-0042-x

[17] Turner BL, Laliberté E. Soil development and nutrient availability along a 2 million-year coastal dune chronosequence under species-rich mediterranean shrubland in southwestern Australia. Ecosystems. 2015;**18**(2):287-309. DOI: 10.1007/s10021-014-9830-0

[18] Vitousek PM, Chadwick OA. Pedogenic thresholds and soil process domains in basalt-derived soils. Ecosystems. 2013;**16**(8):1379-1395. DOI: 10.1007/s10021-013-9690-z

[19] Fichter LS, Pyle EJ, Whitmeiyer SJ. Expanding evolutionary theory beyond darwinism with elaborating, self-organizing, and fractionating complex evolutionary systems. Journal of Geoscience Education. 2010;**58**(2):58-64. DOI: 10.5408/1.3534853

[20] Ryan JG, Ludwig JA, Mcalpine CA. Complex adaptive landscapes (CAL): A conceptual framework of multi-functional, non-linear ecohydrological feedback systems. Ecological Complexity. 2007;**4**(3):113-127. DOI: 10.1016/j.ecocom.2007.03.004

[21] Groffman PM, Baron JS, Blett T, Gold AJ, Goodman I, Gunderson LH, Levinson BM, Palmer MA, Paerl HW, Peterson GD, Poff NL, Rejeski DW, Reynolds JF, Turner MG, Weathers KC, Wiens J. Ecological thresholds: The key to successful environmental management or an important concept with no practical application? Ecosystems. 2006;**9**(1):1-13. DOI: 10.1007/s10021-003-0142-z

[22] Peterson GD. Contagious disturbance, ecological memory, and the emergence of landscape pattern. Ecosystems. 2002;**5**(4):329-338. DOI: 10.1007/s10021-001-0077-1

[23] Rietkerk M, Dekker SC, de Ruiter P, van de Koppel J. Self-organized patchiness and catastrophic shifts in ecosystems. Science. 2004;**305**(5692):1926-1929.DOI: 10.1126/science.1101867

[24] Scheffer M, Carpenter S, Foley JA, Folke C, Walker B. Catastrophic shifts in ecosystems. Nature. 2001;**413**:591-596. DOI: 10.1038/35098000

[25] Scheffer M, Bascompte J, Brock WA, Brovkin V, Carpenter SR, Dakos V, Held H, van Nes EH, Rietkerk M, Sugihara G. Early-warning signals for critical transitions. Nature. 2009;**461**:53-59. DOI: 10.1038/nature08227

[26] Zeng X, Shen SSP, Zeng X, Dickinson RE. Multiple equilibrium states and the abrupt transitions in a dynamical system of soil water interacting with vegetation. Geophysical Research Letters. 2004;**31**:L05501. DOI: 10.1029/2003GL018910

[27] Kefi S, Rietkerk M, Alados CL, Pueyo Y, Papanastasis VP, ElAich A, deRuiter PC. Spatial vegetation patterns and imminent desertification in Mediterranean arid ecosystems. Nature. 2007;**449**:213-217. DOI: 10.1038/nature06111

[28] Hüttl RF, Hinz C, Kögel-Knabner I, Schulin R, Gerwin W, Subke JA. Ecosystems in transition: Interactions and feedbacks with an emphasis on the initial development. Biogeosciences. 2014;**11**:195-200. DOI: 10.5194/bg-11-195-2014

[29] Gerwin W, Schaaf W, Biemelt D, Fischer A, Winter S, Hüttl RF. The artificial catchment "Chicken Creek" (Lusatia, Germany)—A landscape laboratory for interdisciplinary studies of initial ecosystem development. Ecological Engineering. 2009;**35**(12):1786-1796. DOI: 10.1016/j.ecoleng.2009.09.003

[30] Gerwin W, Schaaf W, Biemelt D, Elmer M, Maurer T, Schneider A. The artificial catchment 'Hühnerwasser' (Chicken Creek): Construction and initial properties. Ecosystem Development. 2010;**1**:1-56: urn:nbn:de:kobv:co1-opus-20725

[31] Elmer M, Gerwin W, Schaaf W, Zaplata MK, Hohberg K, Nenov R, Bens O, Hüttl RF. Dynamics of initial ecosystem development at the artificial catchment Chicken Creek, Lusatia, Germany. Environmental Earth Sciences. 2013;**69**(2):491-505. DOI: 10.1007/s12665-013-2330-2

[32] Schaaf W, Elmer M, Fischer A, Gerwin W, Nenov R, Pretzsch H, Seifert S, Winter S, Zaplata MK. Monitoring the formation of structures and patterns during initial development of an artificial catchment. Environmental Monitoring and Assessment. 2013;**185**(7):5965-5986. DOI: 10.1007/s10661-012-2998-x

[33] Gerwin W, Schaaf W, Biemelt D, Winter S, Fischer A, Veste M, Hüttl RF. Overview and first results of ecological monitoring at the artificial watershed Chicken Creek (Germany). Physics and Chemistry of the Earths Parts A/B/C. 2011;**36**(1-4):61-73. DOI: 10.1016/j.pce.2010.11.003

[34] Schaaf W, Elmer M, Fischer A, Gerwin W, Nenov R, Pretzsch H, Zaplata MK. Feedbacks between vegetation, surface structures and hydrology during initial development of the artificial catchment 'Chicken Creek'. Procedia Environmental Sciences. 2013;**19**:86-95. DOI: 10.1016/j.proenv.2013.06.010

[35] Zaplata MK, Winter S, Fischer A, Kollmann J, Ulrich W. Species-driven phases and increasing structure in early-successional plant communities. The American Naturalist. 2013;**181**(1):E17-E27. DOI: 10.5061/dryad.4411b

[36] Schneider A, Gerke HH, Maurer T, Nenov R. Initial hydro-geomorphic development and rill network evolution in an artificial catchment. Earth Surface Processes and Landforms. 2013;**38**(13):1496-1512. DOI: 10.1002/esp.3384

[37] Fischer T, Veste M, Schaaf W, Dümig A, Kögel-Knabner I, Wiehe W, Bens O, Hüttl RF. Initial pedogenesis in a topsoil crust 3 years after construction of an artificial catchment in Brandenburg, NE Germany. Biogeochemistry. 2010;**101**(1/3):165-176. DOI: 10.1007/s10533-010-9464-z

[38] Bormann H, Holländer HM, Blume T, Buytaert W, Chirico GB, Exbrayat JF, Gustafsson D, Hölhel H, Kraft P, Krauße T, Nazemi A, Stamm S, Stoll S, Blöschl G, Flühler H. Comparative discharge prediction from a small artificial catchment without model calibration: Representation of initial hydrological catchment development. Bodenkultur. 2011;**62**(1-4):23-29

[39] Holländer HM, Bormann H, Blume T, Buytaert W, Chirico GB, Exbrayat JF, Gustafsson D, Hölzel H, Krauße T, Kraft P, Stoll S, Blöschl G, Flühler H. Impact of modellers' decisions on hydrological a priori predictions. Hydrology and Earth System Sciences. 2014;**18**:2065-2085. DOI: 10.5194/hess-18-2065-2014

[40] Berdugo M, Soliveres S, Maestre FT. Vascular plants and biocrusts modulate how abiotic factors affect wetting and drying events in Drylands. Ecosystems. 2014;**17**(7):1242-1256. DOI: 10.1007/s10021-014-9790-4

[41] Chamizo S, Cantón Y, Lázaro R, Solé-Benet A, Domingo F. Crust composition and disturbance drive infiltration through biological soil crusts in semiarid ecosystems. Ecosystems. 2012;**15**(1):148-161. DOI: 10.1007/s10021-011-9499-6

[42] Rodríguez-Caballero E, Cantón Y, Chamizo S, Lázaro R, Escudero A. Soil loss and runoff in semiarid ecosystems: A complex interaction between biological soil crusts, micro-topography, and hydrological drivers. Ecosystems. 2013;**16**(4):529-546. DOI: 10.1007/s10021-012-9626-z

[43] Veste M, Littmann T, Breckle SW, Yair A. The Role of Biological Soil Crusts on Desert Sand Dunes in the Northwestern Negev, Israel. Sustainable Land use in Deserts. Heidelberg: Springer; 2001. p. 465. DOI: 10.1007/978-3-642-59560-8_38

[44] Yair A, Almog R, Veste M. Differential hydrological response of biological topsoil crusts along a rainfall gradient in a sandy arid area: Northern Negev desert, Israel. Catena. 2011;**87**(3):326-333. DOI: 10.1016/j.catena.2011.06.015

[45] Belnap J. The potential roles of biological soil crusts in dryland hydrologic cycles. Hydrological Processes. 2006;**20**(15):3159-3178. DOI: 10.1002/hyp.6325

[46] Rodríguez-Caballero E, Cantón Y, Chamizo S, Afana A, Solé-Benet A. Effects of biological soil crusts on surface roughness and implications for runoff and erosion. Geomorphology. 2012;**145-146**(1):81-89. DOI: 10.1016/j.geomorph.2011.12.042

[47] Kidron GJ, Yaalon DH, Vonshak A. Two causes for runoff initiation on microbiotic crusts: Hydrophobicity and pore clogging. Soil Science. 1999;**164**(1):18-27. DOI: 10.1097/00010694-199901000-00004

[48] Fischer T, Gypser S, Subbotina M, Veste M. Synergic hydraulic and nutritional feedback mechanisms control surface patchiness of biological soil crusts on tertiary sands at a post-mining site. Journal of Hydrology and Hydromechanics. 2014;**62**(4):293-302. DOI: 10.2478/johh-2014-0038

[49] Zaplata MK, Winter S, Biemelt D, Fischer A. Immediate shift towards source dynamics: The pioneer species *Conyza canadensis* in an initial ecosystem. Fuel and Energy Abstracts. 2011;**206**(11):928-934. DOI: 10.1016/j.flora.2011.07.001

[50] Biber P, Seifert S, Zaplata MK, Schaaf W, Pretzsch H, Fischer A. Relationships between substrate, surface characteristics, and vegetation in an initial ecosystem. Biogeosciences. 2013;**10**(12):8283-8303. DOI: 10.5194/bg-10-8283-2013

[51] Ulrich W, Piwczynski M, Zaplata MK, Winter S, Schaaf W, Fischer A. Small-scale spatial variability in phylogenetic community structure during early plant succession depends on soil properties. Oecologia. 2014;**175**(3):985-995. DOI: 10.1007/s00442-014-2954-2

[52] Gerull L, Frossar A, Gessner MO, Mutz M. Variability of heterotrophic metabolism in small stream corridors of an early successional watershed. Biogeosciences. 2011;**116**(G2):1-11. DOI: 10.1029/2010JG001516

[53] Gonzales AL, Nonner J, Heijkers J, Uhlenbrook S. Comparison of different base flow separation methods in a lowland catchment. Hydrology and Earth System Sciences. 2009;**13**(11):2055-2068. DOI: 10.5194/hess-13-2055-2009

[54] Hrachowitz M, Savenije HHG, Blöschl G, McDonnell JJ, Sivapalan M, Pomeroy JW, Arheimer B, Blume T, Clark MP, Ehret U, Fenicia F, Freer JE, Gelfan A, Gupta HV, Hughes DA, Hut RW, Montanari A, Pande S, Tetzlaff D, Troch PA, Uhlenbrook S, Wagener T, Winsemius HC, Woods RA, Zehe E, Cudennec C. A decade of predictions in ungauged basins (PUB) – A review. Hydrology Sciences Journal. 2014;**58**(6):1198-1255

[55] McGlynn BL, Blöschl G, Borga M, Bormann H, Hurkmans R, Komma J, Nandagiri L, Uijlenhoet R, Wagener T, editors. A data acquisition framework for predictions of runoff in ungauged basins. In: Blöschl G, Sivapalan M, Wagener T, Viglione A and Savenije H (eds.). Run-Off Prediction in Ungauged Basins: Synthesis across Processes, Places and Scales. Cambrigde: Cambridge University Press; 2013. 490 p. DOI: 10.1017/CBO9781139235761.006

[56] Meiners SJ, Cadotte MW, Fridley JD, Pickett STA, Walker LR. Is successional research nearing its climax? New approaches for understanding dynamic communities. Functional Ecology. 2015;**29**(2):154-164. DOI: 10.1111/1365-2435.12391

[57] Gunderson LH, Holling CS. Panarchy: Understanding Transformations in Human and Natural Systems. 1st ed. Washington, Covelo, London: Island Press; 2002. p. 536

[58] Lovett GM, Burns DA, Driscoll CT, Jenkins JC, Mitchell MJ, Rustad L, Shanley JB, Likens GE, Haeuber R. Who needs environmental monitoring? Frontiers in Ecology and the Environment. 2007;**5**(5):253-260. DOI: 10.1890/1540-9295(2007)5[253:WNEM]2.0.CO;2

Chapter 4

Changes of Water Budget during Ecosystem Development in Post-Mining Sites at Various Spatiotemporal Scales: The Need for Controlled Systems

Jan Frouz

Additional information is available at the end of the chapter

http://dx.doi.org/10.5772/intechopen.74238

Abstract

This chapter describes the development of the soil-water budget at various spatiotemporal scales, including an example of post-mining sites. This includes the formation of soil aggregates and porosity, the development of water retention in the soil profile, and water losses by runoff and evapotranspiration. It is emphasized that the development of soil-water retention is closely linked to carbon storage in post-mining soils, which is strongly affected by litter quality. Plants with a high C:N ratio of litter feature most of the organic matter in Oe and litter layers, which results in a lower soil-water storage, whereas soil covered by vegetation with low litter C:N ratios produces organo-mineral aggregates and deeper A horizons that promote water storage. Moreover, the need for controlled catchment conditions to get a better understanding of how these processes on various spatiotemporal scales interact is emphasized.

Keywords: soil aggregates, soil profile, catchment, water retention, runoff, evapotranspiration

1. Introduction

Water movement in the landscape is a complex process, consisting of many connected and interacting processes at various spatiotemporal scales. These include processes on the level of soil aggregates, which affect infiltration and the ability of soil to hold water. Formation of macropore connectivity and surface channels, which affect surface and subsurface runoff.

Bedrock weathering and transportation processes affect redistribution of clay and nutrients. Organic matter in the soil profile and in the landscape feeds back to water movement and storage in the ecosystem. Many of these processes are affected by plants and soil biota, such as aggregate formation, creation of a porosity network, water intake by plants, or interception of rain by vegetation. Water enters the ecosystem by rainfall and leaves it by runoff and evapotranspiration. In the ecosystem, water can be stored in vegetation and in soil. The previous research shows that the ability of soil to store water is closely related to storage of soil organic matter (SOM) [1]. During ecosystem development, a vegetation cover develops, which reduces water input to the soil by interception and increases water loss by transpiration, but reduces evaporation from open soil surfaces. Vegetation also produces litter and root exudates, which are important for the activity of soil biota. Soil biota, which mostly directly or indirectly feeds on plant products, affects aggregate formation, storage of SOM, and distribution of SOM in the soil profile. Certain types of plants promote the activity of specific assemblies of soil biota, which affect certain patterns of SOM storage and distribution in soil. Plants with a conservative growing strategy promote a soil community that causes no or little bioturbation, which results in a thick litter and Oe layer on the soil surface. On the contrary, fast-growing plants promote intensive bioturbation by soil fauna and the formation of a deep A organo-mineral soil horizon [2–5]. This affects overall SOM storage [6, 2], which very likely affects water storage in soil as well [1]. Plants affect water movement in the system also by other means, such as by a different degree of interception and fate of water trapped by interception or different water consumptions and transpiration rates. Although all of these parameters have been intensively studied, we have only little information about how individual plant traits that affect SOM behavior in the soil relate to various parameters determining water storage and movement in the ecosystem. These interactions have been intensively studied in terms of the relationship between soil development and SOM storage, and although there is a general understanding that SOM may be closely related to soil-water, much less is known about factors and mechanisms affecting the water regime development during ecosystem development.

Mining and open-cast mining cause large disturbances to ecosystems. Most of the affected ecosystems are completely erased, either excavated or buried under overburden, which usually substantially differs from well-developed soils. In addition to texture, the hydrophobicity of the substrate [7], lack of macropores, soil compaction, and sometimes salinity are factors that can affect soil conditions, particularly the soil-water regime [8, 9]. The study of ecosystem development at these sites has a large practical impact. Post-mining sites also represent locations that have a great potential to study these processes. One of the reasons that make these sites suitable for the study of ecosystem development is the presence of sites of various ages, so-called chronosequences, where ecosystem development can be studied by comparing individual parameters or processes on sites of different ages. This approach, called also space-for-time substitution, allows the study of long-term processes in a very short time. Despite its clear advantages, this type of investigation has also its limitations, as each site develops in a specific trajectory, which may differ from the general chronosequence pattern. Another reason why post-mining sites are good systems to study successional processes is that these sites offer a combination of parameters that may not occur elsewhere and are suitable for large-scale landscape manipulations, which would be technically hardly possible or ethically questionable elsewhere.

The aim of this contribution is to describe processes that affect the development of the water regime at post-mining sites after open-cast coal mining near Sokolov based on the extensive study of chronosequences at these sites. In addition, the idea of constructing isolated, controlled micro-catchments that would allow the investigation of these processes on various levels, from soil aggregates to the landscape scale, is presented.

2. Water regime development in chronosequence studies across various spatiotemporal scales

As already explained, the development of a water regime can be divided into two parts: the development of soil, which stores water in the ecosystem, and the development of vegetation, which is an important consumer of water. However, vegetation also determines many transportation processes, which affect movement of soil between ecosystem compartments and surroundings, and directly or indirectly determines soil-forming processes. In this chapter, we will follow the formation of soil aggregates and their role in water retention, followed by the development of the whole soil profile and, finally, the development of vegetation and its role in soil profile development.

2.1. Microscale processes of aggregate formation and porosity

Soil aggregates are secondary particles formed through the rearrangement and cementation of primary mineral particles with SOM [10]. They are often grouped by size into macroaggregates (250–2000 μm) and microaggregates (53–250 μm) [11–13]. A highly influential model on the formation of aggregates was published by Tisdall and Oades [14] that was later modified by Oades [15]. Based on these models, it is now commonly accepted that macroaggregates form first, mainly by the entanglement of particles by fungal hyphae and roots (temporary binding agents) and around fresh particulate organic matter (POM) inputs. When these temporary binding agents and the POM in macroaggregates decompose into fragments, coated with mucilage produced during decomposition, they become encrusted with clay particles and thus form the nucleus for microaggregates within macroaggregates [11]. As a consequence of this "aggregate hierarchy," macroaggregates contain more C and higher amounts of labile C as compared to microaggregates, where SOM is more processed and regarded as relatively stable over longer periods of time. However, because of the higher lability of macroaggregates, the stability of microaggregates (contained within macroaggregates) depends to some extent on the turnover of macroaggregates. Apart from the earlier mentioned roots and fungal hyphae, factors generally positively influencing the formation and stability of aggregates are soil bacteria, producing extracellular polymeric substances, thereby cementing soil particles predominantly in microaggregates [14, 15], and the soil fauna, especially earthworms, forming stable casts and exerting pressure on the walls of their burrows, thereby compacting the soil [13, 16]. At initial post-mining sites, the clay content is another crucial factor as it forms the backbone of stable microaggregates [16]. At post-mining sites, overburden can be formed by clastic material, such as sand or gravel, but often it is deposited in the form of less consolidated material, such as shales, madstones, or stones [17, 18]. Weathering of this material and a gradual increase of the clay content is an important step in soil aggregate formation [19]. At many post-mining sites, aggregates are formed from clay or organic matter that serve as cementing material and fine pieces of weathered consolidated material [20].

Soil fauna can also promote aggregate formation and enhance water-holding capacity by transformation of organic matter. Analyzing soil cores filled with leaf litter and bibionid fly larvae revealed that even litter-feeding fauna can increase the ability of soil to hold water just by turning litter into fecal pellets, which increase water field capacity manifold. Litter transformation into fecal pellets causes fragmentation of the leaf, breaking the cell walls, and opens these capillary surfaces to be accessible to soil-water. Moreover, compaction of material inside the fecal pellet creates additional capillary pores by compacting consumed litter. These holo-organic aggregates can, in some soils, form layers (Oe layer) in several centimeters thick [1, 21–23]. Even more important is the formation of soil aggregates by earthworms [24–27], which can also contribute significantly to the greater ability of soil to hold water [22, 28]. Increased water-holding capacity is, however, in both cases mentioned above accompanied by an increase in wilting point. Nevertheless, in both cases we can conclude that the soil fauna enhanced the ability of soil to hold water, which was then potentially available for plants as increase in water field capacity was in both cases higher than the increase in wilting point (**Figure 1**). Earthworm-created aggregates may also promote formation of larger soil pores between aggregates, which can enhance infiltration and decrease runoff and erosion. However, some other soil biota, namely, ants, might accumulate unconsolidated soil at the surface, which can be easily removed by erosion and in fact increase erosion loss of soil from the plot [29].

Besides aggregate formation, soil macropore development is important in post-mining soils. Several studies indicate a gradual increase of porosity in aging post-mining soils [19, 30]. Besides physical processes and roots, earthworms, ants, and also termites in tropical and subtropical parts of the world are most commonly mentioned in this context [31–35]; however, a much larger diversity of soil invertebrates, such as solitary bees or wasps, spiders [36], coleoptera [37], orthoptera, and even crayfish [38], significantly contribute to biopore formation. These macropores can differ substantially in size and shape and in how they

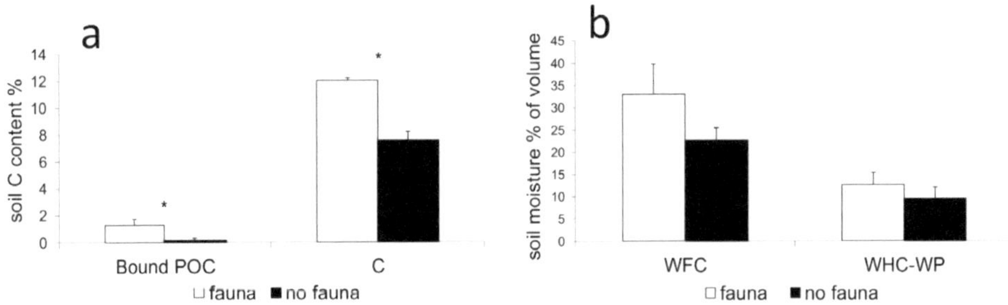

Figure 1. Carbon content in bulk soil C and in proportion of C bound in aggregates from total soil mass (POC bound) in soil aggregates created by soil fauna (specifically by earthworms) and other aggregates (based on data from [20]) (a), water field capacity (WFC), and difference between WFC and wilting point (WFC-WP) (b), for clay post-mining substrate closed in a macrocosm with and without access of soil fauna for 3 years based on data from [22]. *significant difference between fauna-accessible and fauna-non-accessible treatments (t-test, $p < 0.05$).

are formed. Roots basically push themselves in the soil with the help of lubrication effect of root exudates. Similarly, earthworms burrow themselves into the soil by pushing soil to the side, forming a layer of more compacted soil along the burrow wall, which is covered by earthworm mucus. The area in which earthworms actively burrow is called the drilosphere. Ants and most other arthropods commonly dig into the soil, and excavated soil can be transported out from the hole, used to fill unused cavities, or partly spread on the walls to form a ceiling; various other ceilings such as spider net can be used in biopores [33, 36]. Despite the fact that this large variability in biopore construction is well known, a deeper understanding of how individual "construction patterns" affect the function of biopores in soil is still lacking.

2.2. Upscaling small-scale processes to whole soil profile development

Here, we now explore how the small-scale processes described above transfer to the development of the whole soil profile. We use an example of two 20-year-old soils developing de novo at two adjacent locations. The first site is reclaimed leveled and planted with alder (*Alnus glutinosa*) seedlings. Alder is a nitrogen-fixing tree which provides litter with low C:N ratio. Consequently, there were abundant macrofauna community with a high density of earthworms, *Lumbricus rubellus* and *Aporrectodea caliginosa*, in alder site. In contrast, the second site is unreclaimed and keeps its wavelike character created by heaping; it is vegetated by a spontaneous regrowth dominated by willow, birch, and aspen (*Salix caprea*, *Betula pendula*, and *Populus tremula*) and had a less abundant macrofauna, and earthworm species that mix litter with soil are missing. The consequence of this absence of mixing earthworms on spontaneous regrowth was a thick fermentation layer on the soil surface. In the contrary, in the alder plantation, a litter has been immediately fragmented and mixed into the soil, forming an organo-mineral layer [23]. Previous micromorphological observations showed that worm casts can contribute about half of the solid fraction of soil [23]. Intensive mixing of litter into mineral soil can be, under suitable conditions (under trees producing easily available litter such as alder or lime), a reason for fast formation of organo-mineral A layer which can up to 15 cm thick after 40 years of development in some soils. Looking at various soils developing on the same post-mining heap, we can see that the ability to hold water closely connected with organic matter accumulation in these soils (**Figure 2**). As has been shown previously, accumulation of organic matter in mineral soil closely corresponds with bioturbation caused by soil fauna, primarily earthworms [6]. The highest water field capacity was observed at alder stands, which produces easily decomposable litter and is subject to intensive bioturbation by earthworms. In contrast, a lower water field capacity was found in coniferous species with very limited soil faunal activity (**Figure 2**). However, as already mentioned, at the level of individual soil aggregates and also in the level of bulk soil, the wilting point correlates with water field capacity. This means that alder sites have higher water field capacity but also higher wilting point as compared to regrowth sites (**Figure 2**). The water regime at both sites after 30 years of development was basically similar, but higher moisture and consequently higher soil-water storage were found at the alder plantation (**Figure 3**), which is consistent with the already reported high water field capacity at these sites [12, 30, 39].

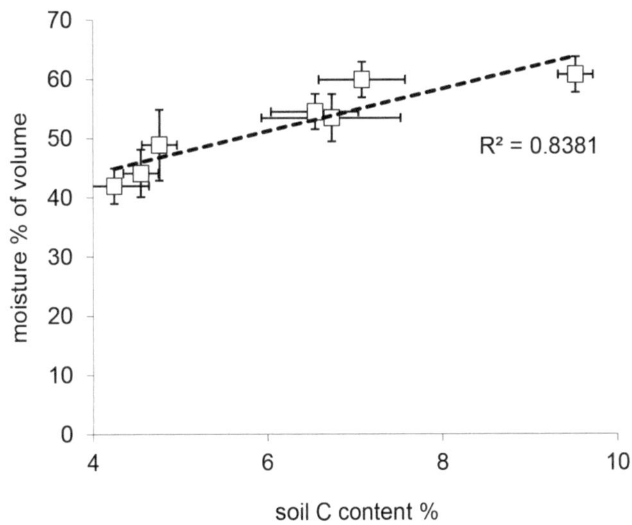

Figure 2. Water field capacity (WFC) of 30-year-old post-mining soils developing on the same clay soils under various tree species in relation to soil carbon content in particular sites based on data from [3].

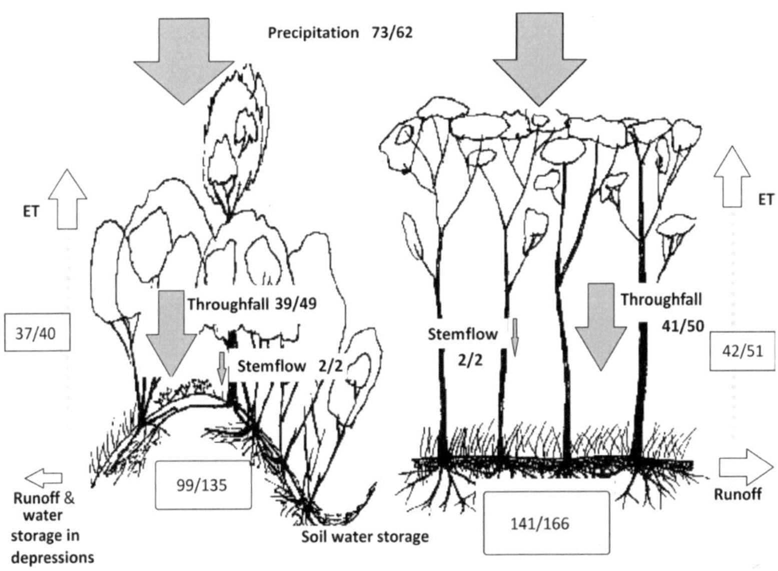

Figure 3. Monthly water budget of unreclaimed mean sites without any technical reclamation spontaneously covered by natural regrowth dominated by *Salix caprea* and *Betula pendula* (left) and reclaimed sites planted by alder (right) both sites about 30 years old, based on data from [39]. Data are monthly mean flows during vegetation season/outside vegetation season in mm or average stock in vegetation season/outside vegetation season.

2.3. Vegetation development and its role in the water regime

With increasing succession age, plant cover increases. Dynamics of this increase may certainly vary between various types of vegetation. For example, Frouz et al. [40] investigating reclaimed alder plantations and unreclaimed post-mining sites, found that biomass of reclaimed sites increased faster early after reclamation, but later on, differences between reclaimed and unreclaimed sites decrease, and older unreclaimed sites show even higher biomass that reclaimed ones. With increasing vegetation cover, water consumption by transpiration also increased [30, 39]. This can be ascribed to an increased temperature buffering. In a study [41] the difference between morning and afternoon temperatures was assumed as a measure of the temperature buffering ability of the ecosystem, which is proportional to the amount of water transpired by evapotranspiration. Buffering increased with increasing plant cover measured as NDVI but also depended on the type of vegetation; reclaimed sites had significantly higher buffering than unreclaimed sites. This indicates a higher water transpiration at reclaimed alder plantations than at unreclaimed woody regrowth sites [41]. A slightly higher water consumption at reclaimed alder plantations compared to unreclaimed plots has been indicated also by a complete water budget on these plots [39].

3. Potential of using controlled catchments in the study of water regime development at post-mining sites

What has been mentioned above opens many potential hypotheses about the development of the water regime in post-mining soils. It seems that soil-water storage is closely linked with SOM storage. Similarly, as proposed by Cejpek et al. [39], plants with a fast-growing strategy, which tend to store more SOM in mineral soil [3, 6], also tend to produce soil aggregates with more bound OM [20] and, consequently, soils store more water [19, 22, 30, 39]. This concept opens many other questions, such as how these parameters relate to the water balance (e.g., to runoff and evapotranspiration), how historical changes in soil carbon storage and water retention affect subsequent ecosystem development, and many others. Answering these questions may be interesting not only for a better understanding of ecosystem development at post-mining sites but also may have more general implication. However, answering these questions faces also many methodical challenges. With common instrumentation we are not able to follow that part of the landscape with woody vegetation that would realistically allow the measurement of all the water movement, including surface and subsurface runoff. In order to answer these and other questions, we plan to build small hydrologically isolated micro-catchments on a heap (similar to rainy hill of Chicken creek catchments [42]) in a way that the installed devices allow comprehensive monitoring of the flow of water and nutrients through the ecosystem as well as the exchange of gases between the ecosystem and the surrounding atmosphere. In particular, we will measure the rain water input, including dry and wet deposition, surface and subsurface runoff, the content of key elements of the discharge, water movement in the soil profile, total radiation, carbon exchange (CO_2) between the atmosphere and the whole ecosystem, and also between soil and atmosphere. The area itself will be divided into four micro-catchments with an area of about 0.25 ha each, two of which will be planted with alder and the other two will be left uncultivated. For each pair of areas (reclaimed and uncultivated), one eddy covariance tower will be located in the direction of the predominant winds. The area will then be equipped with container lysimeters and access shafts allowing for the easy implementation of additional ad hoc experiments.

The main component of the entire experimental catchments will be the monitoring of the water flow. For separation of surface and subsurface runoff, the underground clay layers will be compacted at a depth of 2 m to create an impermeable layer. This impervious layer will lead into a collecting channel fitted with a specific overflow and a subsurface drainage monitoring device. Another specific collecting trough fitted with a further measuring overflow and measuring equipment will then be placed on the surface of the terrain. This will allow a separate observation of surface and subsurface runoff. The flow monitoring devices will also take samples of water to measure flows, which will then allow the calculation of the balance of substances moving with the water.

The water input into the catchment area will be monitored by a set of rain gauges that will both monitor rainfall dynamics and capture rainwater for subsequent chemical analysis.

In addition to the water flow and gas exchange facilities, access shafts will be located in the catchment area. These are plastic shafts equipped with observation windows and preset points where additional accessories can be installed. This will allow the installation of any instrument to monitor the development of soil and nutrient flow during the operation of the experimental basin without the need for further disturbance, which is key to the function of the river basin. The area will also include container lysimeters to monitor soil development and to perform manipulation experiments. Those will be particularly valuable in an experiment aiming at upscaling processes from the soil aggregate level to the whole soil profile.

4. Conclusions

Large-scale hydrological parameters that determine water movement in the landscape level can be tracked in small-scale processes on the level of individual aggregates or soil pores. This tracking indicates that both of these processes can be driven by growth strategy of plants. Experiments in controlled conditions where both macro- and microscopic processes can be studied in more details are needed for better understanding of these interactions.

Acknowledgements

The study was supported by the SoWa Research Infrastructure funded by MEYS CZ grant LM2015075 program "Projects of Large Infrastructure for Research, Development, and Innovations."

Author details

Jan Frouz[1,2*]

*Address all correspondence to: jan.frouz@natur.cuni.cz

1 SoWa Research Infrastructure, Biology Centre CAS, České Budějovice, Czech Republic

2 Institute for Environmental Studies, Faculty of Sciences, Charles University, Prague, Czech Republic

References

[1] Frouz J, Kuráž V. Soil fauna and soil physical properties. In: Frouz J, editor. Soil Biota and Ecosystem Development in Post Mining Sites. Boca Raton: CRC Press; 2014

[2] Frouz J, Thébault E, Pižl V, Adl S, Cajthaml T, Baldrian P, Háněl L, Starý J, Tajovský K, Materna J, Nováková A, de Ruiter PC. Soil food web changes during spontaneous succession at post mining sites: A possible ecosystem engineering effect on food web organization? PLoS One. 2013a;**8**:e79694

[3] Frouz J, Livečková M, Albrechtová J, Chroňáková A, Cajthaml T, Pižl V, Háněl L, Starý J, Baldrian P, Lhotáková Z, Šimáčková H, Cepáková Š. Is the effect of trees on soil properties mediated by soil fauna? A case study from post-mining sites. Forest Ecology and Management. 2013b;**309**:87-95

[4] Ponge J-F. Plant–soil feedbacks mediated by humus forms: A review. Soil Biology and Biochemistry. 2013;**57**:1048-1060

[5] Frouz J. Effects of soil macro- and mesofauna on litter decomposition and soil organic matter stabilization. Geoderma. 2018. DOI: 10.1016/j.geoderma.2017.08.039

[6] Frouz J, Pižl V, Cienciala E, Kalcik J. Carbon storage in post-mining forest soil, the role of tree biomass and soil bioturbation. Biogeochemistry. 2009;**94**:111-121

[7] Doerr SH, Shakesby RA, Dekker LW, Ritsema CJ. Occurrence, prediction and hydrological effects of water repellency amongst major soil and land-use types in a humid temperate climate. European Journal of Soil Science. 2006;**57**:741-754

[8] Bradshaw A. Restoration of mined lands – Using natural processes. Ecological Engineering. 1997;**8**:255-269

[9] Shrestha RK, Lal R. Changes in physical and chemical properties of soil after surface mining and reclamation. Geoderma. 2011;**161**:168-176

[10] Six J, Elliott ET, Paustian K. Soil structure and soil organic matter: II. A normalized stability index and the effect of mineralogy. Soil Science Society of America Journal. 2000a;**64**:1042-1049

[11] Six J, Paustian K, Elliott ET, Combrink C. Soil structure and soil organic matter: I. Distribution of aggregate size classes and aggregate associated carbon. Soil Science Society of America Journal. 2000b;**64**:681-689

[12] Six J, Elliott ET, Paustian K. Soil macroaggregate turnover and microaggregate formation: A mechanism for C sequestration under no-tillage agriculture. Soil Biology and Biochemistry. 2000c;**32**:2099-2103

[13] Bronick CJ, Lal R. Soil structure and management: A review. Geoderma. 2005;**124**:3-22

[14] Tisdall JM, Oades JM. Organic matter and water-stable aggregates in soils. Journal of Soil Science. 1982;**33**:141-163

[15] Oades JM. The role of biology in the formation, stabilization and degradation of soil structure. Geoderma. 1993;**56**:377-400

[16] Six J, Bossuyt H, Degryze S, Denef K. A history of research on the link between (micro) aggregates, soil biota, and soil organic matter dynamics. Soil and Tillage Research. 2004;**79**:7-26

[17] Rojík P. Geological substrates and heaping process of coal mining operations in the Sokolov basin, Czech Republic, implications for reclamation and soil development. In: Frouz J, editor. Soil Biota and Ecosystem Development in Post Mining Sites. Boca Raton, FL: CRC Press; 2014. pp. 1-18

[18] Frouz J, Mudrák O, Reitschmiedová E, Walmsley A, Vachová P, Šimáčková H, Albrechtová J, Moradi J, Kučera J. Rough wave-like heaped overburden promotes establishment of woody vegetation while leveling promotes grasses during unassisted post mining site development. Journal of Environmental Management. 2018;**205**:50-58. DOI: 10.1016/j.jenvman.2017.09.065

[19] Kuraz V, Frouz J, Kuraz M, Mako A, Sustr V, Cejpek J, Romanov OV, Abakumov EV. Changes in some physical properties of soils in the chronosequence of selfovergrown dumps of the Sokolov quarry–dump complex, Czechia. Eurasian Soil Science. 2012;**45**:266-272

[20] Frouz J, Krištůfek V, Livečková M, van Loo D, Jacobs P, Van Hoorebeke L. Microbial properties of soil aggregates created by earthworms and other factors: Spherical and prismatic soil aggregates from unreclaimed post-mining sites. Folia Microbiologica. 2011;**56**:36-43

[21] Frouz J, Keplin B, Pižl V, Tajovský K, Starý J, Lukešová A, Nováková A, Balík V, Háněl L, Materna J, Düker C, Chalupský J, Rusek J, Heinkele T. Soil biota and upper soil layers development in two contrasting post-mining chronosequences. Ecological Engineering. 2001;**17**:275-284

[22] Frouz J, Elhottová D, Kuráž V, Šourková M. Effects of soil macrofauna on other soil biota and soil formation in reclaimed and unreclaimed post mining sites: Results of a field microcosm experiment. Applied Soil Ecology. 2006;**33**:308-320

[23] Frouz J, Pižl V, Tajovský K. The effect of earthworms and other saprophagous macrofauna on soil microstructure in reclaimed and un-reclaimed post-mining sites in Central Europe. European Journal of Soil Biology. 2007a;**43**:184-189

[24] Scullion J, Malik A. Earthworm activity affecting organic matter, aggregation and microbial activity in soils restored after opencast mining for coal. Soil Biology and Biochemistry. 2000;**32**:119-126

[25] Jegou D, Schrader S, Diestel H, Cluzeau D. Morphological, physical and biochemical characteristics of burrow walls formed by earthworms. Applied Soil Ecology. 2001;**17**:165-174

[26] Marashi ARA, Scullion J. Earthworm casts form stable aggregates in physically degraded soils. Biology and Fertility of Soils. 2003;**37**:375-380

[27] Marashi ARA, Scullion J. Porosity and hydrological changes in surface mine soils, paper 405. In: Raine SR, AJW B, Menzies NW, Freebairn DM, Tolmie PE, editors. ISCO 2004 - 13th International Soil Conservation Organisation Conference, Conserving Soil and Water for Society: Sharing Solutions, Brisbane, Australia

[28] Frouz J, Elhottová D, Pižl V, Tajovský K, Šourková M, Picek T, Malý S. The effect of litter quality and soil faunal composition on organic matter dynamics in post-mining soil: A laboratory study. Applied Soil Ecology. 2007b;**37**:72-80

[29] Cerdà A, Jurgensen MF. The influence of ants on soil and water losses from an orange orchard in eastern Spain. Journal of Applied Entomology. 2008;**132**:306-314

[30] Cejpek J, Kuráž V, Frouz J. Hydrological properties of soils in reclaimed and unreclaimed sites after brown-coal mining. Polish Journal of Environmental Studies. 2013;**22**:645-652

[31] Bouché MB, Al-Addan F. Earthworms, water infiltration and soil stability: Some new assessments. Soil Biology and Biochemistry. 1997;**29**:441-452

[32] Lavelle P, Bignell D, Lepage M. Soil function in a changing world: The role of invertebrate ecosystem engineers. European Journal of Soil Biology. 1997;**33**:159-193

[33] Cammeraat ELH, Risch AC. The impact of ants on mineral soil properties and processes at different spatial scales. Journal of Applied Entomology. 2008;**132**:285-294

[34] Baker GH, Brown G, Butt K, Curry JP, Scullion J. Introduced earthworms in agricultural and reclaimed land: Their ecology and influences on soil properties, plant production and other soil biota. Biological Invasions. 2006;**8**:1301-1316

[35] Ouellet G, Lapen DR, Topp E, Sawada M, Edwards M. A heuristic model to predict earthworm biomass in agroecosystems based on selected management and soil properties. Applied Soil Ecology. 2008;**39**:35-45

[36] Colloff MJ, Pullen KR, Cunningham SA. Restoration of an ecosystem function to revegetation communities: The role of invertebrate macropores in enhancing soil water infiltration. Restoration Ecology. 2010;**18**:65-72

[37] Nichols E, Spector S, Louzada J, Larsen T, Amequita S, Favila ME. Ecological functions and ecosystem services provided by Scarabaeinae dung beetles. Biological Conservation. 2008;**141**:1461-1474

[38] Robertson KM, Johnson DL. Vertical redistribution of pebbles by crayfish in Mollisol catenas of central Illinois. Soil Science. 2004;**169**:776-786

[39] Cejpek J, Kuráž V, Vindušková O, Frouz J. Water regime of reclaimed and unreclaimed post-mining sites. Ecohydrology. 2017:e1911

[40] Frouz J, Vobořilová V, Janoušová I, Kadochová Š, Matějíček L. Spontaneous establishment of late successional tree species English oak (*Quercus robur*) and European beech (*Fagus sylvatica*) at reclaimed alder plantation and unreclaimed post mining sites. Ecological Engineering. 2015;**77**:1-8

[41] Bujalský L, Jirka V, Zemek F, Frouz J. Relationships between the normalised difference vegetation index and temperature fluctuations in post-mining sites. International Journal of Mining, Reclamation and Environment. 2017. DOI: 10.1080/17480930.2017.1278659

[42] Gerwin W, Schaaf W, Biemelt D, Fischer A, Winter S, Hüttle RF. The artificial catchment "Chicken Creek" (Lusatia, Germany) a landscape laboratory for interdisciplinary studies of initial ecosystem development. Ecological Engineering. 2009;**35**:1786-1796

Chapter 5

Water Cycle Process Research: Experiments and Observations

Lihu Yang, Xianfang Song and Yunfeng Qiao

Additional information is available at the end of the chapter

http://dx.doi.org/10.5772/intechopen.70545

Abstract

The evaluation of methods and instrumentation for measuring water cycle parameters and for monitoring the status of hydrological process will assist governmental personnel, researchers, and water resources practitioners in determining strategies for field and laboratory measurements. This chapter aims to specify the instruments and techniques developed during the long-term monitoring phase of field experimental stations and the establishment phase of indoor experimental laboratory in the Key Laboratory of Water Cycle and Related Land Surface Processes, Institute of Geographic Sciences and Natural Resources Research, Chinese Academy of Sciences. The two field experimental stations, Dongtaigou and Chongling, have been initiated to observe and quantify the water cycle process for more than 10 years, which has formed a complete set of observing and experimental methods in watershed. The experimental laboratory is a new integrated water cycle experiment platform, based on the new technology integrated control, measurement, sensors, and information processing. It includes artificial rainfall system, experimental sink of runoff and erosion, river simulation system, and transformation dynamical processes experimental device among precipitation, vegetation water, surface water, soil water, and groundwater. The continued instrumentation development and advanced experimental strategies will serve as a first port of call for professionals studying the behavior of water footprint.

Keywords: field station, experimental laboratory, instruments and techniques, water cycle, experiments and observations

1. Introduction

An understanding of water cycle processes is essential for assessing water resources as well as the changes to the resources caused by changes in the land use or climate. Experimentation and observation are central activities within the water cycle process research. The range of measuring and monitoring instrumentation and techniques for defining water cycle process variables is immense. Hence, this chapter is not intended to be a comprehensive overview of such instrumentation and techniques, but to present the specific instruments and techniques developed during the long-term monitoring phase of field experimental stations and the establishment phase of indoor experimental laboratory in the Key Laboratory of Water Cycle and Related Land Surface Processes (KLWCRLSP), Institute of Geographic Sciences and Natural Resources Research (IGSNRR), Chinese Academy of Sciences (CAS).

To investigate the dynamic changes in land surface water cycle and the related geographic processes, the KLWCRLSP established two field experimental stations and one indoor experimental laboratory. One station is Dongtaigou Field Station and the other is Chongling Field Station. The experimental laboratory is experimental hall of water and soil process. To clearly demonstrate the water cycle principle, we present a chapter introducing experimental approaches. This chapter is valuable for studying mechanism of water cycle processes *in situ* as well as in the laboratory. The goal of this chapter is to illustrate the development and application of innovative experimental techniques broadly across the areas of subsurface and surface hydrology and hydrometeorology.

2. Field experimental stations

2.1. Dongtaigou Field Experimental Station

Dongtaigou Field Experimental Station is one of the field stations of IGSNRR, CAS, in Dongtaigou catchment (**Figure 1**). It was established in 2003 and designed for the long-term observation of water cycle processes impacted by soil and water conservation projects.

Dongtaigou catchment is a part of the Baihe watershed in the northern part of Yanshan Mountain in Northern China (40° 45′N, 116° 37′E). A detailed water cycle experiment has been initiated in a 0.64 km^2 research catchment [1].

The slopes in the catchment are steep, with an average of 30° and altitude of 290–530 m. The south-north oriented catchment is in the temperate zone and a semi-humid monsoon climate, with average annual temperature of 9–9.5°C and precipitation of 511 mm. The precipitation occurs mainly in June–September with the type of storms which take up 81.2% of the total amount of the year. The bedrock of the region is mainly andesite, covered with meager cinnamonic soil. Constrained by the natural conditions, the catchment has a single and simple vegetation with the coverage of 70%. It is covered with perennial shrubs and herbs such as twigs of the chaste tree, axillaries, and apricot and a small amount of other economic trees such as hawthorn and pear.

Detailed observations of surface and subsurface water dynamics have been made at the catchment. The instruments hydrometrically observing the dynamics of soil water, groundwater, and stream flow response to rainfall and evaporation are described as follows:

http://dx.doi.org/10.5772/intechopen.70545

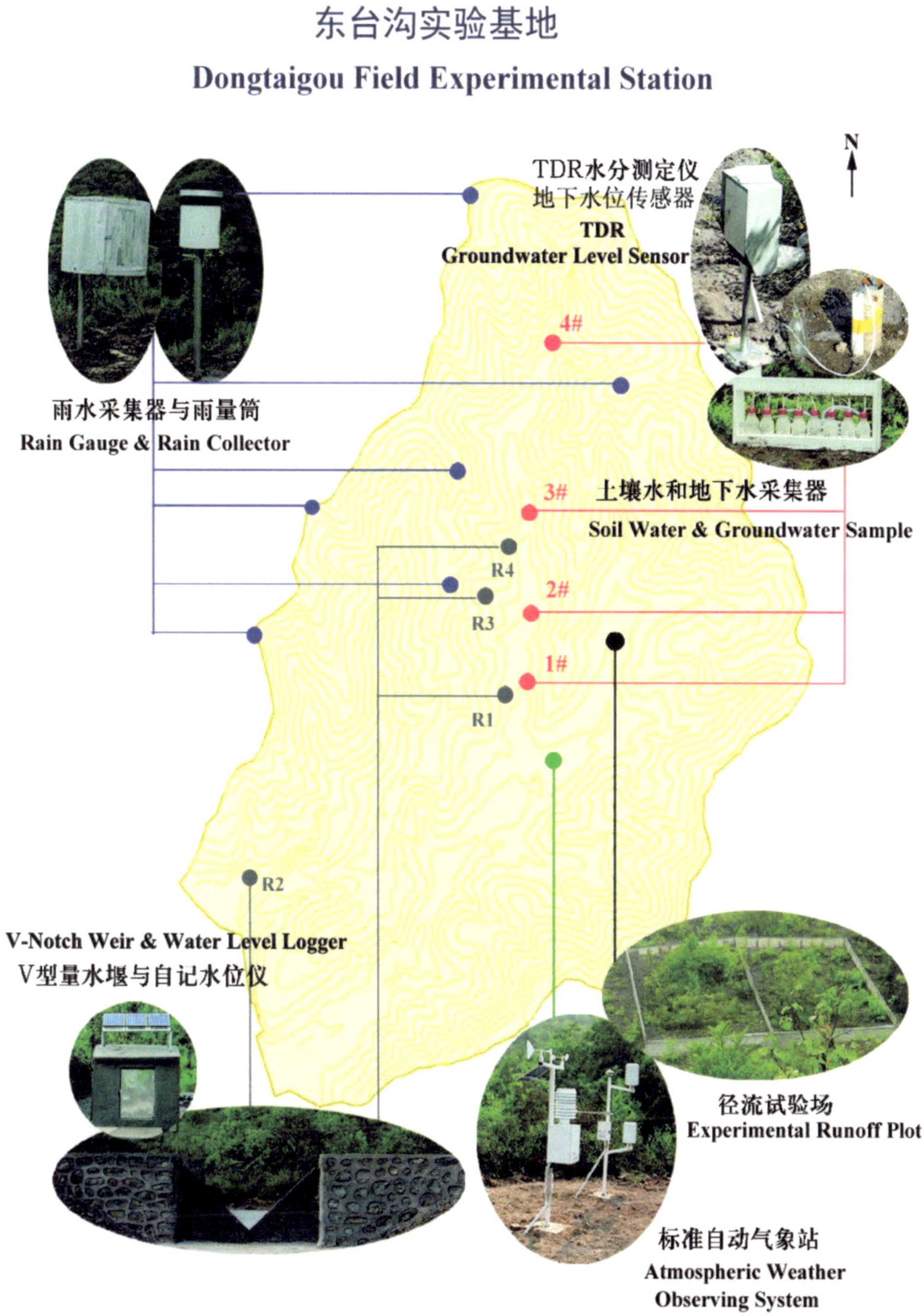

Figure 1. Layout of the Dongtaigou experiment catchment, showing measurement instrumentation and structures.

2.1.1. Meteorological observation

The Atmospheric Weather Observing System is located at the center of meteorological field at outlet of catchment (**Figure 1**). It is made by monitor sensor company (www.monitorsensors.com) to make observations of atmospheric conditions [2]. The measurements taken include air temperature, humidity, wind speed, wind direction, barometric pressure, solar radiation, net radiation, precipitation, evaporation, photoactive radiation, soil temperature, soil moisture, and soil heat flux. The data are automatically recorded every 5 min except the rain gauge (every 1 min) and evaporation sensor (every 30 min). The evaporation is measured as the depth of water (in mm) from the Class A Pan. For the accurate measurement of soil profiles, soil moisture sensors are installed at 10, 15, 30, and 40 cm depths, temperature sensors at 0, 10, 20, and 40 cm depths, and soil heat flux sensors at 15 cm depth.

2.1.2. Precipitation observation

There are six HOBO Logging Rain Gauges (Onset Computer Corporation) monitoring the precipitation of catchment (**Figure 1**). The rain gauge is a self-contained device that includes a high-quality tipping-bucket with an integrated data logger. The tipping-bucket mechanism is designed such that one tip of the bucket occurs for each 0.2 mm of rainfall. Each bucket tip is detected when a magnet attached to the tipping bucket actuates a magnetic switch as the bucket tips, thus effecting a momentary switch closure for each tip. The switch is connected to a HOBO Event Data Logger, which records the time of each tip.

In the catchment, precipitation is sampled for hydrogen and oxygen stable isotopes and main ions analysis. A rain collector consisting of a polyethylene bottle and funnel is placed outside and a ping pong ball is positioned at the funnel mouth to prevent evaporation during rainfall. After each rainfall event, rainwater is collected and immediately transferred to a bottle and sealed and stored [3].

2.1.3. Soil water moisture and groundwater level observation

Along the channel region, four observing pots are arranged namely 1#, 2#, 3#, and 4#. The DataTaker DT50 (Thermo Fisher Scientific Australia Pty Ltd, http://www.datataker.com/) is employed to sample the Time-domain reflectometer (TDR) probes and groundwater level sensor (DS30, SEBA Hydrometrie GmbH & Co. KG). The frequencies of data collection are 10-min interval in the flood season and 30-min interval in non-flood period. A total of four TDR probes are installed horizontally at depths of 20, 40, 70, and 100 cm to continuously monitor soil moisture throughout the 110 cm soil profile. The groundwater level sensor is fixed in a depth in which all possible water levels are within the measuring range. The pressure sensor always measures the water above the sensor. To measure the distance between the ground and the water table or the height of water table above sea level, the parameters of the recording device must be changed correspondingly.

Soil water samples are collected after rainfall. Soil water is sampled at the depth same as that of sensor, using a suction lysimeter designed by IGSNRR, which was composed of a Teflon pipe and porous ceramic tube. A vacuum pump of about −0.8 MPa is applied to the suction lysimeter for 12 h of equilibrium to collect soil water [4]. The ground water sample is collected directly from pump discharge.

2.1.4. Surface water runoff monitor

Two small catchments are carefully selected as the contrastive study site. One is named Donggou in which 22 stonemasonry dams were constructed (R1), and the other is Xigou with conservation of the natural environment (R2). There are also two little similar branches in Donggou (R3 and R4). Four V-notch weirs mounted into the concrete are built at the outlet (**Figure 1**). Odyssey Capacitance Water Level Logger (Dataflow Systems PTY Ltd, New Zealand) is used to monitor the surface water level. The logger automatically records the water level at a 1-min interval. The surface water is sampled from the stream when flooding. The sampling container is completely filled and then capped and properly stored.

2.1.5. Experimental runoff plot

Two experimental runoff plots (5 × 10 m) have been established on mid-slopes as indicated in **Figure 2**. The two plots are covered with shrubs mainly consisting of Vitex negundo var. heterophylla, wild jujube, wild grass, and two cypresses [5]. The plot borders are made of concrete. The edges of the runoff plots are about 50 cm above the soil surface to prevent input from splashes entering the plot from the surrounding areas and are sufficiently embedded into the bedrock so that the water insides and outsides will not exchange. In order to examine water horizontal processes on the overland, in the shallow soil, and the soil bedrock interface, three layers of water movement are monitored at the down slope end of the plot. A pipe is positioned at very layers, and a V-Notch (Triangular) Weir tank (410 mm long, 210 mm wide, and 305 mm deep) was installed to collect the runoff which would then be piped into a collecting cylinder [6]. The weirs from top to bottom are 210 mm high with 30, 20, and 20° V-notch thin-plate, respectively. The collecting cylinder is made of a metal sheet and covered with a sheet metal to prevent direct entry of rainfall. Runoff volume is calculated by measuring the head of water over the V-notch crest. The total volume of runoff is measured by volumetric method to calibrate the runoff calculated from the wire [7].

Meanwhile, a few measurements are also made to observe rainfall, soil water content, and soil water potential dynamics. An automatic recording tipping bucket rain gauge is installed some 3 m from side plot. In assessing the relationship between water content and capillary pressure head (tension), a comparison is made between the prevailing capillary pressure heads recorded by the automatic tension meters at the time of the TDR probe automatic measurements. The probes are installed at 10, 20, and 30 cm depths in two separate profiles up and down of plot. Campbell Scientific CR10X data loggers (Campbell Scientific Inc., USA) are employed to sample the TDR probes (20-min step), tension meter sensors (20-min step), and water level sensors of wires (1-min step).

2.2. Chongling Field Experimental Station

Chongling Field Experimental Station was constructed in 2004 by Key Laboratory of Water Cycle and Related Land Surface Processes (KLWCRLSP), cooperated with Baoding Soil and

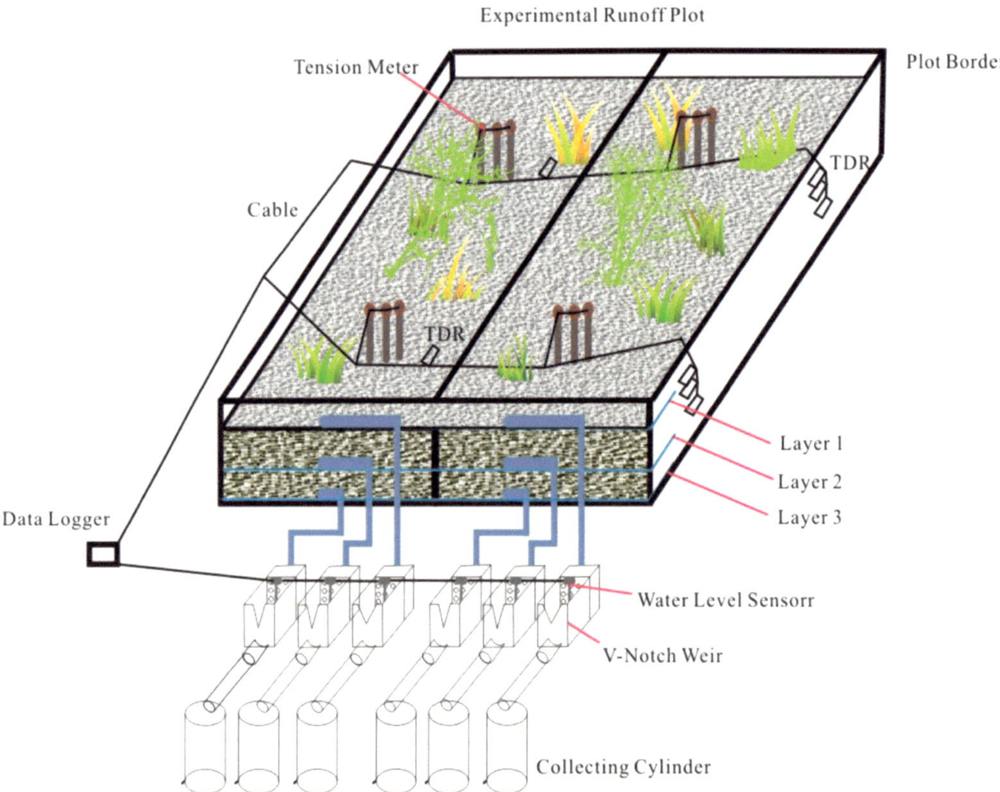

Figure 2. Schematic representation of two experimental runoff plots.

Water Conservation Station. It is also one of the field stations of IGSNRR, CAS and has been developed into a prime research location involving many institutions.

It is located in Chongling catchment in Yixian County, Hebei province of China (115° 21′E, 39° 23′N). Chongling catchment is in the north of Taihang mountainous region, which is a tributary of north Yishui River. It covers a total area of approximately 6 km^2, with a length of 4.4 km and a width of 1.5 km at an altitude of approximately 70–300 m above the sea level. The research catchment was chosen for studying on hydrological process affected by typical vegetation in North China.

The area has a temperate climate with average annual air temperature of 11.6°C, and the maximum and minimum temperature is observed in July and January with extreme values of 40 and −23.4°C. There are dry season (from September to May) and rainy season (from June to August). The annual precipitation ranges from 217.0 to 1004.3, on an average of 641.2 mm. The mean annual evaporation is 1906 mm by E601B Evaporator. The soil type in the catchment is predominantly sandy loam soil and loess, which is mainly cumulated in the valley, with depth of 1–2 m. The vegetation coverage in this area is diversified, including woody plants (Acacia, Arborvitae, Poplar, Pine), shrubs (*Vitex negundovar*, *Ziziphus jujuba* Mill. var. spinosa), and herbage (*Carex humilis*, *Carex lanceolata*, *Bothriochloa ischaemum*) [8].

The catchment drains in a southeasterly direction and the contributing hillslopes are each divided into five gully channels from east to west of the watershed, that is, Yangshugou, Wanmulingou, Yanghuaigou, Langweibagou, and Huyaogou. There are also two little gully channels (Chenglingou and Langyaogou) in the southeast of the watershed.

A summary of the instruments observing the precipitation, surface water, soil water, groundwater, and so on is presented in **Figure 3**.

2.2.1. *Comprehensive meteorological observation*

A meteorological field (25 × 2 5 m) has been made at Chongling catchment since 2004, providing a continuous, very high-quality record as shown in **Figure 3**. The instruments include Vantage Pro2 Weather Station (Davis Instruments Corp., USA), E601B evaporation pan (China), Φ20 cm evaporation pan (China), eddy covariance (EC), and flux observation system.

The Vantage Pro2 station includes a console and a versatile integrated sensor suite. The console displays and records the station's weather data every 10 min, provides graph and alarm functions, and interfaces to a computer using the WeatherLink software. The sensor suite to the console is used to monitor wind speed and direction, temperature and humidity, wind chill, dew point, barometric pressure, ultraviolet radiation, heat index, temperature humidity sun wind (THSW) index, rain rate, and solar radiation.

Evaporation is determined by the water levels, which are monitored by automatic logger (E601B pan) and by manual measurement (Φ20 cm pan). For automatic measurement of the evaporation, use is made of an Odyssey Capacitance Water Level Logger (Dataflow Systems PTY Ltd, New Zealand). The sensor is supplied with Teflon-covered measuring element and the logger stores measurements at 1-min interval. Manual evaporation pan measurements are made by measuring the volume of water in the evaporation pan. Manual measurements are made twice a day, at 8:00 a.m. and 6:00 p.m., to see the difference between day evaporation and night evaporation.

There is an eddy covariance and flux observation system in the northeast corner of the field [9]. The measurements of eddy covariance (EC) are made from the tower at 20 m height, and the setup consists of three-dimensional sonic anemometer (CSAT3, Campbell Scientific, Inc., USA), an open-path CO_2/H_2O analyzer (Li-7500A, LI-COR, Inc., USA). The frequency of data acquisition is 10 Hz. In addition to the EC measurements, several meteorological variables, such as air temperature/relative humidity (HMP155A-L, Vaisala Oyj, Finland) and wind speed (010C-1, Met One Instruments, Inc. USA), are measured at about 2, 6, 12, and 20 m heights above the ground. Furthermore, incoming and outgoing short- and long-wave radiation (CNR4 Net Radiometer, Kipp & Zonen B.V., The Netherlands), photosynthetically active radiation (LI190SB, LI-COR, Inc., USA), and wind direction (020C-1, Met One Instruments, Inc., USA) are measured from the tower at 20 m height. To deal with water movement in soils, the soil profiles are selected to measure the soil water content (CS616, Campbell Scientific, Inc., USA), temperature (109-L, Campbell Scientific, Inc., USA) at 10, 20, 40, and 80 cm depths, and heat flux (HFP01, Hukseflux Thermal Sensors B.V., The

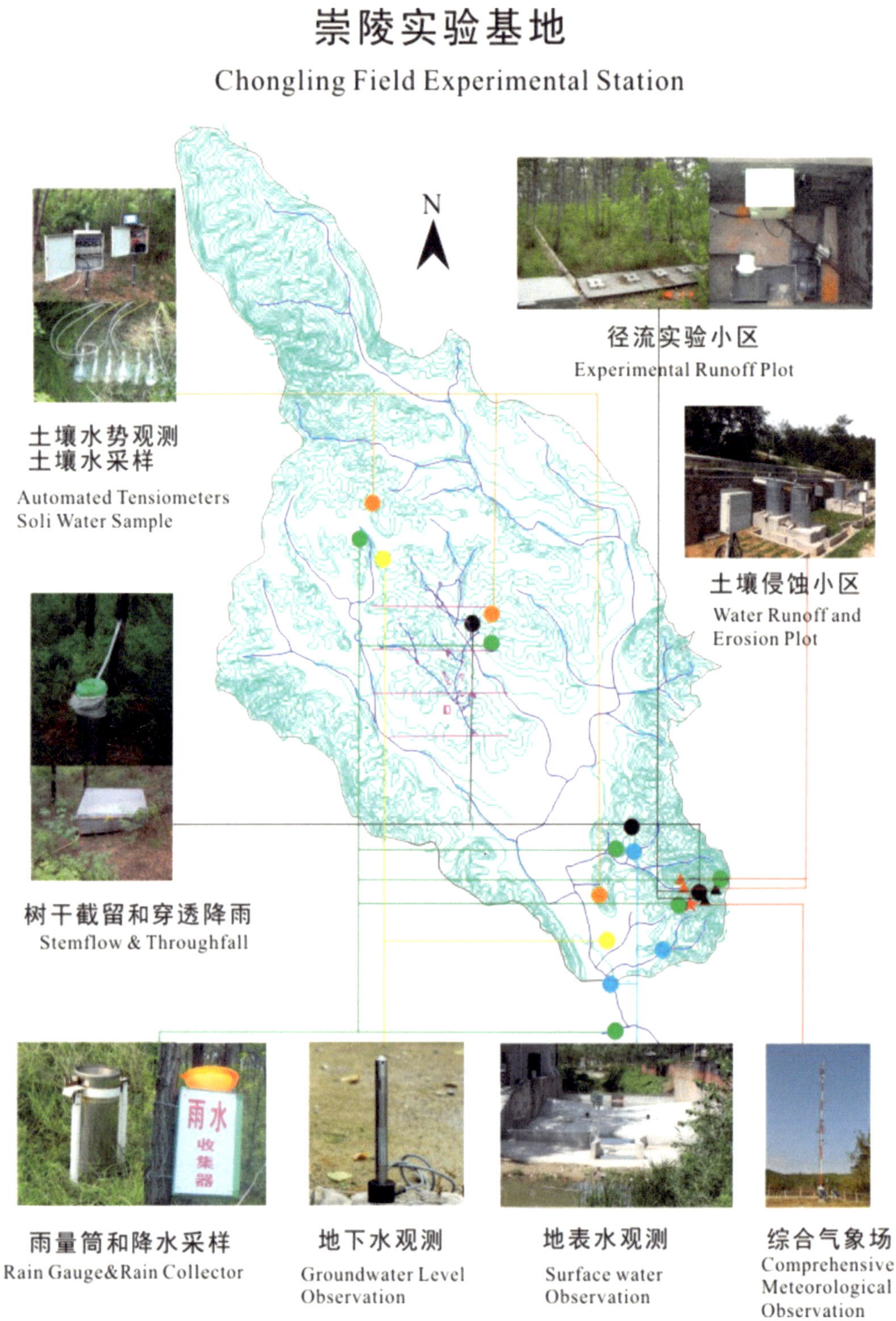

Figure 3. Layout of Chongling catchment showing the location of instruments.

Netherlands) at 5 cm depth. The data are automatically recorded in the CR3000 data logger (Campbell Scientific, Inc., USA) at a 30-min interval.

2.2.2. Precipitation observation

There are six rain gauges monitoring the precipitation of catchment (**Figure 3**). The rain gauge is the tipping bucket rain gauge type (SL3-1) made by Shanghai Meteorological Instrument Factory Co., Ltd. The two buckets in a tipping bucket rain gauge rest on a pivot so that when one bucket has received 0.1 mm of rain, it tips by gravity, empties the rainwater, and allows the other bucket to start collection. During the tip, an electrical switch is closed and triggers the HOBO Event Data Logger (Onset Computer Corporation, USA) to register each "tilt," thus giving a fairly continuous record of precipitation.

Six rain collectors are placed near the rain gauges to sample the rain water. After each rainfall event, rainwater is collected and immediately transferred to a bottle and then sealed and stored.

2.2.3. Soil water potential observation

Soil water potential has been monitored at four deferent sites consisting mainly of grassland, one under acacia and two under old arborvitae. Soil water potential is observed by automated tensiometer in the catchment [10]. Each site has one profile, and the depths of observation are 10, 20, 30, 40, 50, 70, and 100 cm. The data are automatically recorded at 2-min intervals and averaged every 30 min by data logger (CR800, Campbell Scientific, Inc., USA).

Suction lysimeters designed by IGSNRR are used to collect pore water from unsaturated soil at four sites. After installation below ground level, vacuum is applied to the porous ceramic tube through Teflon pipe from bottle. The negative air pressure created inside the tube draws pore water into the tube through the porous ceramic tub. The pore water is transported to the bottle through Teflon pipe. Suction lysimeters perform best in moist soil and below the water table and work as long as the soil water potential is in the 0/-800 mbar range.

2.2.4. Surface water observation

Runoff is monitored at Youlingou (R1) and Langyaogou (R2) by the V-notch weir and outlet of catchment (R3) by the compound weirs (**Figure 3**). The compound weirs (**Figure 4**) are composed of V-notch weir (up), Flat V weir (middle), and Parshall flume (down) [11]. The weirs are instrumented with Odyssey Capacitance Water Level Logger (Dataflow Systems PTY Ltd, New Zealand). Water level-discharge relationships can be applied and meet accuracy requirements for the weirs. The runoff can be calculated according to water level-discharge relationship.

2.2.5. Groundwater level observation

Existing wells are selected and used for long-term water level monitoring. It currently includes 10 active observation wells located across the catchment. The locations of the wells are shown in **Figure 3**. Six wells (G1–G6) are manually monitored in the dry season (from November to April) every 10 days and in wet season (from May to October) every 5 days. The rest four

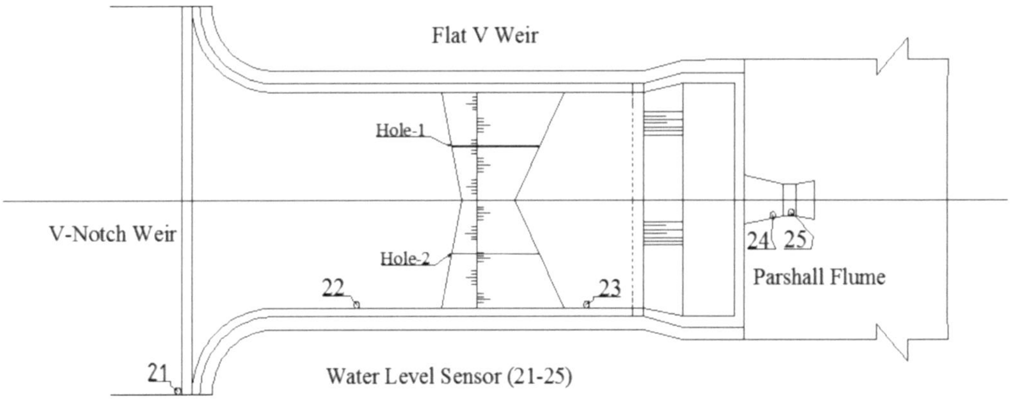

Figure 4. Schematic representation of the compound weirs.

wells are equipped with electronic data loggers that record water levels every 30 min. To date, a multi-parameter groundwater data logger (CTD-Diver, Eijkelkamp Soil & Water, The Netherlands) has been installed to monitor water level, conductivity, and temperature at G7. The KADEC-MIZU II groundwater monitoring data loggers (North One Co., LTD, Japan) are used to monitor the groundwater level at G8, G9, and G10. The wells are visited approximately every 1 month for field verification, and water samples are collected periodically to test groundwater chemistry. The ground water is directly pumped from the well and stored.

2.2.6. Throughfall and stemflow

To investigate the effect of different types of tree (Acacia, Arborvitae, and Pine) on the spatial variability of throughfall, throughfall collectors are partitioned into three zones as shown in **Figure 3** [12]. In each zone, one collector 1 m long, 1 m wide, and 20 cm high is made of a metal sheet and located at a fixed position. The collector is connected by polyethylene hose into the tipping-bucket flow gauge designed by IGSNRR (**Figure 3**). The gauge had been previously calibrated and recorded the by HOBO Event Data Logger (Onset Computer Corporation, USA).

The stemflow is measured simultaneously with throughfall. Three sets of stemflow collars are fitted to trees in each zone. The rubber collar is encircled at the trunk at an angle of approximately 45° to the horizontal at the level of the breast height (about 1.3 m above ground) and tightened closely with silicone sealant to avoid the leaching of water, as shown in **Figure 3**. The stemflow is diverted from the rubber collars to tipping bucket gauge similar to that used for gross rainfall via a PVC suction hose, 3 cm in diameter. The outputs from that are accumulated on HOBO Event Data Logger (Onset Computer Corporation, USA).

Gross rainfall amounts are measured in the neighboring open area, outside the forest, using the tipping bucket rain gauges (SL3-1, Shanghai Meteorological Instrument Factory Co., Ltd., China).

2.2.7. Experimental runoff plot

Two experimental runoff plots (5 × 15 m) have been set up on 15° slopes consisting mainly of pine and shrub, respectively. The plot borders 20 cm above the soil surface are made of concrete to prevent water loss and input from splash and are sealed with the bedrock to prevent the water exchange inside and outside of the plot (**Figure 5**). A pipe, connecting the collecting trough at two layers, is positioned at the downslope end of the plot to monitor the overland flow and base flow at soil-bedrock interface. From this pipe, the runoff flows into the collecting tanks (400 mm long, 200 mm wide, and 300 mm deep). The tanks are performed with a 210 mm high 30 and 20° V-notch thin-plate weir to measure the overland flow and base flow, respectively. The flow from collecting tank is piped into a collecting cylinder which is made of a metal sheet and covered with a sheet metal to prevent direct entry of rainfall. The water level sensor (L304S-3-B-F, Beijing Hua Yi Ao Feng Automation Equipment Co., Ltd, China) is installed in the tank. Runoff volume is calculated by weir equation. The total volume in collecting cylinder is measured by volumetric method to calibrate the runoff calculated from the wire.

The measurements, including soil water tension [10] and volumetric water content (EC5, Decagon Devices, Inc., USA), are also conducted at upper and lower sites in every plot. The probes are installed at 10, 20, 30, and 40 cm depths in two separate profiles. A Campbell Scientific CR1000 data logger is developed to record signals from soil water tension and volumetric water content sensors at a common 30-min time step and from water level sensor at

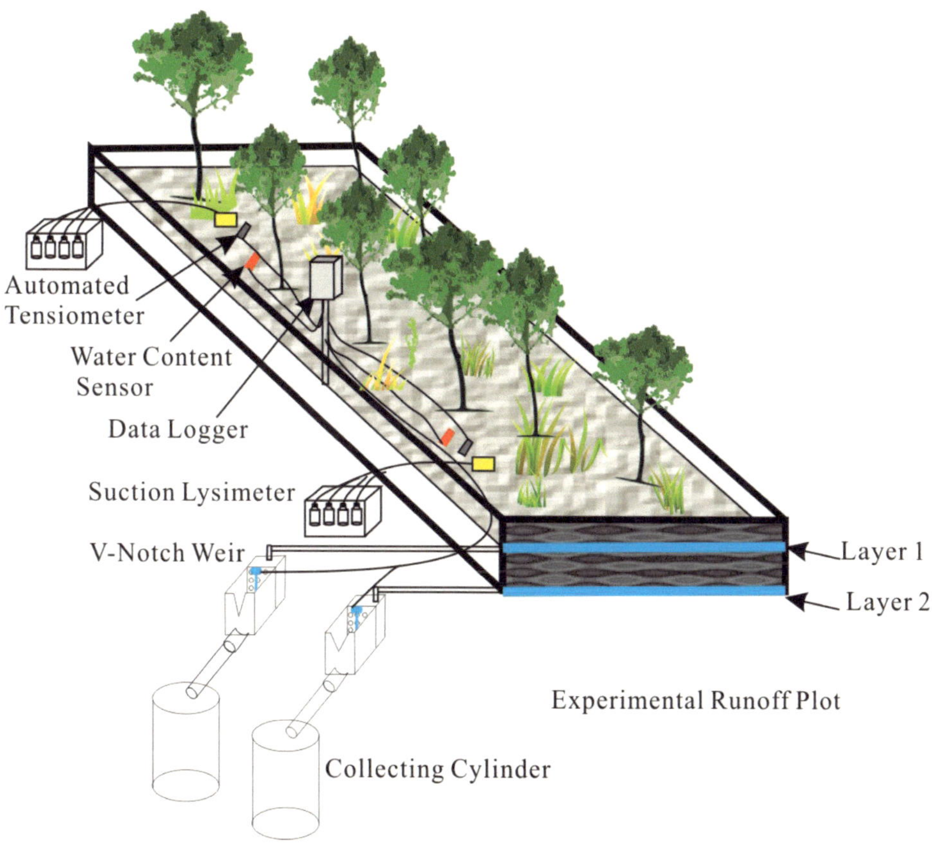

Figure 5. Layout of experimental runoff plot.

1-min time step. Suction lysimeters designed by IGSNRR are placed at the same depth of water potential sensors to collect pore water from unsaturated soil. The pore water is transported to the bottle through Teflon pipe.

2.2.8. *Water runoff and erosion plot*

In the catchment, a total of 13 water runoff and erosion plots were set up at both sides of the channel representing different type of vegetation. Eight plots (5 × 20 m) from P1 to P8 are located at southern hill slope of the channel. The plots are characterized by different vegetations, which are corns/wheat, bare, grasses, shrubs, paper mulberries, peanuts, peaches, apricots, *Phyllanthus urinarias*, and *Angelica keiskei*. Each runoff plot is located at a slope of 15°. The plot borders 50 cm above the soil surface are made of concrete and sufficiently embedded into the soil. At the downslope end of each plot is a trough, connected to a drum for storage of runoff. Two collecting tanks of the same size are used for each runoff plot.

In addition, a rainfall simulation system designed by IGSNRR is set up on the P1 and P2 plots. The system includes a submersible pump, electromagnetic flowmeters, sprinkler nozzles, and spray pipes (**Figure 6**). The sprinkler nozzle is installed at a height of 6.0 m so that the drops could reach a horizontal distance of at least 10 m to cover the whole 2 plots. Three rain gauges are positioned at every plot to monitor the simulated and natural rainfalls. A

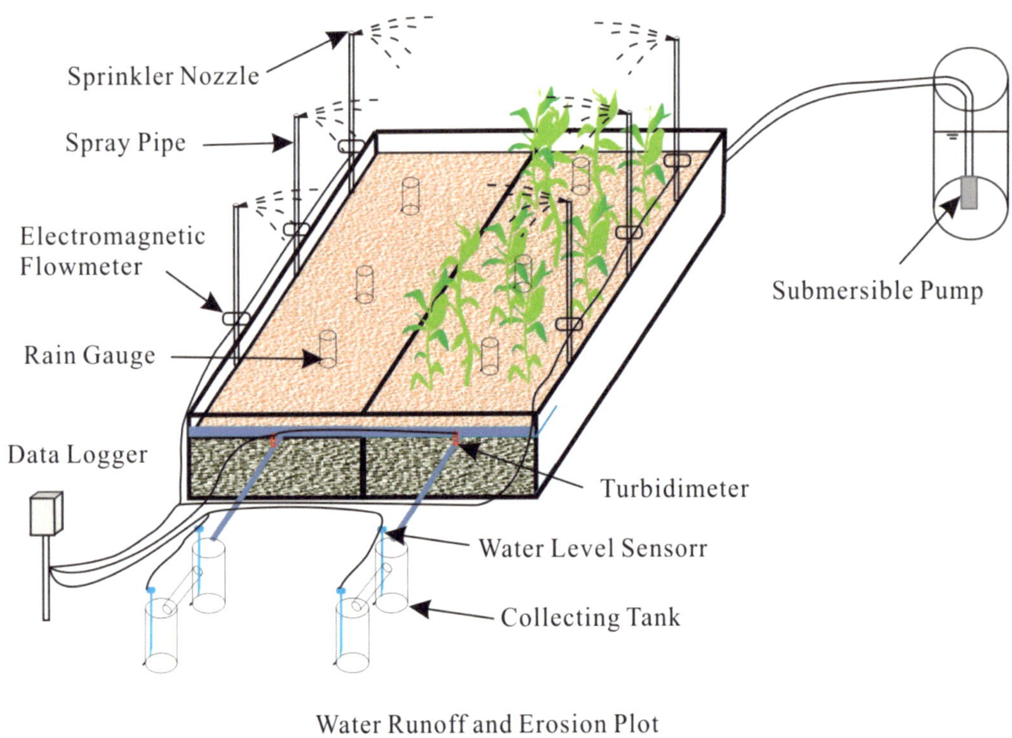

Figure 6. Schematic layout of water runoff and erosion plots.

turbidimeter (Ananlite NEP180, Mcvan Instruments PTY Ltd., Australia) and two water level sensors (L304S-3-B-F, Beijing Hua Yi Ao Feng Automation Equipment Co., Ltd, China) are installed in transferring pipe and collecting tanks, respectively, in plots 1 and 2. Volume of surface runoff was calculated by measuring the height of the water in the first and second collecting tanks. The sediment concentration is also calculated from the correlation and relationship between sediment concentration and turbidity. A Campbell Scientific CR1000 data logger (Campbell Scientific Inc., USA) is employed to automatically monitor the precipitation, pipe flow, runoff, and sediment concentration in plots 1 and 2 each min. Runoff collected from plots 3 to plot 8 is manually measured within a day after each runoff event. Meanwhile, a sediment sample is collected. Afterward, samples are dried and weighed to estimate their sediment concentration.

Five plots are located at northern hill slope of the channel. The size of the runoff plot is different due to the difficulty in finding the location with 20 m slope length. A collecting trough is positioned at the downslope end of the plot. Sediment and surface runoff from this collecting trough enter the first collecting tank, which splits overflow into five equal parts and passes one part, as a sample, into the second collecting tank. For each rainfall event, runoff volume and sediment loss from the plot are calculated.

3. Experimental laboratory

Experimental Hall of Water and Soil Process is located in the geographical museum of IGSNRR. It is 80 m long, 18 m wide, and 22 m high. It is a new integrated water cycle experiment platform, based on the new technology integrated control, measurement, sensors, information processing, developed from China's first artificial rainfall runoff laboratory, slope erosion laboratory, and fluvial geomorphology laboratory in the 1950s. It includes artificial rainfall system, experimental sink of runoff and erosion, river simulation system, and transformation dynamical processes experimental device among precipitation, vegetation water, surface water, soil water, and groundwater.

3.1. Artificial rainfall system

The artificial rainfall system finished in December 2015 is set up at the height of 18 m in the hall. It includes three rainfall zones: Z1, Z2, and Z3 (**Figure 7**). The total area is 370 m^2. The rainfall can be achieved in each separate zone or in all three zones at the same time. The system consists of variable speed pumps, stainless pipes, control center, laser rainfall monitor, and sets of solenoid valves and spraying nozzles (**Figure 7**). Every set of solenoid valves and spraying includes three valves and three nozzles which can be combined to produce 12–300 mm/h rain and mobile storm. A pressure-compensated flow control valve and a pressure gauge are located at the same altitude of the nozzle allowing a precise control of water pressure and consequently the constancy of rain kinetic energy. The artificial rainfall system is automatically regulated in the control center. The calibration tests showed that the uniformity of the rainfall intensities was greater than 85%.

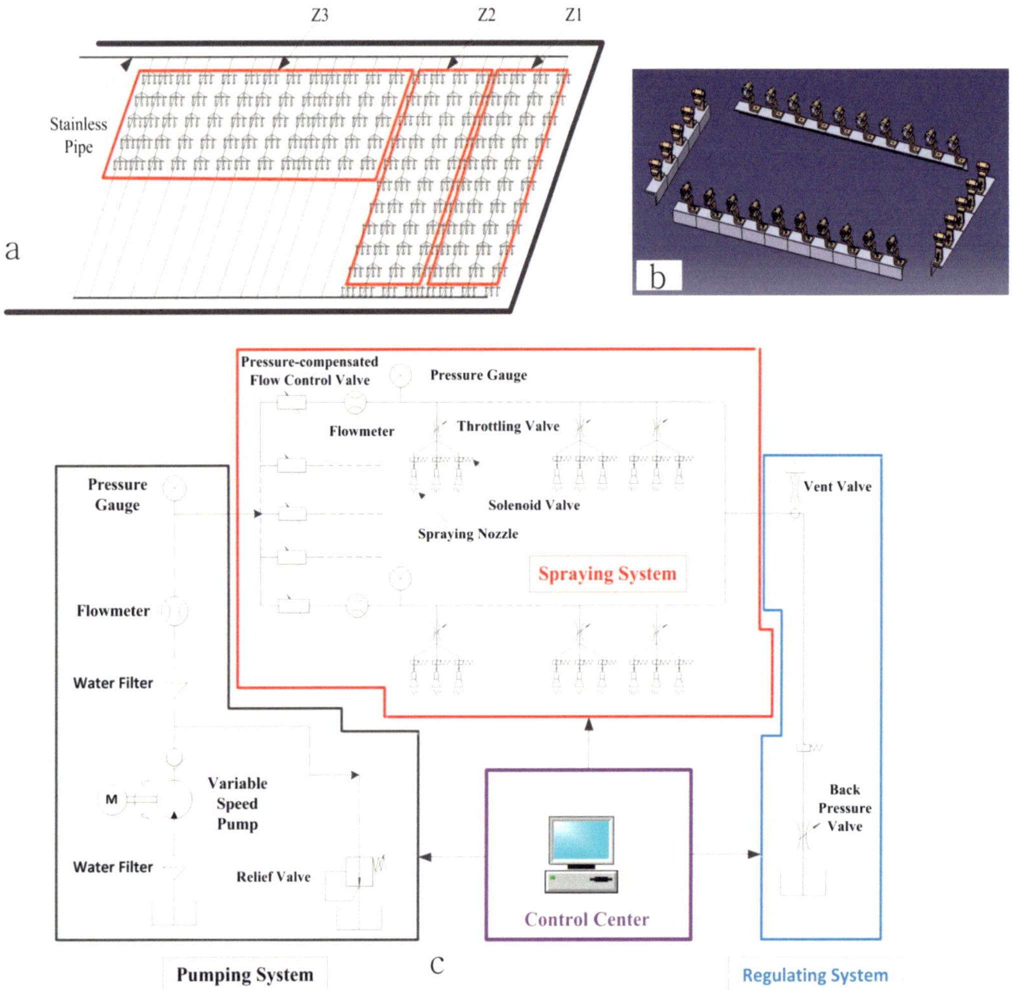

Figure 7. Schematic layout of artificial rainfall system: (a) three zones, (b) laser rainfall intensity monitor, and (c) four component sections.

Laser rainfall intensity monitor is installed at the mid-height of nozzles. It is composed of an array of laser transmitters and receivers (**Figure 7**). It achieves the rain non-touch measurement using orthogonally multiplexed laser beams according to the light attenuation law. The measurement error is less than 2%.

3.2. River simulation system

The river simulation system is 38 long, 6 m wide, and 1 m high (**Figure 8**). The borders are made of concrete and sealed with the ground to prevent water leaching from them. The crustal lifting simulation system is installed in the middle [13]. It is composed of 12 square steel blocks (2 × 2 m). Each block is supported by four stainless steel-threaded rods, which can be adjusted up and down. The 12 square blocks can be automatically motioned to form 82 types of crustal shape.

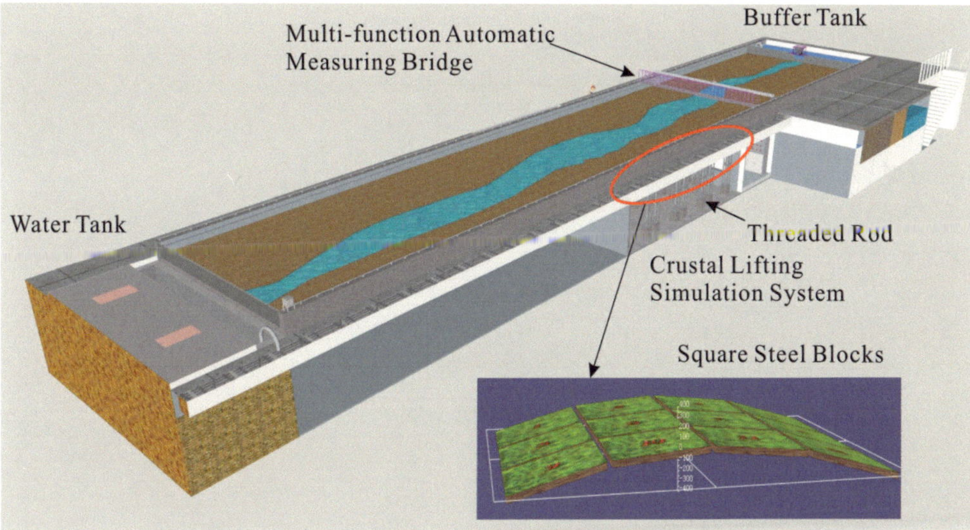

Figure 8. Schematic layout of the river simulation system (top) and one type of crustal shape (bottom).

The rate of motion is as slow as 30–70 mm/day. The multi-function automatic measuring bridge is placed above the system to move from the upstream to the downstream. It can automatically measure water flow, water depth, and cross-section of the modeled river. At the end, there is a big tank where the recycled water can be pumped to the upstream.

3.3. Experimental sink of runoff and erosion

It consists of two metal rectangular boxes, 10 m long, 3 m wide, and 0.8 m high, and each one is located under artificial rainfall zone 1 and zone 2 (**Figure 9**). The interval area, 1 m wide, is kept between the two boxes in order to easily assemble them into a bigger one. The slope of the experimental sink could be adjusted automatically from 0 to 35°. One 5 cm hole is cut into the downslope end of each plot. A short metal stub pipe is welded on to the hole to form an outlet. Two water flow monitors [14] are horizontally set up in front of the each box for the measurement of the runoff. The box outlet and flow monitor are fitted together with a flexible PVC pipe. The monitor should have lids to prevent direct rainfall from entering them. For simulated rainfalls, runoff volume measurements and sediment sample collection are performed every 5 s and 5 min, respectively.

3.4. Transformation dynamical processes experimental device among precipitation, vegetation water, surface water, soil water, and groundwater

The transformation dynamical processes experimental device among precipitation, vegetation water, surface water, soil water, and groundwater (TDPEDPVSSG) finished in July 2014 is a complex equipment to study the water process among the five different types of water (**Figure 10**). It is hermetically sealed in the house (7 m long, 5 m wide, and 7.5 m high) and consists of two sections joined together, the up section and the down section.

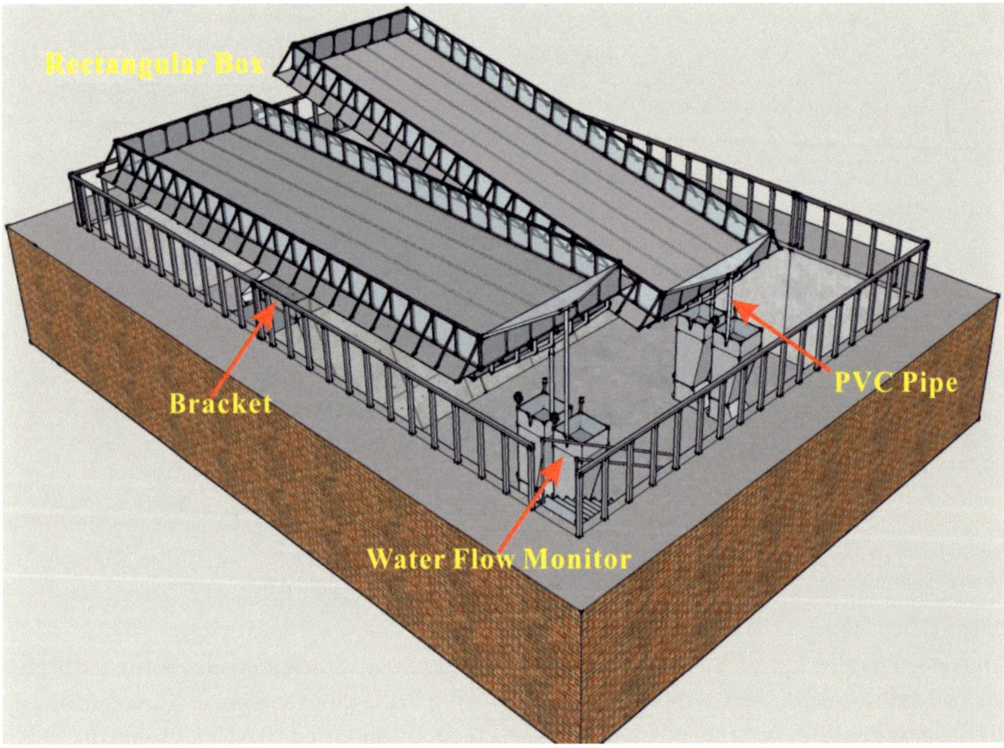

Figure 9. Schematic layout of experimental sink of runoff and erosion.

The down section has two weighable lysimeters. Each lysimeter has a rectangle stainless steel tank with a surface area of 6 m^2 (3 m long, 2 m wide) and a depth of 3 m. It is designed to have enough depth to accommodate the rooting depth of most plants and control the groundwater level. A drainage discharge and water supply system at the bottom is designed to facilitate the fluctuation of groundwater level. The gap between the concrete wall and the stainless container is less than 2 cm to avoid alteration of the energy balance of the system. This gap has been covered with a flexible and impermeable rubber film in the surface. Each lysimeter tank rests on a base frame that transmits the weight through a lever system with a counterweight to an electronic load cell. The lever arm reduces the majority of the total mass of tank and soil to a small fraction of some kilograms that are measured by load cell. It measures those soil mass with an accuracy of 60 g which corresponds to a precipitation or water column of 0.01 mm. The output signal of the sensor is transmitted to a computer located in the control room. The frequency of data collection is 30-min interval.

In the northern lysimeter, the silt loam is homogeneously placed. However, in the southern lysimeter, three horizons of soil (silt loam and silty sandy loam) are placed, and each horizon depth is 1 m. The type of soil structure is prevalent in this region of North China. Fourteen sets of soil moisture, temperature, and electrical conductivity sensor (5TE, Decagon Devices,

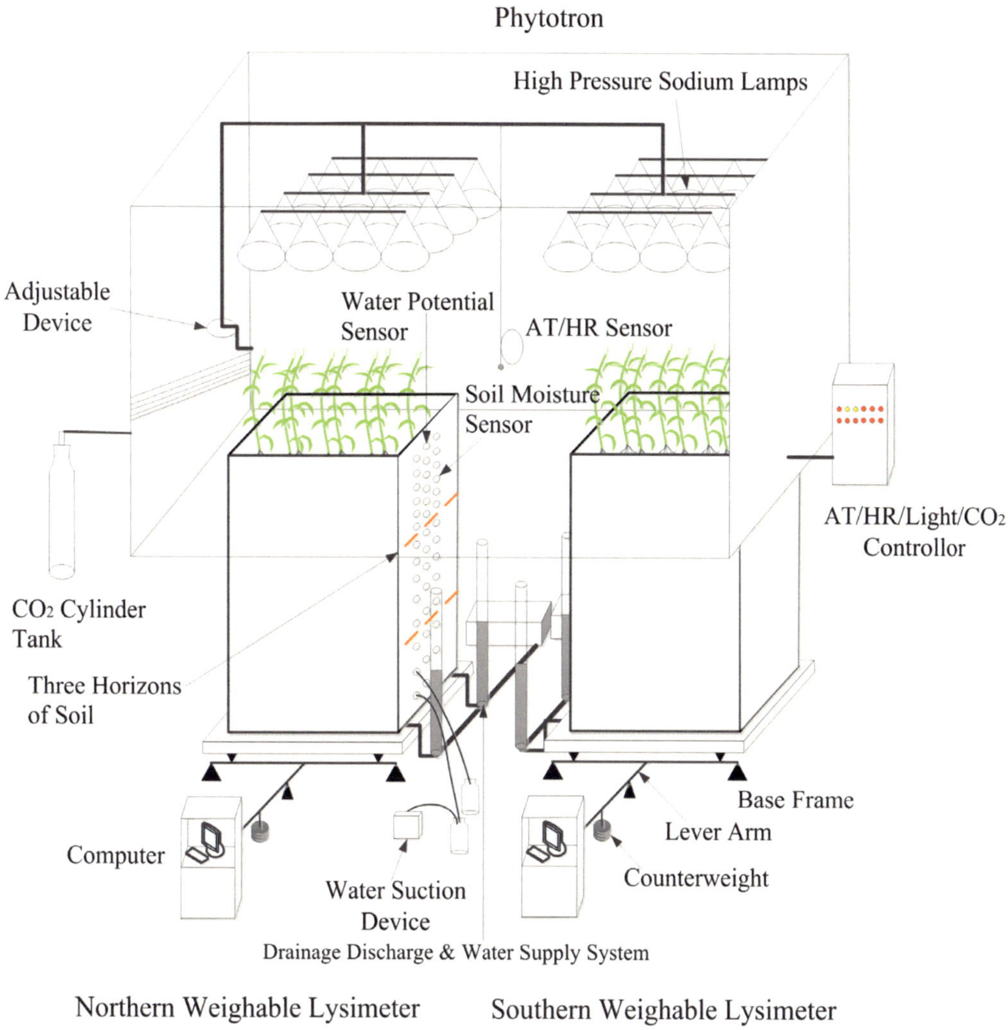

Figure 10. Schematic layout of the transformation dynamical processes experimental device among precipitation, vegetation water, surface water, soil water, and groundwater.

Inc., USA), dielectric water potential (MPS-2, Decagon Devices, Inc., USA), and suction lysimeter designed by IGSNRR are installed in side of the tank at the depths of 20, 43, 53, 63, 73, 83, 92.5, 110.5, 130.5, 150.5, 180.5, 210.5, 240.5, and 275.5 cm.

The up section is a phytotron where the temperature, humidity, light, and CO_2 can be automatically controlled. The temperature and humidity are generally controlled by air conditioner and humidifier. The light (0–30,000 LUX) is produced by 12 high-pressure sodium lamps above the lysimeter, which can be adjusted up and down manually. CO_2 is emitted from the steel CO_2 cylinder tank. The precision of controlling temperature, humidity, and CO_2 are ±1°C, ±5%, and ±100 ppm, respectively.

4. Conclusions

The instruments and techniques developed during the long-term monitoring phase of field experimental stations and the establishment phase of indoor experimental laboratory have been specifically described. The methods in the filed observations will enhance the quantitative research about the hydrology process. The new integrated water cycle experimental hall can be used to characterize the water movement among precipitation, vegetation water, surface water, soil water, and groundwater. These characterizations will improve the parameterization in numerical models. In addition, the continued instrumentation development and various techniques are recommended. It is particularly important to develop measurement and predictive techniques for *in situ* or indoor hydrological characteristics in the many instances.

Acknowledgements

This work was financially supported by Key Program of National Natural Science Foundation of China (Grant No.41730749), CAS Key Technology Talent and Key Research Program of the Chinese Academy of Sciences (grant no. KJZD-EW-TZ-G10). The authors like to thank the personnel of the Key Laboratory of Water Cycle and Related Land Surface Processes, Chinese Academy of Sciences, for providing important ideas in construction of field experimental stations and indoor experimental laboratory.

Author details

Lihu Yang[1,2]*, Xianfang Song[1,2] and Yunfeng Qiao[1,2]

*Address all correspondence to: yanglihu@igsnrr.ac.cn

1 Key Laboratory of Water Cycle and Related Land Surface Processes, Institute of Geographic Sciences and Natural Resources Research, Chinese Academy of Sciences, Beijing, China

2 University of Chinese Academy of Sciences, Beijing, China

References

[1] Yang C. Experiments of Water Cycle Process in Typical Catchment, Mountainous Area, North China—Case Study in Dongtaigou Catchment. Beijing: Graduate University of Chinese Academy of Sciences; 2006. p. 134 (in Chinese)

[2] Yang C, Yu J, Song X, Xia J, Liu C. Reference crop evapotranspiration calculation in short interval of mountainous area in North China. Progress in Geography. 2004;**23**(6):71-80 (in Chinese)

[3] Liu X, Song X, Xia J, Yu J, Yang C, Li F. A study on oxygen isotope in precipitation of Dongtaigou basin in Chao and Bai river basin. Geographic Research. 2005;**24**(2):196-205 (in Chinese)

[4] Song X, Wang P, Yu J, Liu X, Liu J, Yuan R. Relationships between precipitation, soil water and groundwater at Chongling catchment with the typical vegetation cover in the Taihang mountainous region, China. Environmental Earth Sciences. 2011;**62**(4):787-796. DOI: 10.1007/s12665-010-0566-7

[5] Yu J, Yang C, Liu C, Song X, Hu S, Li F, Tang C. Slope runoff study in situ using rainfall simulator in mountainous area of North China. Journal of Geographical Sciences. 2009;**19**(4):461-470. DOI: 10.1007/s11442-009-0461-x

[6] Li F, Song X, Liu C, Yu J, Yang C, Liu X, Hu K, Tang C. Discharge recession from runoff plots in representative mountain area in North China. Journal of Beijing Forestry University. 2006;**28**(2):79-84 (in Chinese)

[7] Li F, Song X, Liu C, Yu J, Yang C, Liu X, Hu K, Tang C. Automatic approach to the measurement of low flow in the slope runoff processes. Geographic Research. 2006;**25**(4):666-672

[8] Hu K. Experiment and Mathematical Simulation Study on Slope Hydrological Processes in Lithoid Mountainous Areas in North China—Case Study at Chongling Catchment. Beijing: Graduate University of Chinese Academy of Sciences; 2006. p. 88 (in Chinese)

[9] Tan L, Liu S, Mo X, Yang L, Lin Z. Environmental controls over energy, water and carbon fluxes in a plantation in Northern China. Chinese Journal of Plant Ecology. 2015;**39**(8):773-784. DOI: 10.17521/cjpe.2015.0074 (in Chinese)

[10] Yang L, Song X, Ma Y, Qiao Y, Chen Q. An automatic tensiometer. Patent No. ZL201320 638893.3. China; 2014

[11] Yang L, Song X, Cao L, Zhang Y, Ma Y, Zhang G, Wang S. A automatic device measuring the flow in the catchment. Patent No. ZL201110116288.5. China; 2015

[12] Hu S, Yu J, Hu K, Jin M. Impacts of Chinese Pine (*Pinus tabulaeformis*) plantations on rainfall redistribution processes: A case study for the mountainous area of North China. Acta Ecologica Sinica. 2010;**30**(7):1751-1757 (in Chinese)

[13] Qiao Y, Song X, Yang L, Jin D. The crustal lifting simulation system. Patent No. ZL 201520834298.6. China; 2015

[14] Yang L, Song X, Qiao Y. The device measuring the runoff. Patent No. ZL201521130097.4. China; 2016

Chapter 6

Experimental Study on the Mechanisms of Soil Water-Solute-Heat Transport and Nutrient Loss Control

Quanjiu Wang, Beibei Zhou, Lijun Su and Yuyang Shan

Additional information is available at the end of the chapter

http://dx.doi.org/10.5772/intechopen.76280

Abstract

The release and migration of nutrients, pesticides, and other chemicals in the runoff from agricultural lands is not only an economic loss but a threat to the quality of our surface and groundwater. In contrast to pollution from point sources, pollution from non-point sources is often low in intensity but high in volume. The development of a physically based model to simulate the transport of soil solutes would provide a better understanding of transport mechanisms and assist in the development of effective methods to control the loss of nutrients from soils and the pollution of waterways. As a result, numerous studies have been conducted in this area. But due to the soil genesis and human activity, the process is very complex, which can have a great impact on soil water movement, solute transport, as well as nutrient loss. In this study, we determined water movement and solute and heat transport through columns of disturbed soil samples. We also carried out simulated rainfall experiments on an artificial slope to study the nutrient loss.

Keywords: water movement, soil heat transport, nutrient loss

1. Introduction

Due to the fast development of agriculture and industry, water resource scarcity and nutrient loss became more and more serious in China. In order to solve those problems, it is necessary to present some high-efficiency technical methods and theories to understand the whole process of water-solute-heat transport and nutrient loss.

Water infiltration process is a complex process, and it is necessary to know the rules of water movement well and establish formulas and models to describe the whole process. Darcy's law was presented in 1856 [1]. In 1907, Edgar Buckingham applied "capillary potential" to soil water for the first time, which showed the energy state of soil water. Green and Ampt proposed an infiltration model based on capillary theory [2]. Richards introduced Darcy's law to describe soil water flow [3]. Philip presented a basic equation to describe the water movement in a one-dimensional vertical soil column [4]. In addition, the formula for infiltration of the Kostiakov infiltration model, Horton infiltration model, and Holtan infiltration model is also used [5–7]. Shu puts forward the model of capillary bundle infiltration [8]. Ghosh combines the one-dimensional vertical infiltration formula and Kostiakov empirical formula to obtain the new infiltration formula [9]. Many experts and scholars have proposed some new methods because it is difficult to find out the results in accordance with the actual results. Parlange presented an approximate solution for Richards equation [10], and Hogarth and Parlange improved the solution [11]. The finite difference method and finite element method were also used to solve the solution of the water eqaution [12]. Yang and Lei established a numerical model for the one-dimensional saturated water flow in the FORTRAN language and verified it in laboratory [13]. There are a large number of basic formulas and empirical formulas to describe the process of one-dimensional soil water movement. However, both the classical infiltration model and empirical model have different parameters which make them difficult and time-consuming. In this study, we want to find a simple and feasible method to determine soil hydraulic parameters.

Soil thermal conductivity is not only one of the important indexes of soil thermal properties but also an important parameter for simulating the soil hydro-thermal-solute-coupled model. How to estimate soil thermal conductivity quickly and accurately is one of main contents of studying soil thermal properties [14]. At present, a number of indirect estimation models to describe the relationship between thermal conductivity and soil texture, bulk density, water content, and organic matter were proposed by domestic and foreign scholars [15–23]. There are two types of indirect estimation models: empirical models [15, 16] and semi-theoretical models [17–20]. The empirical model mainly established the relationship between thermal conductivity and soil moisture content, such as the Chung-Horton model and Campbell model. These models are simple to calculate, but the model parameters are uncertain which will lead to large errors between the calculated data and measured values due to the difference of soil qualities in different regions [21]. The semi-theoretical model showed the relationship between thermal conductivity and soil saturation, such as the Johansen model, Côté-Konrad model, and Lu-Ren model [22]. These models have a theoretical basis and have given the model parameters for different soil textures. However, the model parameter values varied greatly with different soil particles and organic matter content, which limited the application of this model. In general, different models have their own advantages and disadvantages, but the effect of particle composition and organic matter content on the parameters of different types of soils needs further study. In this chapter, the thermal conductivity of undisturbed soil was measured by heat pulse methods. So, in this study, a new method based on analyzing the influence of soil particle composition on thermal conductivity, the relationship between thermal conductivity, saturation, bulk density, soil particle composition, and organic matter, was established. The improved Côté-Konrad and Lu-Ren models were also proposed to provide a reference method for obtaining soil thermal conductivity in a simple and rapid manner.

http://dx.doi.org/10.5772/intechopen.76280

The process of soil-dissolved chemical transfer to the runoff and transport to the field outlet was complex. Modeling the large number of processes involved and their interactions requires the solution of relatively complicated, coupled linear and nonlinear partial differential equations subject to time-dependent boundary conditions [24]. To reduce mathematical complexity, we applied the refined model [25] to data from our experiments in this study, in which the presumed exchange layer is replaced by a mixing zone, which can be regarded as an extension of the deposited layer or "shield" [26]. Assuming that the exchange rate was controlled by raindrop splash and that the effects of diffusion could be neglected, we replaced the exchange rate, km, with the variable er, the raindrop-induced water transfer rate, developed by Gao et al. [27]. This modification obviated the need to calibrate km. Laboratory experiments were performed to assess the accuracy of the new model's predictions. So, in this study, we carried out simulated rainfall experiments on an artificial slope to study the nutrient loss and test our new theory.

2. Materials and methods

2.1. Experimental soils

In this study, four soils were collected, and the soil's physical characteristics were listed in **Table 1**.

2.2. Experimental measurement

2.2.1. *Horizonal infiltration experiment*

Four soils in **Table 1** were collected for the infiltration experiments, and the negative hydraulic heads were designed as −2.5, −6, −9, −12, −15, and − 18 cm. Soil samples were filled in the column, every 10 cm, and the bulk density of soil was designed as 1.4 g/cm^3. The length of h_2 and h_3 were measured, and the values of h_1 were calculated by the formula $p = h_3 - h_1 + h_2$. The

Soil textural	<0.002 mm	0.002 mm < d < 0.02 mm	>0.02 mm	Saturated hydraulic conductivity (cm/min)	Initial water content (cm^3/cm^3)	Saturated water content (cm^3/cm^3)
Loessal soil	2.7	12.96	84.34	0.0416	0.01	0.47
Red glue soil	14.89	38.88	47.66	0.0155	0.04	0.63
Dark loessial soil	20.82	41.49	35.6	0.0047	0.04	0.54
Lou soil	16.65	44.76	38.22	0.0058	0.04	0.53

Table 1. The soil's physical characteristics.

water head in pressure regulator pipe values was adjusted by the values of h_1 (**Figure 1**). The standpipe was filled with distilled water before the experiment. Opening the right valve, the experiment continued until the bubble was emptied. For each infiltration measurement, cumulative infiltration was recorded every minute until it reached a steady state.

2.2.2. Soil thermal experiment

The test equipment uses a three-probe heat pulse probe (**Figure 2**) which was connected to the data collector, and sensor probes were used on two sides to observe and monitor the changing

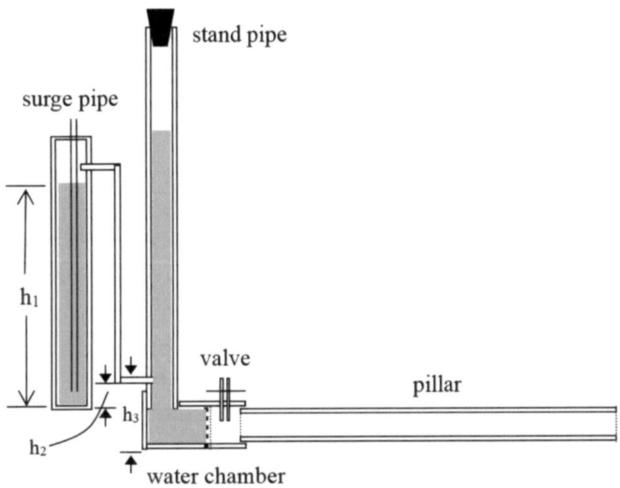

Figure 1. The sketch map of experimental equipment for horizonal soil column.

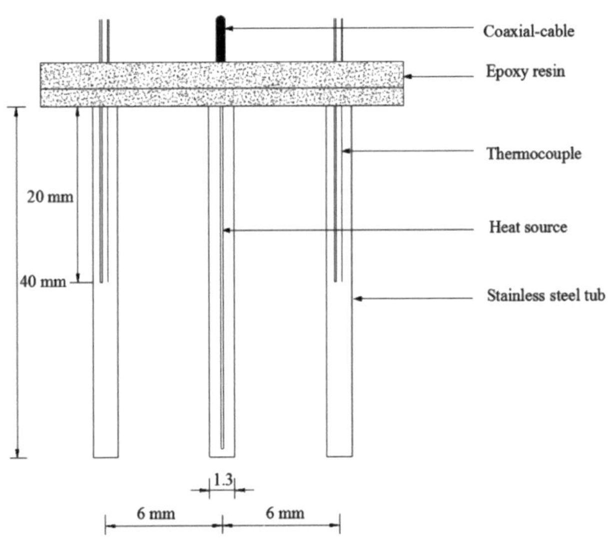

Figure 2. Schematic diagram of the heat pulse probe.

http://dx.doi.org/10.5772/intechopen.76280

temperature in the process with time after the middle probe sent the heat pulse (**Figure 3**). The diameter, length, and space distance of the three probes were 1.3, 40, and 6 mm, respectively (as shown in **Figure 1**). The 5–6 gL^{-1} agar solution was used to demarcate in advance in actual, which was to prevent natural convection of water when heated. The Data Collector (US CR1000 Data Collector) controls the heated input via a relay, and the electric current was determined by a precise resistance (10 Ω) of the assigned voltage. The data collector also recorded the temperature change of the sensing probe at intervals of 1 s. The volumetric heat capacity of the agar solution is $4.18 MJm^{-3}{}^{\circ}C^{-1}$. The distance r was obtained by a non-linear fitting temperature–time curve and averaged by repeating the calibration process 10 times.

To study the variation characteristics of the undisturbed soil thermal conductivity, the ring knife was used to take samples in the experimental ground. Each measurement point was arranged as follows: 10 measuring points per column, step length of 3 m between 2 points, and setting in 2 columns; 4 kinds of water contents were given to measure soil thermal conductivity in each measurement point, and the actual water contents were determined by the measured value at the end of the measurement (that means that the actual moisture content of the soil sample is supposed to be equal to that in the ring knife after finishing the measurement).

2.2.3. *Nutrient runoff experiment*

The basic component of the experiments was a rain simulator, which could generate a variable intensity of rainfall. The nozzles used to simulate rainfall were 15 m from the soil surface. We used six steel soil flumes with the following dimensions: 1 m in length × 0.40 m in width × 0.50 m in height. The flumes were filled with soil to a depth of 0.35 m; this depth allowed

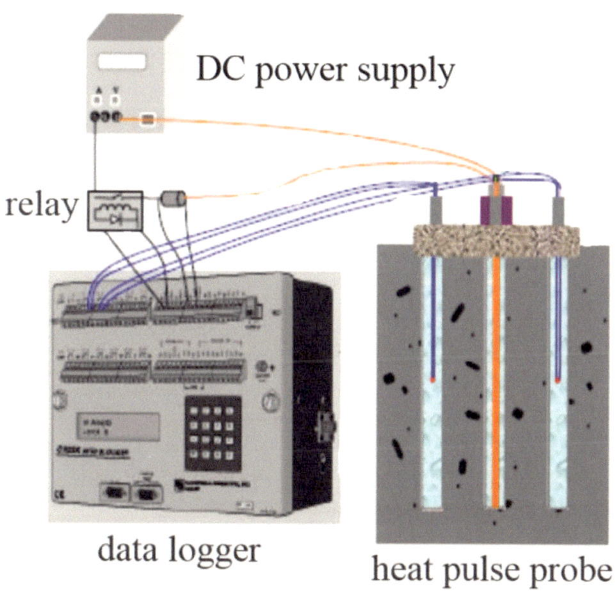

Figure 3. Schematic diagram of experimental apparatus.

infiltration without causing the bottoms of the flumes to become dank and left a 0.15 cm "lip" above the soil level to prevent water losses from splashing. The flumes' angle of inclination could be varied between 0° and 30° (**Figure 4**). The experiments were performed from April to September 2010 in a laboratory for simulating artificial rainfall at the Institute for Soil and Water Conservation, Chinese Academy of Sciences, Shaanxi Province, China.

Three treatments were established to test our model. In treatment 1, three initial levels of soil moisture (5, 10, and 20%, measured gravimetrically) were used to study the influence of the soil's initial water content on our model. The rainfall rate was 90 mm/h and the slope gradient was 5°. Treatment 2 was designed to investigate the influence of variation in the rainfall intensity on our model. Three different rainfall intensities (60, 96, and 129 mm/h) were examined, with an initial soil moisture content of 10% and a slope gradient of 5°. Treatment 3 was designed to assess the influence of the slope gradient. Slopes of 5, 15, and 25° were investigated, with a rainfall intensity of 90 mm/h and an initial (gravimetric) soil moisture content of 10%. All treatments were run three times.

The soil samples were sieved (0.004 m in aperture) to remove coarse rock and debris and then air dried (to about 2%, gravimetrically). Potassium, used as a tracer, was dissolved in water and added to the test soils based on their designed soil water contents and potassium concentrations; the soil was then thoroughly mixed. The soil flume was filled with the prepared soil sample layer by layer to achieve a dry bulk density of 1.35 g/cm^3. To obtain a flat surface, a sharp-edged straight blade was used to remove excess soil. The soil surface was covered with plastic for approximately 24 h before the beginning of the experiments. During the experiments, the outflow from one of the holes in the flume was collected into plastic containers every minute to measure the amount of runoff and its sediment and soil concentrations. We directly measured the depth of the exchange layers along a vertical section. The potassium content in the runoff was measured with an atomic absorption spectrophotometer (Perkin-Elmer 5100ZL). The soil water content was measured by drying, and the sediment was isolated by filtration on filter paper and weighed after drying.

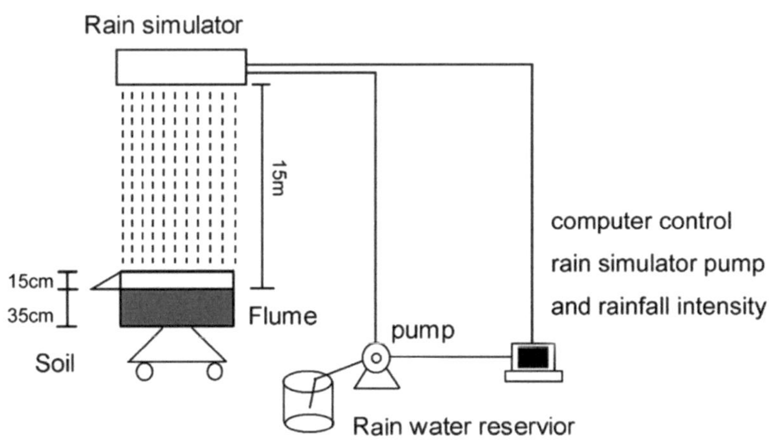

Figure 4. Experimental setup of artificial rainfall.

http://dx.doi.org/10.5772/intechopen.76280

2.3. Theory

2.3.1. Models of infiltration process

2.3.1.1. Kostiakov model

The Kostiakov model [5] was presented by large amount of experiments and can be expressed as:

$$I = at^b \tag{1}$$

where I is cumulative infiltration (cm), t is the infiltration time (min), a and b are empirical constants. The empirical constants a and b have no physical meaning and are obtained by experimental data fitting.

2.3.1.2. Philip model

Philip [4] proposed an infiltration equation based on Boltzmann transformation as power series:

$$I = St^{0.5} + At + Bt^{1.5} + \cdots \tag{2}$$

where S is the soil sorptivity (cm/min$^{1/2}$) and A is stable infiltration rate (cm/min). The result is exactly enough with two terms.

A horizontal one-dimensional infiltration equation, neglecting gravity action, can be expressed as:

$$I = St^{0.5} \tag{3}$$

A vertical one-dimensional infiltration equation, neglecting gravity action, can be expressed as:

$$I = St^{0.5} + At \tag{4}$$

2.3.1.3. Wang's model

Wang et al. [28] use $k(h) = k_s(h_d/h)^m$ and $(\theta_0 - \theta_r)/(\theta_s - \theta_r) = (h_d/h)^n$ to calculate unsaturated hydraulic conductivity and soil moisture characteristic curves. The relationship between wetting front and cumulative infiltration time can be expressed as:

$$I = x_f(\theta_s - \theta_i)\left(1 - \frac{n}{m+n-1}\right) \tag{5}$$

$$q = \frac{1}{x_f}\frac{h_d}{m-1} \tag{6}$$

$$x_f = \sqrt{\frac{2h_d k_s(m+n-1)}{(m^2-1)(\theta_s - \theta_i)}t^{\frac{1}{2}}} \tag{7}$$

Eq. (5), (6), and (7) also can be expressed as:

$$I = A_1 x_f, \qquad i = A_2/x_f, \qquad x_f = A_3 t^{1/2} \tag{8}$$

$$\begin{aligned} A_1 &= \frac{k_s h_d^m h_s^{1-m}}{m-1} \\ A_2 &= (\theta_s - \theta_i)\left(1 - \frac{n h_d^{m-1} h_s^{m-1}}{m+n-1}\right) \\ A_3 &= \frac{2k_s h_d^m h_s^{1-m}(m+n-1)}{(m-1)(\theta_s - \theta)\left(m+n-1-n h_d^{m-1} h_s^{m-1}\right)} \end{aligned} \tag{9}$$

Therefore, parameters n, m, and h_{d} can be expressed as:

$$\begin{aligned} n &= \sqrt{\frac{\theta_s - \theta_r}{A_1 + \theta_i - \theta_r}} - 1 \\ h_d &=^d \frac{A_2}{anK_s} \\ m &= \frac{A_3(\theta_s - \theta_r)}{aK_s h_d} + \frac{1+n}{n} \end{aligned} \tag{10}$$

where θ_{i} is the soil water content (cm^3/cm^3), θ_{s} is the saturated soil water content (cm^3/cm^3), θ_{r} is the residual water content (cm^3/cm^3), h_{d} is air entry suction (cm), Ks is saturated hydraulic conductivity (cm/min), and n, m is the parameter.

2.3.1.4. Green-Ampt model

Green and Ampt [2] proposed the model. The equation is expressed as:

$$i = K_s\left(1 + \frac{h_0 + h_f}{z_f}\right) \tag{11}$$

$$I = (\theta_s - \theta_i) z_f \tag{12}$$

where i is infiltration rate (cm/min), Ks is saturate hydraulic conductivity (cm/min), h_0 is ponder depth (cm), h_{f} is wetting front suction (cm), I is cumulative infiltration (cm), and θ_{s}, θ_{i} are saturated water content (cm^3/cm^3) and initial water content (cm^3/cm^3), respectively.

2.3.2. Models of soil heat conductivity

1. Thermal conductivity empirical model by Campbell [16].

Campbell proposed an empirical formula for calculating soil thermal conductivity based on soil texture, bulk density, and volume moisture content, which can be specifically expressed as

$$\lambda = A + B\theta - (A - D)\exp\left[-(C\theta)^E\right] \tag{13}$$

where θ is the volume of water content (cm^3/cm^3), and the parameters *A, B, C, D,* and *E* can be calculated according to soil bulk density, clay content, quartz, and other mineral volume ratios as follows:

$$\begin{cases} A = 0.65 - 0.78\rho_b + 0.60\rho_b^2 \qquad B = 1.06\rho_b \\ C = 1 + \dfrac{2.6}{m_c^{0.5}} \qquad D = 0.03 + 0.1\rho_b^2 \qquad E = 4 \end{cases} \tag{14}$$

where m_c is the clay content and ρ_b is the soil bulk density.

2. Semi-theoretical model of thermal conductivity by Johansen [18].

For the unsaturated soil, the relationship between λ and K_e (Kersten) is established based on thermal conductivity λ_{dry} (W/(m·K)) of dry soil and thermal conductivity λ_{sat} (W/(m·K)) of saturated soil.

$$\lambda = \left(\lambda_{sat} - \lambda_{dry}\right)K_e + \lambda_{dry} \tag{15}$$

And the relationship between K_e and conventional soil moisture content or saturation S_r ($S_r = \theta/\theta_s$, where θ_s is saturated water content) is established:

$$K_e = \begin{cases} 0.7\log S_r + 1.0 & 0.05 < S_r \le 0.1 \\ \log S_r + 1.0 & S_r > 0.1 \end{cases} \tag{16}$$

$$\lambda_{sat} = \lambda_s^{1-n}\lambda_w^n \tag{17}$$

where λ_w = 0.594 W/(m·K) under the condition 20°C, *n* is the soil porosity, and λ_s is obtained by the quartz content (*q*) of the whole solid, its thermal conductivity being λ_q = 7.7 W/(m·K), and thermal conductivity (λ_0) of the other minerals is $\lambda_s = \lambda_q^q\lambda_0^{1-q}$, where, λ_0 = 2.0 W/(m·K) ($q > 0.2$), λ_0 = 3.0 W/(m·K) ($q \le 0.2$).

$$\lambda_{dry} = \frac{0.135\rho_b + 64.7}{2700 - 0.947\rho_b} \tag{18}$$

3. Improved Johansen model by Côté and Konrad [19].

In order to simplify the calculation of the logarithmic function formula in the Johansen model, Côté and Konrad proposed a new relationship between K_e and S_r based on the parameter *k*:

$$K_e = \frac{kS_r}{1 + (k-1)S_r} \tag{19}$$

where *k* is an independent parameter related to the soil texture and its values for coarse sand, small sand, clay, and higher organic matter content are 4.60, 3.25, 1.40, and 1.20, respectively. And a new formula to estimate λ_{dry} is given as follows:

$$\lambda_{dry} = \chi 10^{-\eta n} \tag{20}$$

where χ (W/(m·K)) and η are parameters that are affected by particle traits. The χ and η values for crushed rock, mineral soil, and soil with high organic matter were 1.70 and 1.80, 0.75 and 1.2, and 0.30 and 0.87, respectively.

4. Improved Johansen model by Lu and Ren [20].

In order to make the Johansen model more suitable for calculating the thermal conductivity under the condition of low soil water content, Lu and Ren proposed a new exponential function expression of K_e about S_r:

$$K_e = \exp\{\alpha[1 - S_r^{\alpha-1.33}]\} \tag{21}$$

where α is the parameter determined by the soil texture, and for coarse soils with sand content larger than 40% and fine soils with sand content of less than 40%, α = 0.96, 0.27. 1.33 refers to the shape parameters. A new formula is given for mineral soil as follows:

$$\lambda_{dry} = -an + b \tag{22}$$

where a and b are the empirical coefficients; when $0.2 < n < 0.6$, the value is: 0.56, 0.51.

2.3.3. *Models of nutrient runoff on the slope*

To better understand the factors affecting the loss of solutes to the runoff, we applied our experimental data to the model developed by Wang et al. [25]. This model is described and justified in full detail in the publication cited above and is only briefly outlined here. The model is based on a soil water system that is divided into three vertically distributed horizontal layers: runoff or water ponding on the surface; an exchange layer below that; and the underlying soil. The variation in solute mass in the exchange layer changes over time and can be modeled using a power function. The transport of solutes from the exchange layer to the surface runoff is assumed to be dependent on the mass exchange rate. The model can be expressed as:

$$c(t) = k_m \frac{C_o \rho_b H_o}{r(t)(Ptp + \rho_b \theta_o H_o)} t^b \tag{23}$$

where $c(t)$ represents the solute concentration (mg/L) in the runoff, k_m is the exchange rate, c_o denotes the initial solute concentration in the surface soil (g/g), ρ_b is the soil's dry bulk density (g/cm), H_o is the depth of the exchange layer (cm), $r(t)$ is the runoff volume (L), p is the rainfall intensity (cm/min), t_p is the time between the initiation of rainfall and the formation of the runoff (min), θ_o is the initial soil moisture content (%), t is time (min), and b is an empirical parameter.

We adopted a new model, in which the presumed exchange layer is replaced by a mixing zone, which can be regarded as an extension of the deposited layer or "shield" concept presented by Hairsine and Rose [26]. Assuming that the exchange rate is controlled by raindrop splash and that the effects of diffusion can be neglected, we replaced the exchange rate k_m with the variable e_r, developed by Gao et al. [27]. This substitution obviates the need to calibrate k_m. This new variable is the rate at which soil water is ejected from the soil during rainfall:

$$e_r = \frac{ap}{\rho_b}\theta \tag{24}$$

$$a = \frac{e}{p^{\rho}} \tag{25}$$

where e_r is the raindrop-induced water transfer rate, a is the detachability of the bare soil (g/cm) [29], e is the rainfall-induced soil detachment per unit soil area (g/cm), ρ is a constant parameter, ρ_b is the bulk density of the dry soil (g/cm), p is the rainfall intensity (mm/min), and θ is the soil water content (%) [30, 31]. Eqs. (23) and (24) can be combined to give:

$$c(t) = \frac{ap\theta C_o H_o}{r(t)(Ptp + \rho_b \theta_o H_o)} t^b \tag{26}$$

3. Results and discussions

3.1. The infiltration in the horizonal soil column

3.1.1. Cumulative infiltration with times

In the horizontal one-dimensional suction process, the soil water content increases with times, along with cumulative infiltration. However, the cumulative infiltration amount is different at different negative hydraulic heads and at various soil textures in the same infiltration time. From **Figure 5**, it can be seen that under different negative hydraulic head conditions, the change of cumulative is the largest in Loessal soil, followed by red glue soil and black loessial soil, with Lou soil showing the smallest cumulative change in the same infiltration time.

It can be seen from **Figure 6** that the cumulative infiltration capacity with infiltration time is almost the same for each soil. The regulation in infiltration gradually decreases with a negative hydraulic head increase.. Among them, the most significant change was observed in Loessial soil, and there is a large difference between −9 and − 12 cm; the reason is the soil porosity ratio. Lou soil also shows a large difference. From −2.5 to −18 cm, red glue soil, black Loessial soil, and Lou soil show no significant difference, because no difference in porosity was observed between these two hydraulic heads. As for various soil textures, there are great differences in soil moisture absorption characteristics at different negative hydraulic head conditions, and the lighter the soil texture, the more is the difference.

In order to obtain a negative pressure suction effect on soil infiltration characteristics of quantitative analysis, we use the Kostiakov infiltration equation to fit the measured data. The Kostiakov model could fit the cumulative infiltration and infiltration time very well, which were shown **Table 2**.

From **Table 2**, it could be found that the correlation coefficient R^2 are all lager than 0.99 which indicated that the relationship between cumulative infiltration and time all have followed a power function under different negative hydraulic heads. For different soil textures, the order of coefficient a is Loessial soil >red glue soil>dark loessial soil>Lou soil, and index b had no

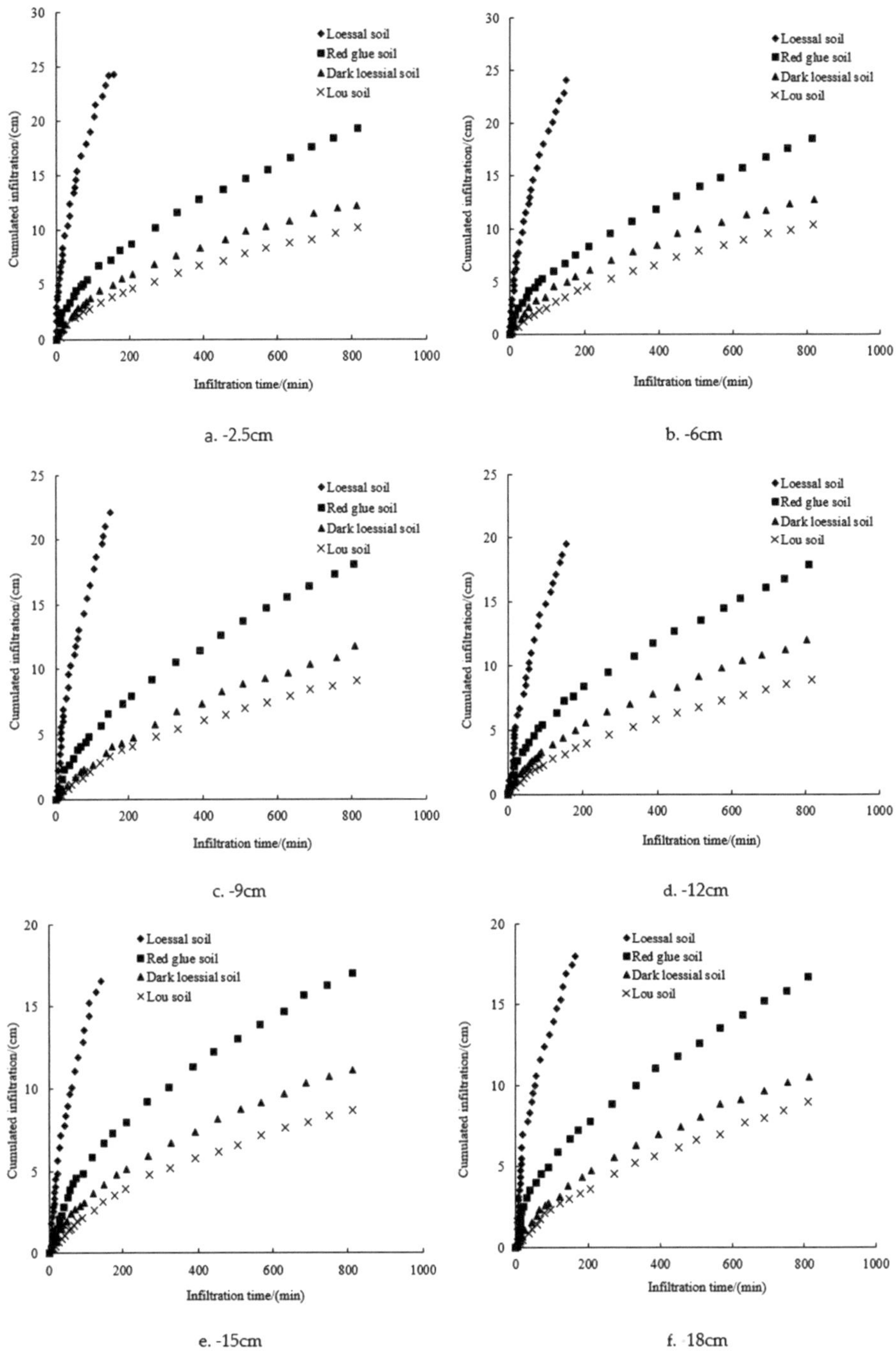

Figure 5. Relationship between cumulated infiltration and measured infiltration time at different hydraulic heads with various soils. (a) −2.5 cm; (b) −6 cm; (c) −9 cm; (d) −12 cm; (e) −15 cm; (f) −18 cm.

http://dx.doi.org/10.5772/intechopen.76280

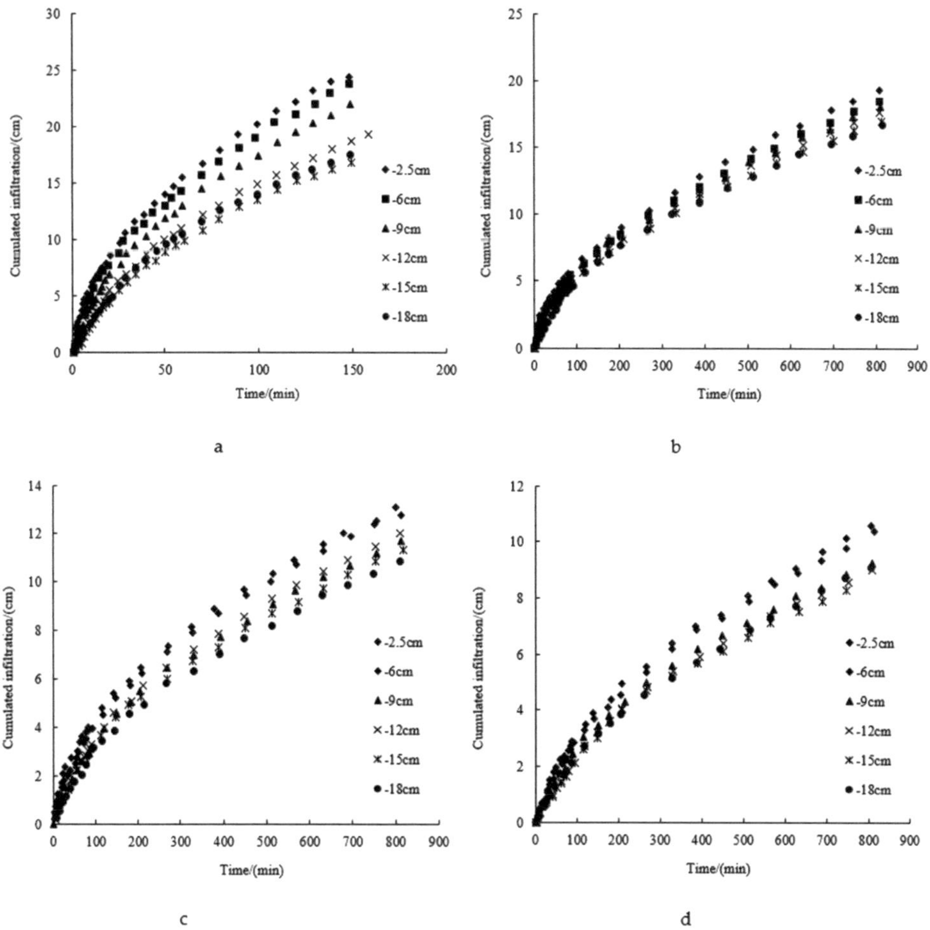

Figure 6. Relationship between cumulated infiltration and measured infiltration time at different negative hydraulic heads. (a) Loessal soil; (b) red glue soil; (c) dark loessial soil; (d) Lou soil.

significant changing tendency. For each soil, the parameter increased with the increase of negative hydraulic heads, while for parameter b, the opposite is true.

3.1.2. Determining the soil sorptivity based on horizonal one-dimensional experiments

According to horizontal one-dimensional experiments, we can easily obtain the soil sorptivity with the analysis to cumulative infiltration change with $t^{1/2}$. In **Figure 7**, it shows the change processes of cumulative infiltration with $t^{1/2}$ under four negative hydraulic heads. Sub-graphs a–f are Loessal soils, sub-graphs g–l are red glue soils, sub-graphs m–r are dark Loessial soils, sub-graph s–x are Lou soils.

As shown in **Figure 7**, the cumulative infiltrations of four kinds of soils had a linear relationship with $t^{1/2}$. We use a linear function to describe the curves, and the results were shown in **Table 3**.

Soil textural	Parameters	Hydraulic head (cm)					
		−2.5	−6	−9	−12	−15	−18
Loessal soil	*A*	1.3548	1.1979	0.9380	0.5566	0.4585	0.7439
	B	0.5949	0.6086	0.6502	0.7239	0.7507	0.6453
	R^2	0.9939	0.9937	0.9951	0.9899	0.9847	0.9922
Red glue soil	*A*	0.353	0.3167	0.2844	0.3820	0.2095	0.2906
	B	0.6025	0.6105	0.6255	0.5787	0.6727	0.6129
	R^2	0.9954	0.9956	0.9956	0.9921	0.9857	0.9925
Dark loessial soil	*A*	0.2343	0.1942	0.1559	0.1631	0.1421	0.1017
	B	0.6106	0.6418	0.6551	0.6529	0.6666	0.7125
	R^2	0.9941	0.9926	0.9964	0.9889	0.9938	0.9923
Lou soil	*A*	0.1718	0.1272	0.1244	0.1149	0.073	0.0934
	B	0.6198	0.6672	0.6561	0.6579	0.744	0.6903
	R^2	0.9976	0.9966	0.9951	0.9985	0.9947	0.9976

Table 2. Fitted value for parameters of I-t at different negative hydraulic heads.

The regression coefficients R^2 in **Table 3** were all above 0.95, which indicated that Philip equation can describe the infiltration rule very well under different negative hydraulic heads. Meanwhile, the soil sorptivity decreased with increasing soil viscidity (Loessal soil > red glue soil >dark Loessal soil > Lou soil). And beyond that, soil sorptivity decreases with the increasing negative hydraulic head.

3.1.3. Determining the parameters using Wang's proposed equation

We used three kinds of textured soils (red glue soil, dark Loessial soil, and Lou soil) for horizontal one-dimensional infiltration experiments. The length of the soil column is 50 cm. The upper boundary was a constant hydraulic head (i.e., when x = 0, the hydraulic head was designed as a different negative hydraulic head). The hydraulic heads of red glue soil were − 21 cm and − 30 cm. The hydraulic heads of dark Loessial soil were − 18 cm and − 24 cm. The hydraulic heads of Lou soil were − 21 cm and − 34 cm. The lower boundary condition was free discharge. The duration of the experiment was 810 min. The saturated water content, the retention water content, and the initial water content were measured, noting down the changes of the cumulative infiltration with time. The parameters in the Brook-Corey model can be determined by MATLAB programming based on the experimental data. **Figure 8** shows the relationship between cumulative infiltration and wetting front under different negative hydraulic head conditions.

As shown in **Table 4**, there is a good linear relationship between cumulative infiltration and wetting front which is in agreement with the theoretical derivation.

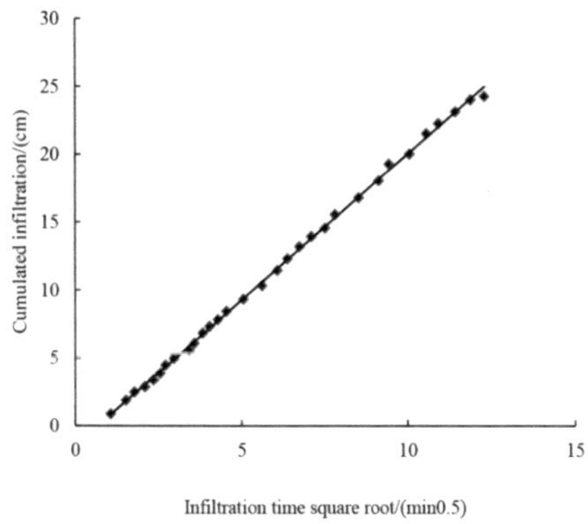

a. -2.5cm

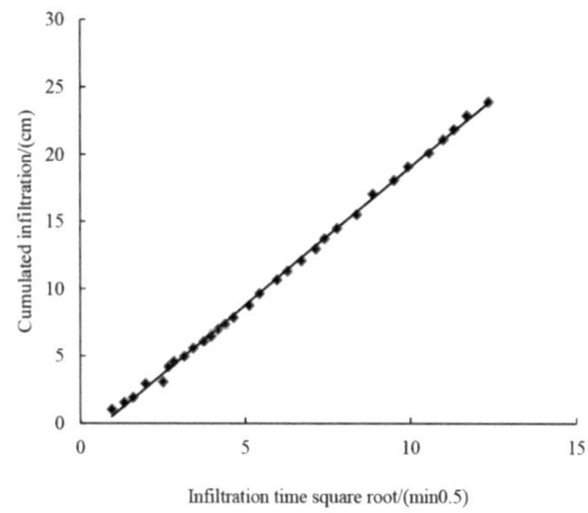

b. -6 cm

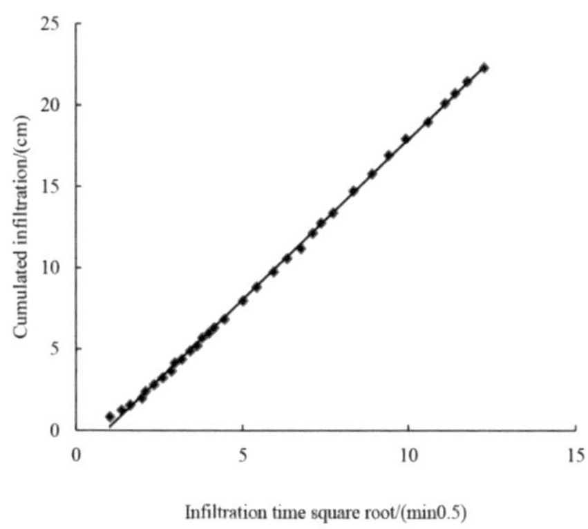

c. -9.0 cm

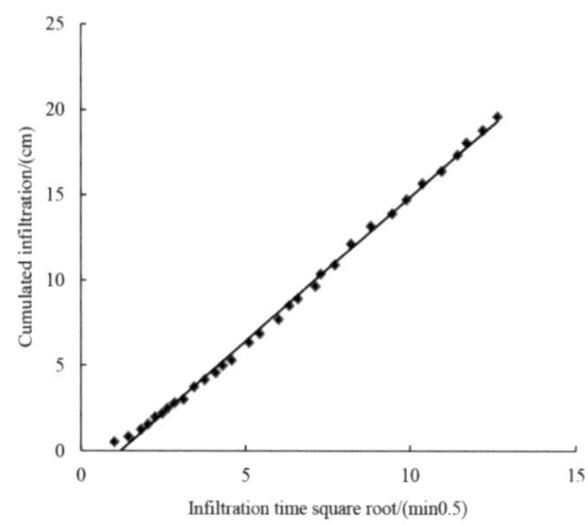

d. -12 cm

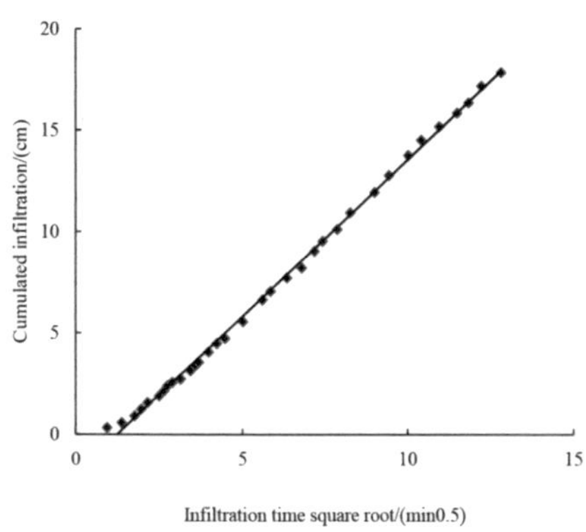

e. -15 cm

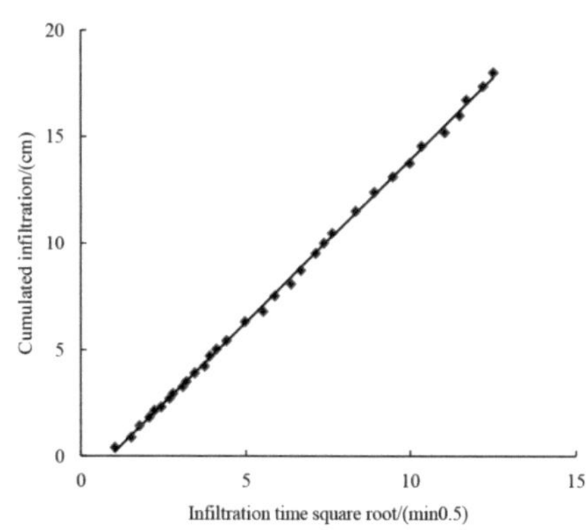

f. -18 cm

A

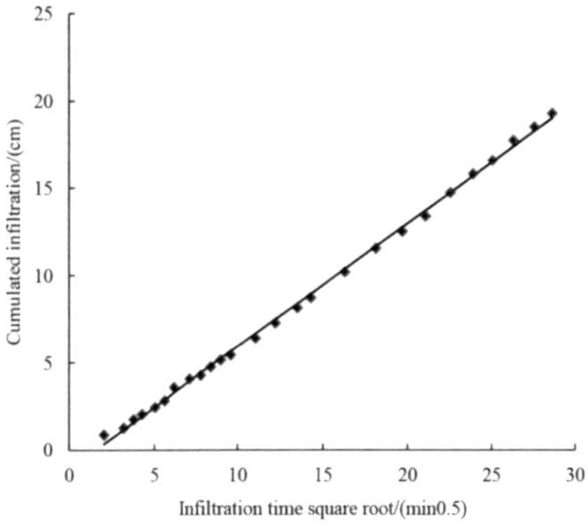

g. -2.5 cm

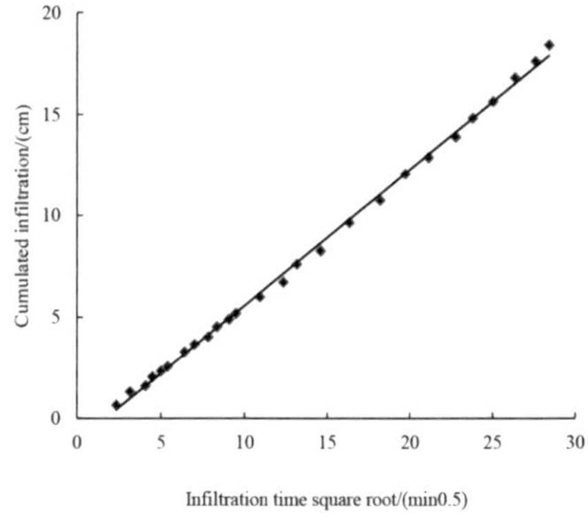

h. -6 cm

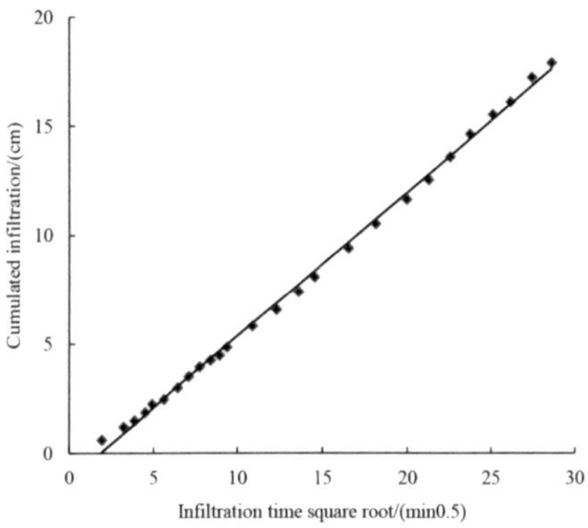

i. -9 cm

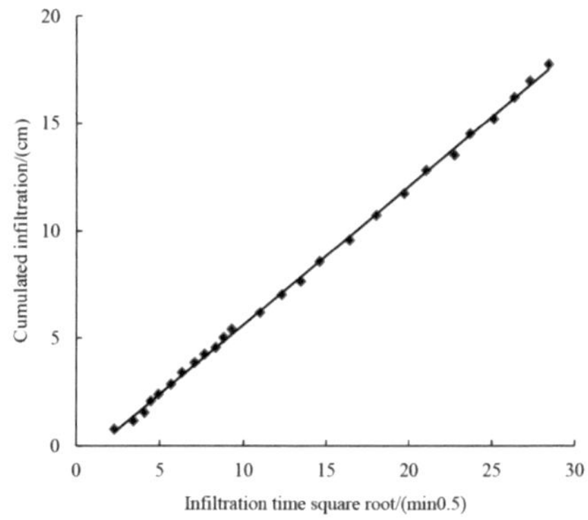

j. -12 cm

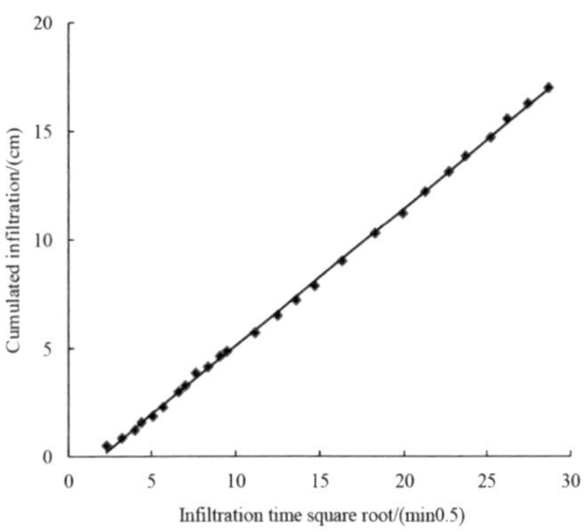

k. -15 cm

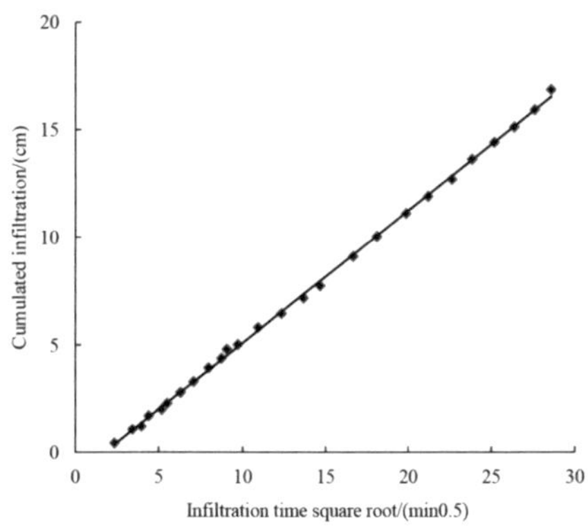

l. -18 cm

B

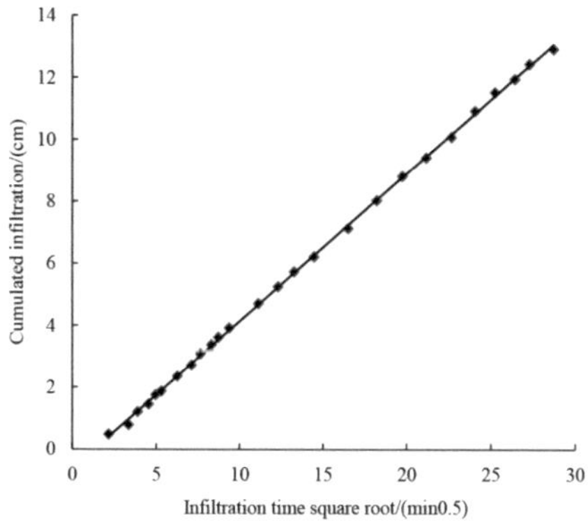

m. -2.5 cm

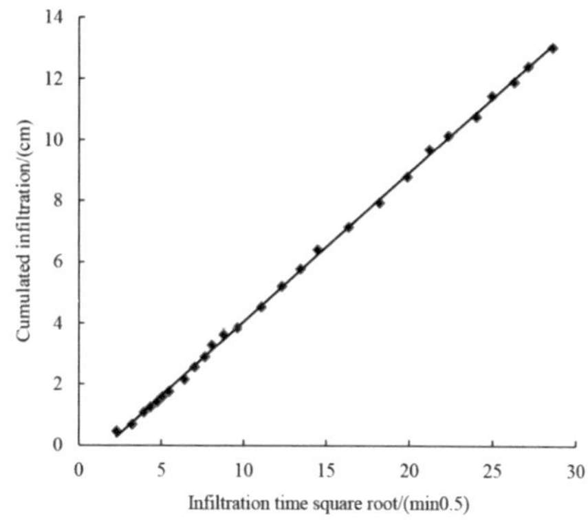

n. -6 cm

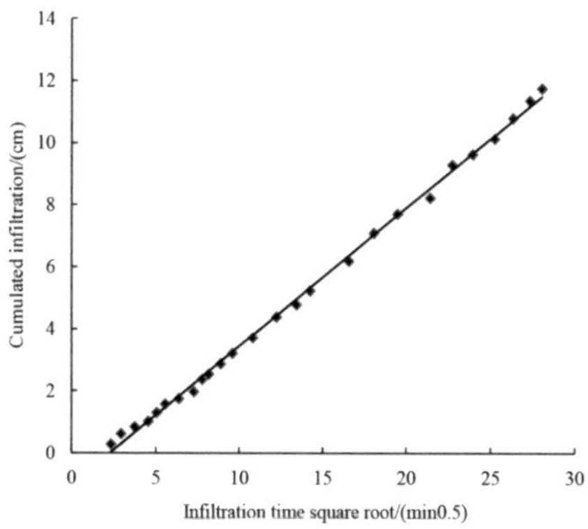

o. -9 cm

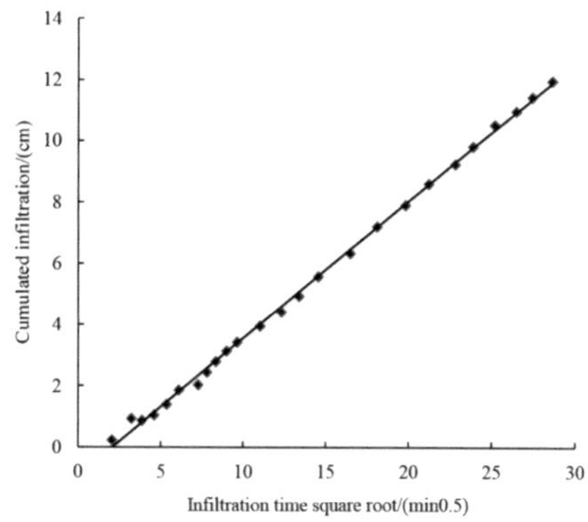

p. -12 cm

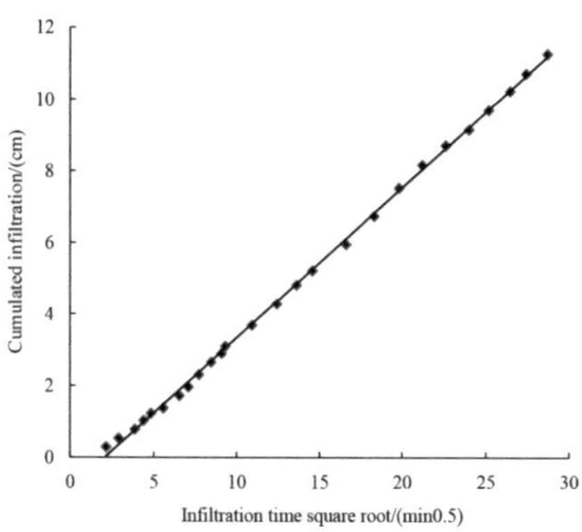

q. -15 cm

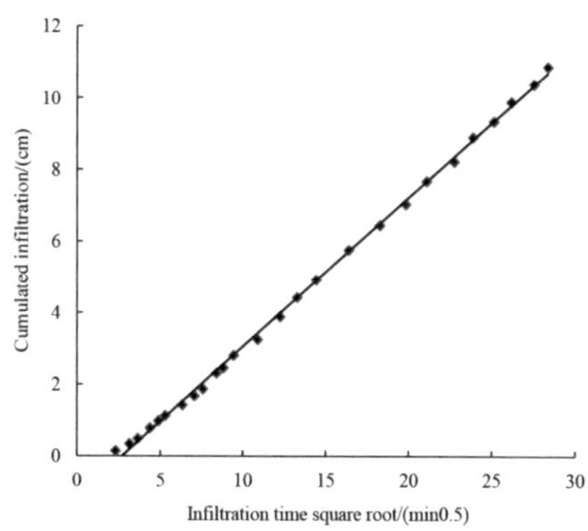

r. -18 cm

C

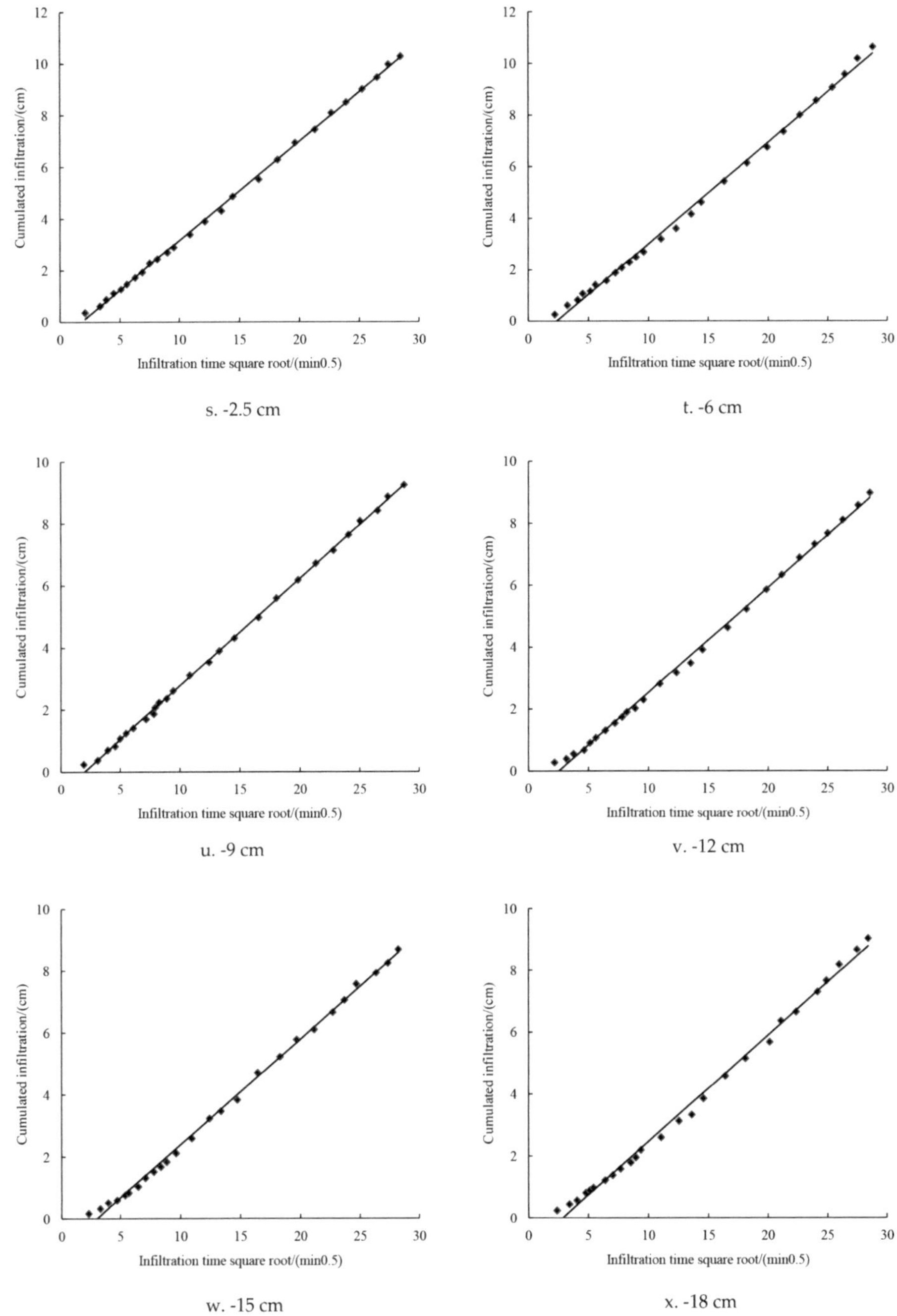

D

Figure 7. Relationship between cumulated infiltration and square root of measured infiltration time at different negative heads. A. Loessal soil; B. Red glue soil; C. Dark loessial soil; D. Lou soil.

Hydraulic head (cm)	Loessal soil		Red glue soil		Dark loessial soil		Lou soil	
	S	R^2	*S*	R^2	*S*	R^2	*S*	R^2
−2.5	1.9904	0.994	0.6448	0.9903	0.4436	0.993	0.3466	0.9882
−6	1.8747	0.9903	0.6083	0.9878	0.4423	0.9893	0.344	0.9759
−9	1.7482	0.9836	0.5968	0.9856	0.3902	0.9814	0.3112	0.9843
−12	1.4331	0.9706	0.6008	0.9959	0.4012	0.9838	0.2947	0.9777
−15	1.3145	0.9686	0.5695	0.986	0.3776	0.9832	0.2875	0.9688
−18	1.3654	0.9866	0.5597	0.9918	0.3552	0.9746	0.2918	0.9684

Table 3. The suction of four soil at different negative hydraulic heads.

Substitution of A1 and A2 (listed in **Table 5**) into Eq. (10) yields the hydrodynamic parameters (in the Brooks-Core model). The results are listed in **Table 5**.

The soil water characteristic curves can be easily obtained by the values in **Table 5**. Comparing the calculated results and the experimental results (determined by centrifuge), the results were listed in **Figure 9**. As shown in **Figure 9**, the calculated data concur with experimental data. The results indicated that the parameters in the Brooks-Core model can be accurately and easy computed by the new method.

3.2. Analysis of soil thermal conductivity characteristics

Soil thermal conductivity reflects the size of soil thermal conductivity, and soil texture has a certain impact on thermal conductivity. According to the principle of thermal pulse probe, soil thermal parameters were measured and soil thermal conductivity was calculated. **Figure 10** shows the curves of soil thermal conductivity with soil moisture content in four experimental sites of Shenmu (sand), Ansai, Yichuan, and Changwu. It can be seen from **Figure 10** that soil thermal conductivity increases rapidly with the increase of water content when the soil water content is lower than 0.13 cm^3/cm^3. When the soil water content is higher than 0.13 cm^3/cm^3, the increasing trend of soil thermal conductivity is relatively reduced. Under the same moisture content, the trend of soil thermal conductivity is as follows: Shenmu sand soil>Ansai sandy loam soil>Yichuan clay loam soil> Changwu silty loam soil. So we can see that the higher the sand content, the lower the silt content, and the greater the soil thermal conductivity [22].

A study by Lu and Ren et al. showed that the soil can be divided into two categories according to sand content of the soil: It is coarse soil when the sand content is more than 40%, and $S_r = 0.3$, K_e-S_r relationship curve of the coarse soil is divided into two linear ranges; when the sand content is less than 40%, it is fine soil, it is composed of $S_r = 0.13$ and $S_r = 0.30$, and the K_e-S_r relationship curve of the fine soil is divided into three linear intervals; **Figure 11** shows the K_e-S_r curves of the normal form of soil thermal conductivity. The soil samples of Changwu and Ankang belong to fine soil, Shenmu, Mizhi, Ansai, Shangnan, Yichuan, Luochuan, and

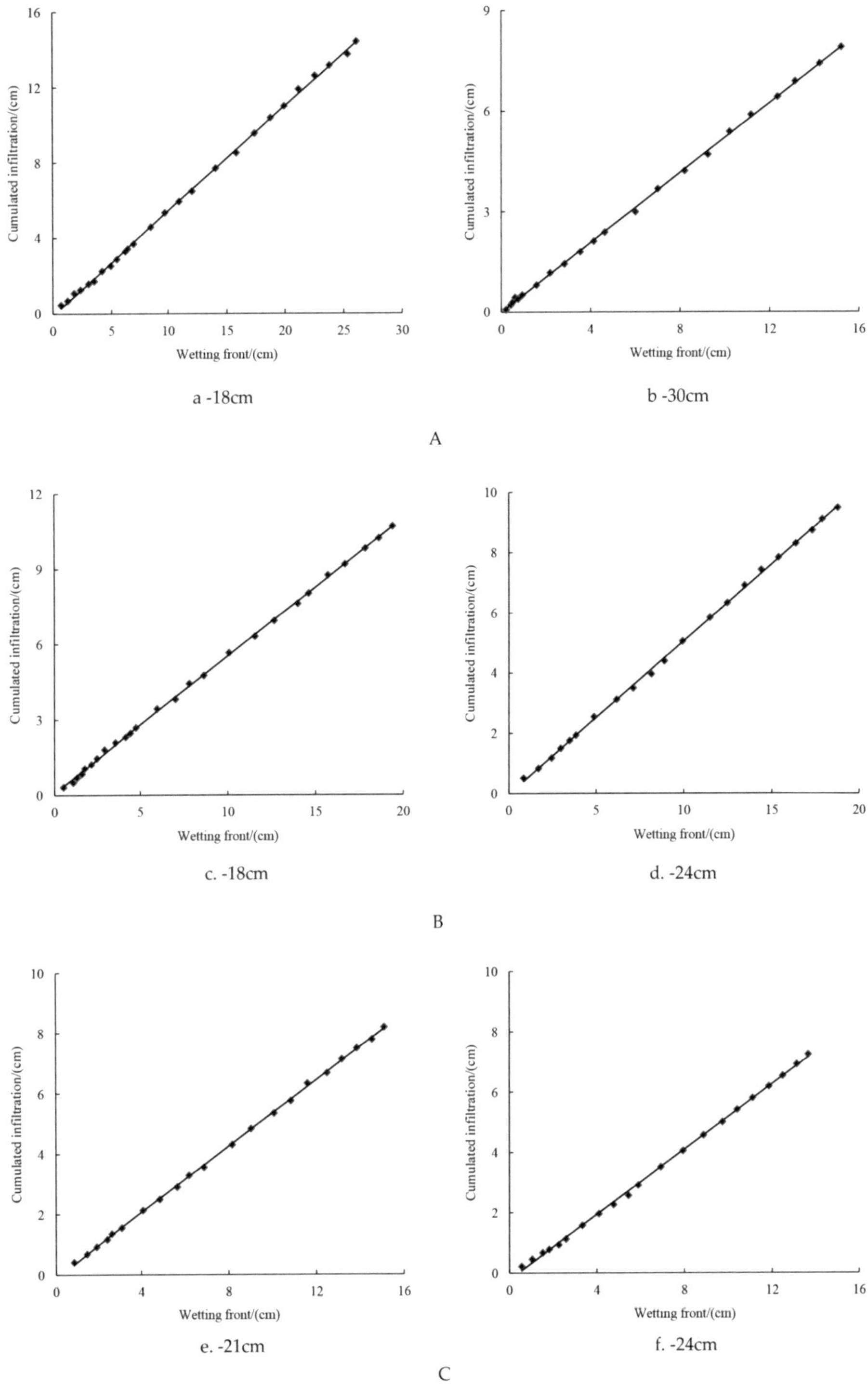

Figure 8. Relation curve between cumulated infiltration and wetting front. A. Red glue soil; B. Dark loessial soil; C. Lou soil.

http://dx.doi.org/10.5772/intechopen.76280

Soil textural	Hydraulic head (cm)	A1	R^2	Pressure head (cm)	A1	R^2
Red glue soil	−21	0.552	0.9997	−30	0.5192	0.9995
Dark loessial soil	−18	0.5539	0.9997	−24	0.5059	0.9994
Lou soil	−21	0.5356	0.9995	−24	0.5195	0.9966

Table 4. Relation fitting values of cumulated infiltration and wetting front.

Soil textural	n	M	h_d	m/n
Red glue soil	0.17	2.51	26	14.65
Dark loessial soil	0.32	2.94	40.11	9.18
Lou soil	0.23	2.69	54.06	11.69

Table 5. Parameters calculation of three soils.

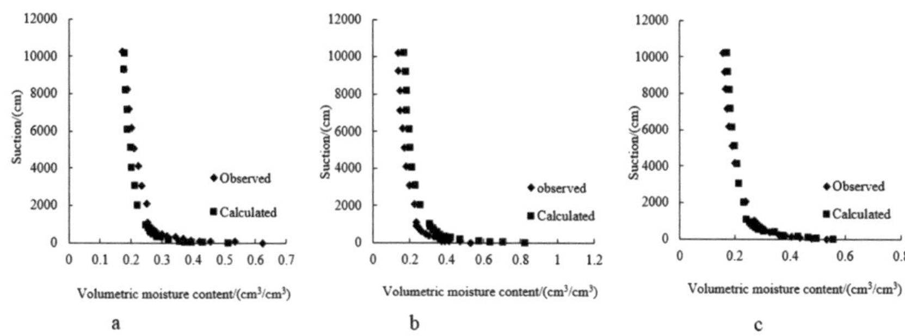

Figure 9. Compared between observed and calculated soil water characteristic curve. a. Red glue soil; b. Dark loessial soil; c. Lou soil.

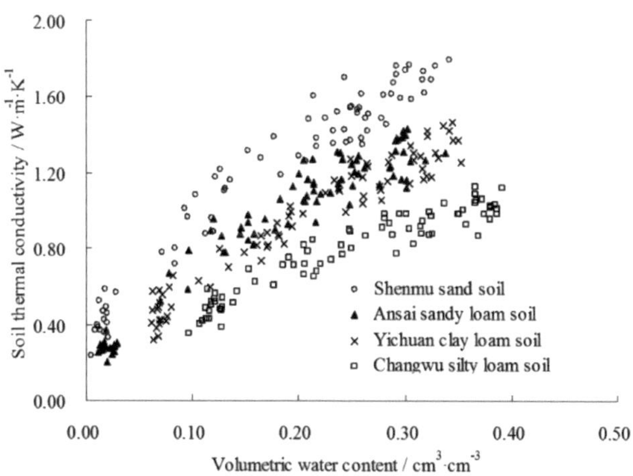

Figure 10. Trend of soil thermal conductivity with water content.

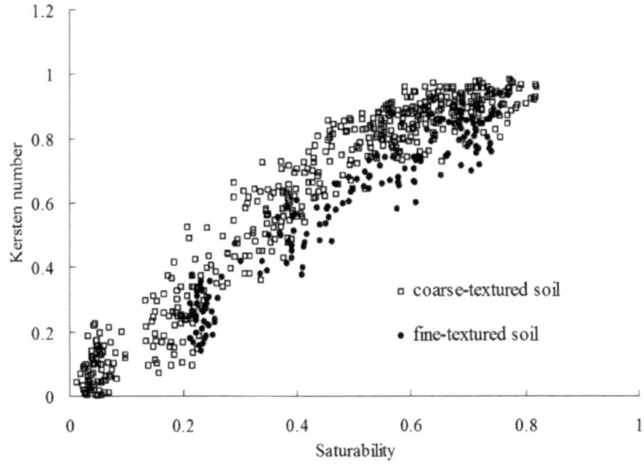

Figure 11. Thermal conductivity as a function of water content as indicated by the normalized form (the Kersten number K_e - the degree of saturation S_r).

Zhangye kind of matter is coarse soil. It can be seen from the figure that the K_e value of fine soil is obviously smaller than the K_e value of coarse soil when $0.2 < S_r < 0.6$.

3.2.1. Accuracy analysis of the soil thermal conductivity model

3.2.1.1. Campbell model

The sandy soil and sandy loam soil of coarse soil in Shenmu and Ansai, silty loam soil and silty clay loam soil of fine soil in ChangWu and Ankang, respectively, were selected. The thermal conductivities of these four soils were calculated by the Campbell model, and the results were shown in **Figure 12**. According to the statistical analysis, it can be seen that the difference between the calculated value and the measured value of the heat pulse is small when the water content of the soil is less than 0.20 cm^3/cm^3, and the relative error (R_e) of Shenmu sand soil and Ansai sandy loam soil are 13.51 and 9.56%, respectively; When the soil water content is higher than 0.20 cm^3/cm^3, the measured value of the heat pulse is larger than the calculated value, and the R_e of Shenmu sand soil and Ansai sandy loam soil are 19.40 and 13.38%, respectively; the larger the volume of moisture content, the greater the difference; relative to the thermal pulse's measured value, the calculation of the coarse soil model is too small for the coarse soil. For the fine soil, when the soil moisture content is less than 0.25 cm^3/cm^3, the R_e are 26.29 and 21.19%, respectively, and the measured value of the heat pulse is larger than the calculated value of the model. When the soil moisture content is higher than 0.25 cm^3/cm^3, and the R_e of Shenmu sand soil and Ansai sandy loam soil are 14.15 and 6.60%, respectively, the difference between the calculated value and the measured value of the heat pulse is small. Therefore, the model needs to be improved when calculating the thermal conductivity using the Campbell model [8, 9].

3.2.1.2. Johansen model, Côté-Konrad model and Lu-Ren model

Côté-Konrad model and Lu-Ren model are all semi-theoretical models of thermal conductivity based on Johansen model. The three models are used to calculate soil thermal conductivity for

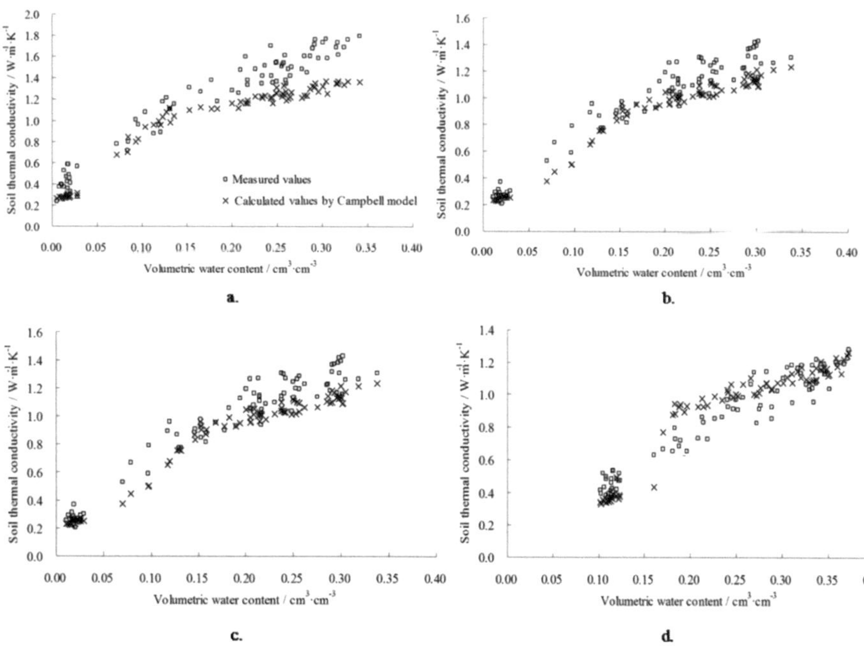

Figure 12. Comparison of soil thermal conductivity values calculated by Campbell model with measured. (a) Shenmu sand soil, (b) Ansai sandy loam soil, (c) Changwu silty loam soil and (d) Ankang silty clay loam soil.

the following four soils: Shenmu sand soil, Ansai sandy loam soil, Changwu silty loam soil, and Ankang silty clay loam soil. The model's calculated values and measured values are shown in **Figure 13**. It can be seen from the figure that the calculated values of Johansen model are significantly smaller than the measured values, the calculation error is larger, the coefficient of determination R^2 is in the range of 0.656–0.827, the root mean square error (RMSE) is in the range of 0.0848–0.2548, and the relative error R_e is in the range of 10.32–20.41%. For fine soil, the Côté-Konrad model and Lu-Ren model have a good fitting effect on soil thermal conductivity and the precision is high. Where the variation coefficient of R^2 is in the range from 0.842 to 0.940, the variation range of RMSE is from 0.0810 to 0.1208, the relative error R_e is in the range from 9.67 to 10.57%. The coefficient of determination R^2 of the Lu-Ren model is in the range from 0.874 to 0.937, RMSE varied from 0.0725 to 0.1238, and the relative error R_e varied from 8.28% to 9.91%. For coarse soil (sand content greater than 40%), the Côté-Konrad model and Lu-Ren model can still well fit soil thermal conductivity when the saturation $S_r < 50\%$, but the prediction accuracy of the model is poor, and the calculated value is obviously smaller than the measured value when the saturation $S_r > 50\%$. This phenomenon may be due to the large soil voids and the weak water-holding capacity, resulting in the measured value of water content being lower.

3.2.2. *The improved Côté-Konrad model and the improved Lu-Ren model*

The comparison between the calculated values of Campbell model, Johansen model, Côté-Konrad model, and Lu-Ren model and the measured values of the thermal pulses show that the soil thermal conductivity is closely related to the soil particle composition, organic matter

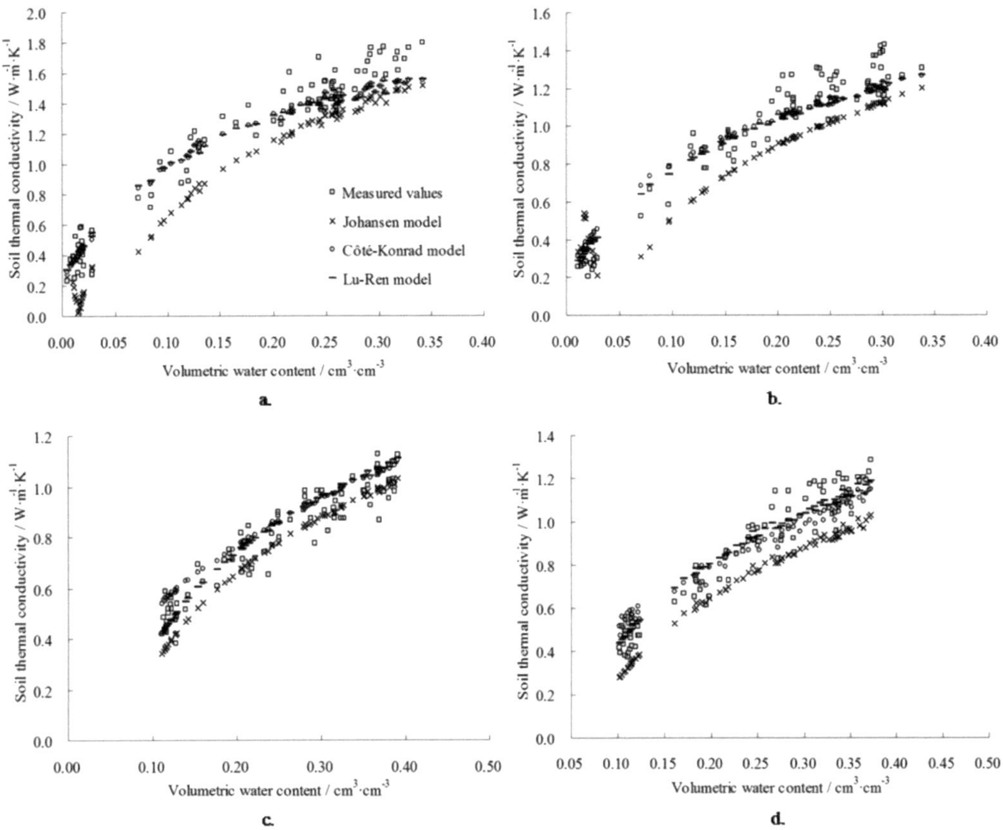

Figure 13. Comparison of soil thermal conductivity values calculated by different models (Johansen model, Côté-Konrad model and Lu-Ren model) with measured. (a) Shenmu sand soil, (b) Ansai sandy loam soil, (c) Changwu silty loam soil and (d) Ankang silty clay loam soil.

content, and bulk density. For different soils with different textures, model parameters are also different. In Johansen model, the parameter λ_s is related to the quartz content of the whole solid, and the thermal conductivity λ_{dry} of the dry soil is related to soil bulk density. In Côté-Konrad model, the parameter k is related to the content of coarse sand, small sand, clay, and organic matter. In Ren model, the parameter α is related to the soil sand content, and the soil is divided into coarse soil and fine soil according to the sand content. Under certain conditions, these models can calculate the soil thermal conductivity more accurately but cannot reflect the effect of soil particle composition and organic matter content on soil thermal conductivity. The improved Côté-Konrad model and the improved Lu-Ren model established the relationship between the model parameters, the composition of the particles, and the content of organic matter, respectively, and can describe the relationship between soil texture and soil thermal conductivity in detail.

In this chapter, the data of the five sites (466 sample points) of Mizhi, Shenmu (sandy loam), Ansai, Yichuan, and Changwu combined with R^2, R_e, and **Figure 13**, the relationship between the soil texture and the thermal conductivity, is fitting for the improved model, and the results of the parameter fitting are shown in **Table 6**. The comparison between soil thermal conductivity and the measured values is shown in **Figure 14**. From the fitting error, it can be seen that the accuracy of the two improved models is not very different, and they have high accuracy. However, it can be seen from **Figure 14** that the fitting value is larger than the measured value, and the soil thermal conductivity is more than 1.1 W/(m·K), the RMSE, R^2, and *Re* are 0.0964, 0.9274, and 9.62% for the improved Côté-Konrad model when soil thermal conductivity is less than 0.6 W/(m·K), respectively. For the improved Lu-Ren model, although the fitting value and the measured value are also different, the discrete points in the figure are evenly distributed near the 1:1 line; RMSE, R^2,and *Re* are 0.0961, 0.9278, and 9.59%, respectively.

According to R^2, R_e, soil thermal conductivity of four samples of Shenmu (sand), Shangluo, Luochuan, and Ankang combined with the model parameter fitting values in **Table 6** was predicted. The comparison between the predicted values and the measured values about the different models is shown in **Figure 15a–d** and **Table 7**, where, the sand content in four experimental sites was as follows: Shenmu> Shangluo> Luochuan> Ankang; clay content: Shenmu <Shangluo <Luochuan <Ankang; silt content: Shenmu <Shangluo <Luochuan <Ankang. Analysis of the simulation error shows that two improved models can be used to simulate soil thermal conductivity of different soils. For Shenmu sand soil and Ankang silty clay loam soil, the RMSE of the improved Côté-Konrad model is less than 0.1183, the R^2 is greater than 0.9259, and the R_e is less than 9.47%, which is better than the Côté-Konrad model,

Model	Model parameters				RMSE	R^2	*Re*/%
	a_1 / b_1	a_2 / b_2	a_3 / b_3	a_4 / b_4			
Improved Côté-Konrad model	4.1381	−0.8413	4.1506	−0.2200	0.0964	0.9274	9.62
Improved Lu-Ren model	−0.5863	0.9451	0.1080	0.0567	0.0961	0.9278	9.59

Note: a_i is the parameter in the improved Côté-Konrad model; b_i is the parameter in the improved Lu-Ren model; $i = 1, 2, 3, 4$.

Table 6. Parameters fitted values and errors by improved Côté-Konrad model and improved Lu-Ren model.

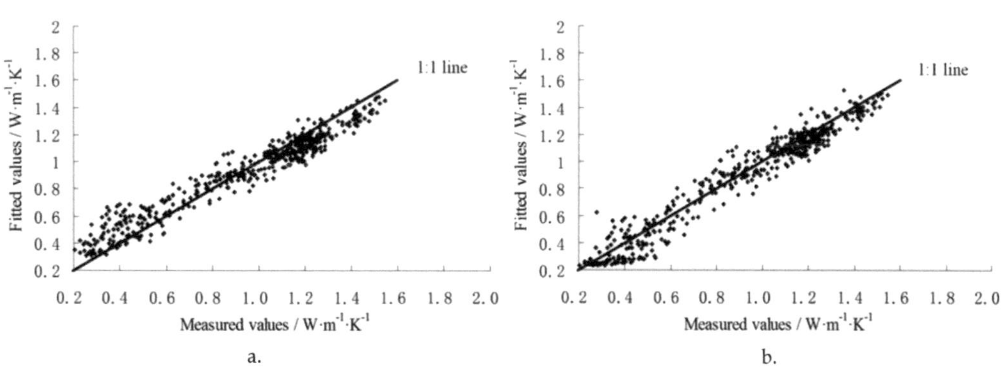

Figure 14. Fitted values of soil thermal conductivity by improved Côté-Konrad model and improved Lu-Ren model. (a) Improved Côté-Konrad model and (b) Improved Lu-Ren model.

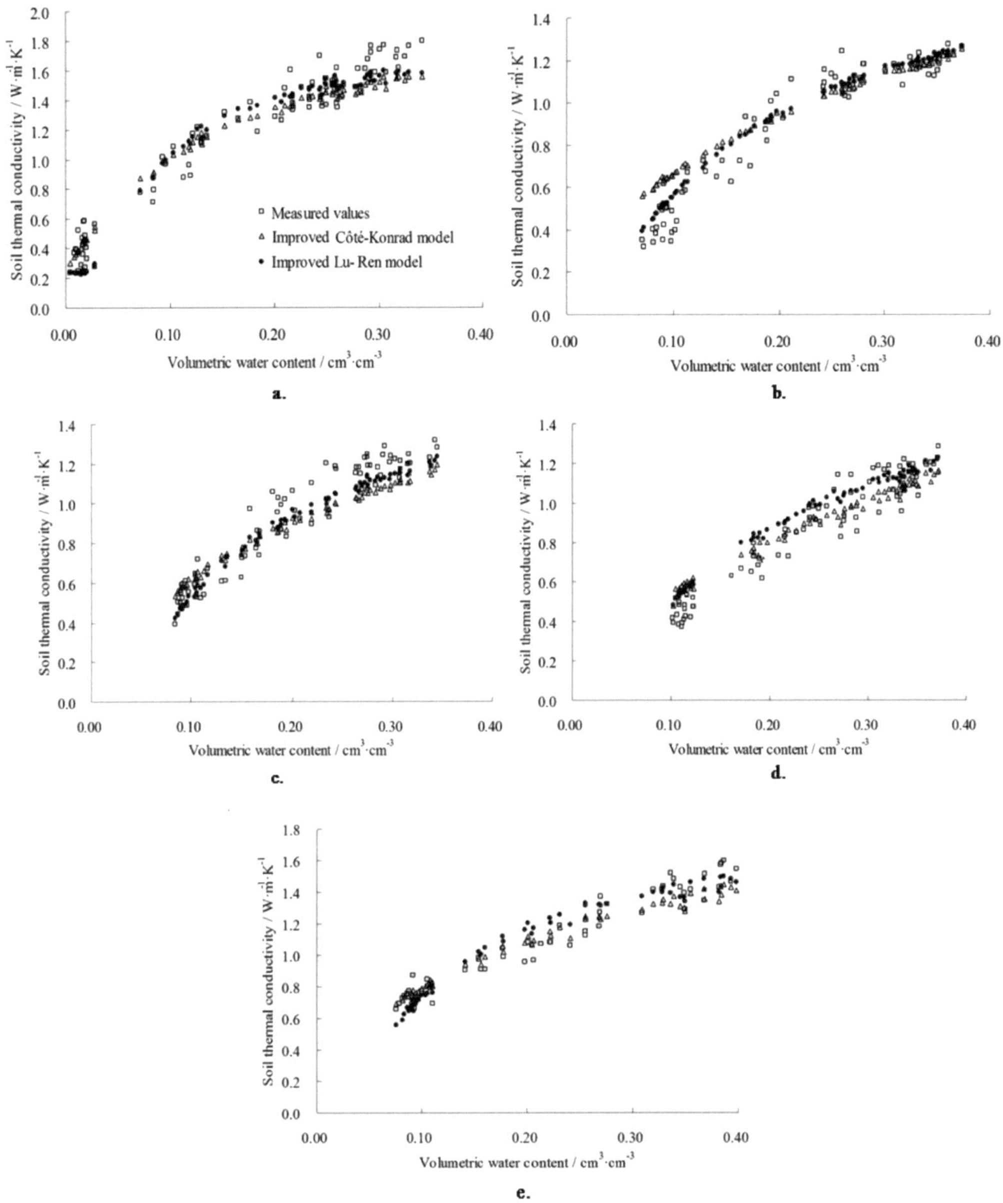

Figure 15. Comparison of soil thermal conductivity values predicted by improved Côté-Konrad model and improved Lu-Ren model with measured. (a) Shenmu sand soil, (b) Shangluo loam soil, (c) Luochuan clay loam soil, (d) Ankang silty clay loam soil and (e) Zhangye sandy clay loam soil.

Lu-Ren model, and improved Lu-Ren model. In other words, the improved Côté-Konrad model can be used to simulate the soil thermal conductivity for the soil with high sand content or high silt content. For the Shangnan loam soil and Luochuan clay loam soil, the RMSE of improved Lu-Ren model is less than 0.0815, R^2 is greater than 0.9326, and R_e is less than 8.11%, which are obviously better than the other three models. In other words, the improved Lu-Ren model can be used to simulate soil thermal conductivity.

Sampling area	Côté-Konrad model			Lu-Ren model			Improved Côté-Konrad model			Improved Lu-Ren model		
	RMSE	R^2	R_e/%	RMSE	R^2	R_e/%	RMSE	R^2	R_e/%	RMSE	R^2	R_e/%
Shenmu	0.1208	0.9401	9.67	0.1238	0.937	9.91	0.1183	0.9425	9.47	0.1366	0.9234	10.94
Ankang	0.1088	0.9062	10.57	0.1014	0.9185	9.85	0.0951	0.9259	9.55	0.0986	0.8775	10.94
Shngluo	0.081	0.8422	9.87	0.0725	0.8736	8.83	0.1243	0.8451	13.17	0.0766	0.9412	8.11
Luochuan	0.0946	0.8872	10.5	0.0747	0.9298	8.28	0.1063	0.8514	10.97	0.0815	0.9326	8.21
Zhangye	0.1216	0.8985	8.68	0.1349	0.8911	8.81	0.1026	0.9069	8.15	0.1034	0.9053	8.22

Table 7. Soil thermal conductivity simulated values and errors by different soil thermal conductivity models in sampling area.

In order to further verify whether the improved model can be extended to other soils, soil thermal conductivity of Zhangye samples in Gansu Province is predicted by the improved model. As the soil samples are sandy clay loam soil, the sand content is 60.13%. From the above model comparison analysis, we can see that the improved Côté-Konrad model is better for soil thermal conductivity with higher sand content. **Figure 15e** and **Table 7** show the prediction results of thermal conductivity and the measured values and the simulation error, respectively. Through the error analysis, we can see that the results show that the improved Côté-Konrad model is slightly higher than other three models where the RMSE and R^2 of the improved Côté-Konrad model are 0.1026 and 0.9069, respectively, which is slightly higher than other three models, R_e is 8.15%, slightly lower than the other three models. Therefore, by selecting the appropriate improved model, soil thermal conductivity for different soil textures can be calculated accurately.

3.3. Solute transport in runoff by raindrops

Most of the parameters in our model were measured directly. The depths of the exchange layers, H_o, were measured directly in the soil profiles and are shown in **Table 8**. Assuming that the exchange layer was saturated when the runoff was generated, the water content θ was assumed to be equal to that in saturated soil, that is, 0.42 cm^3/cm^3. The initial soil moisture content, θ_o, was 0.1 g/g except in the experiments where the initial soil moisture content was varied. The rainfall intensity, p, was controlled by a computer. In all experiments except those investigating the influence of varying this parameter, p was fixed at 90 mm/h. The ponding times, t_p, are shown in **Table 8**. ρ_b has a value of 1.35 g/cm^3, and $r(t)$ was measured directly.

To determine the relationship between the rainfall-induced soil detachment per unit area, e, and the rainfall intensity, p, we performed a series of rainfall experiments at rainfall intensities of 36 and 90 mm/h with a slope gradient of 5° and a second series of experiments at rainfall intensities of 60 and 84 mm/h with a slope gradient of 15°.

The measured and simulated (assuming $\rho = 2$) relationships between the rainfall-induced soil detachment per unit area, e, and the rainfall intensity, p, for a slope gradient of 5° are shown in **Figure 16**. The model's predictions agree well with the experimental results ($R^2 > 0.90$); this

Treatment		t_p (min)	H_o (cm)	θ (%)
Initial soil moisture content	5%	2.83	0.2	5
	15%	1.83	0.23	15
	20%	1.21	0.24	20
Rainfall intensity	60 mm h^{-1}	4.00	0.18	10
	96 mm h^{-1}	2.40	0.23	10
	129 mm h^{-1}	1.70	0.27	10
Slope gradient	5°	3.9	0.16	10
	15°	3.3	0.18	10
	25°	2.3	0.3	10

*Values shown represent the average of three runs.

Table 8. Experimental parameters used in the numerical model.

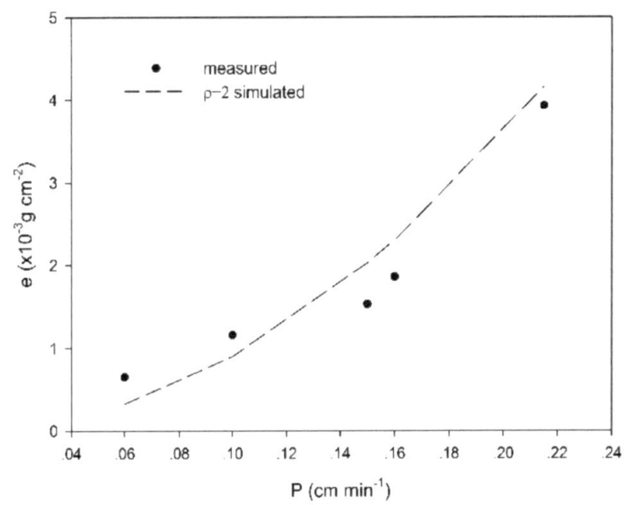

Figure 16. The relationship between the rainfall-induced soil detachment per unit area, e, and rainfall intensity, p.

result is consistent with the findings of Meyer [32], Foster [33], and Liebenow et al. [34], whose results were incorporated into the USDA WEPP model [35]. When the slope gradient is 15° or greater, ρ = 1, which is consistent with the results of Gao et al. [29] and corroborates the conclusions drawn by Sharma et al. [36, 37] and Jayawardena and Bhuiyan [38]. The relationships under these conditions between the rainfall-induced soil detachment per unit area, e, and rainfall intensity, p, at slope gradients of 15 and 25°, are shown in **Figure 17**. The values of the bare-soil detachability parameter, a, were calculated using Eq. (25) and are shown in **Table 9**. The values of the bare-soil detachability parameter, a, shown in **Table 9**, were quite similar, which may be attributed to the fact that the soil detachability is constant when the ponding depth is below a critical or breakpoint depth [30, 31, 38, 39]. Parameter b was estimated using

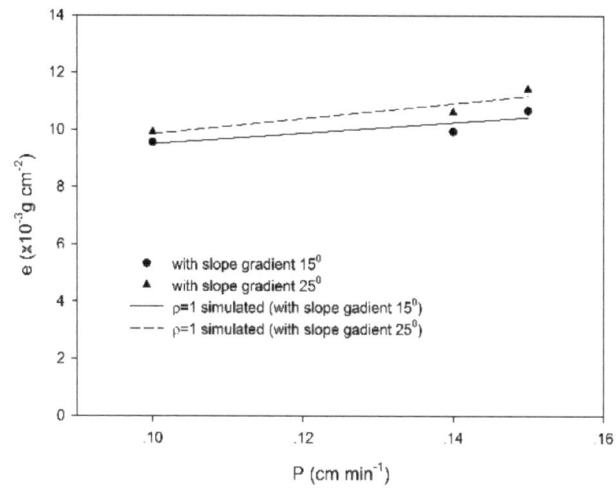

Figure 17. The relationship between the rainfall-induced soil detachment per unit area, e, and rainfall intensity, p, at slope gradients of 15 and 25°.

Treatment		e (×10^{-3} g cm^{-2})	p (cm min^{-1})	ρ	a (g cm^{-3})	b^*
Initial soil moisture content (%)	5	2.18	0.15	2	−0.4	0.094
	15	1.9	0.15	2	−0.4	0.083
	20	1.68	0.15	2	−0.4	0.075
Rainfall intensity (mm h^{-1})	60	1.16	0.1	2	−0.2	0.116
	96	1.86	0.16	2	−0.2	0.082
	129	3.93	0.215	2	−0.2	0.085
Slope gradient (°)	5	1.85	0.15	2	−0.35	0.082
	15	10.11	0.15	1	−0.35	0.072
	25	11.93	0.15	1	−0.35	0.082

*Calibrated to best fit the runoff solute data.

Table 9. Data used in calculating the bare-soil detachability, a.

the best fit to the experimental data listed in **Table 9**. The predictions made using Eq. (26) are compared to our experimental results in **Figure 18**.

The simulated data agreed well with the experimental results for all three treatments except for the experiment where the rainfall intensity was 129 mm/h, suggesting that the use of the raindrop-induced water transfer rate, e_r, in place of the exchange rate, k_m, is reasonable for conditions involving relatively natural rainfall. The model, however, did not appear to accurately predict the solute concentrations in the runoff observed under conditions that give rise to severe soil erosion.

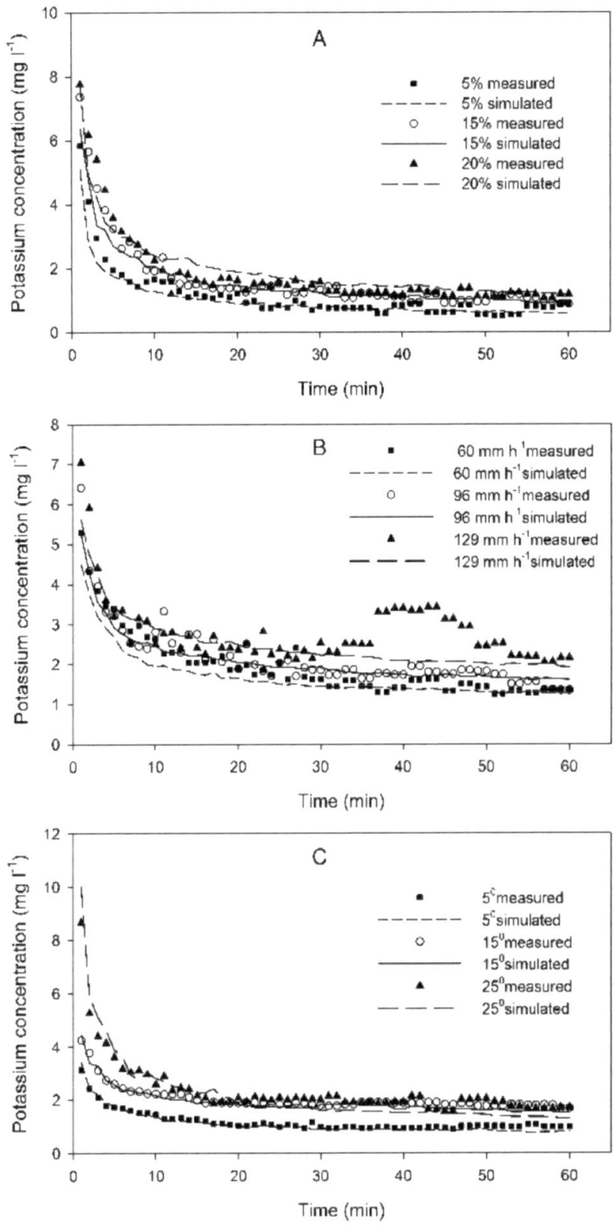

Figure 18. Simulated and measured runoff concentrations of potassium under different initial soil moisture contents (A), rainfall intensities (B), and slope gradients (C).

Our results support the conclusion drawn by Walter et al. [40], who argued that the depth of the exchange layer decreases as the rate of infiltration increases. The initial soil moisture content, rainfall intensity, and slope gradient influence the solute concentration of the runoff solution by virtue of their effects on the depth of the exchange layer, the infiltration rate, and the length of time between the initiation of rainfall and the formation of the runoff.

The agreement of the simulated results with the measured data was quantified by calculating the root mean square error (RMSE) [31]. RMSE can be expressed as:

$$RMSE = \sqrt{\frac{\sum_{i=1}^{N} (p_i - o_i)^2}{N}} \tag{27}$$

where N is the total number of data points, p_i is a given simulated data point, and o_i is the corresponding experimental data point. The RMSEs are shown in **Table 10**. **Table 10** and **Figure 18** illustrate that the model has correctly captured the temporal behavior of the solute concentration in the runoff under all conditions investigated.

Figure 18A shows that the measured and simulated solute concentrations for different initial soil moisture contents changed with time. The results indicated that the refined model [25] could predict the movement of solutes in the overland flow under different initial soil moisture contents. Also, the higher initial soil moisture contents were associated with higher solute

Treatment		RMSE (mg l^{-1})	R^2
Initial soil moisture content (%)	5	0.227	0.90
	15	0.245	0.93
	20	0.308	0.91
Rainfall intensity (mm/h)	60	0.295	0.80
	96	0.229	0.88
	129	0.508	0.46
Slope gradient (°)	5	0.081	0.94
	15	0.127	0.88
	25	0.336	0.86

Table 10. The root means square errors (RMSEs) and R^2 between the measured and simulated data.

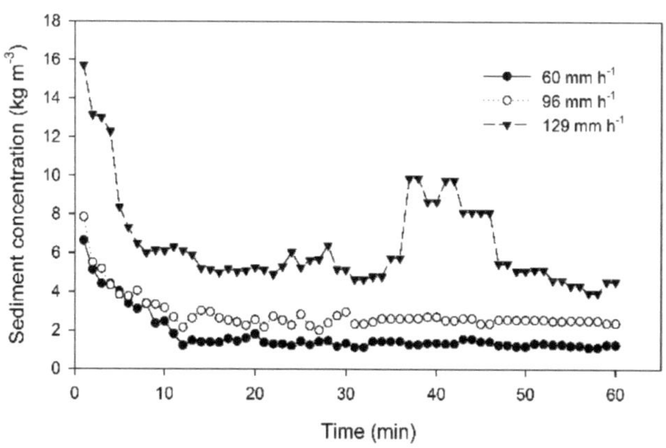

Figure 19. Variation of the sediment concentrations over time at different rainfall intensities.

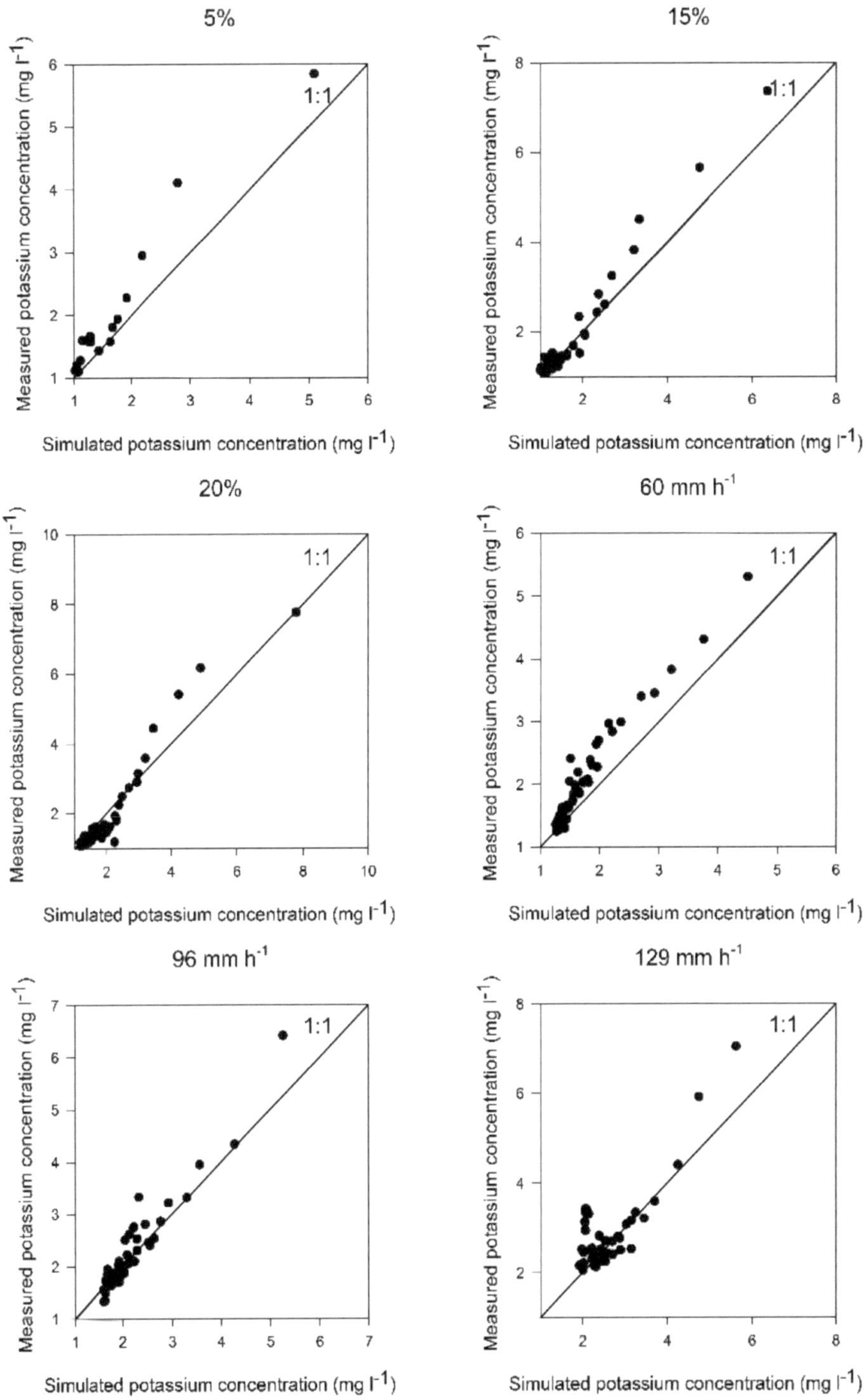

Figure 20. Graph of potassium concentration calculated by Eq. (26) against potassium concentration measured (under different initial soil moisture content, rainfall intensity conditions).

concentrations per unit time. **Figure 18A** and **Table 10** indicate that the differences between the measured and simulated solute concentrations under an initial soil moisture content of 20% were more distinct than those under the other two initial soil moisture contents, which implied

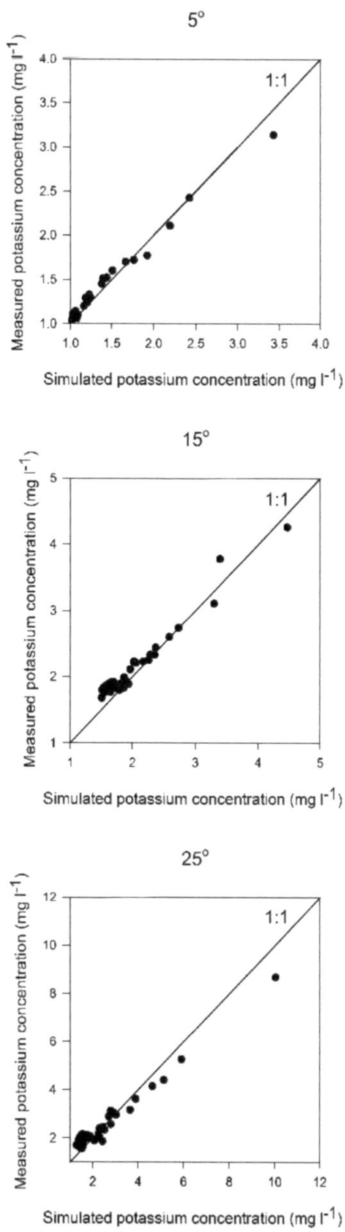

Figure 21. Graph of potassium concentration calculated by Eq. (26) against potassium concentration measured (under different slope gradient conditions).

that the model did not accurately predict the solute concentrations of the runoff in conditions of severe soil erosion.

Comparisons between the simulated and the experimental solute concentrations for the different rainfall intensities over 60 min are shown in **Figure 18B**. At a rainfall intensity of 129 mm/h, the solute concentration of the runoff increased substantially between 37 and 49 min after the initial generation of the runoff (**Figure 18B**). The mass of sediment in the runoff between 37 and 43–49 min showed a corresponding spike (**Figure 19**), which indicated that solute loss is closely related to sediment loss [41–45]. These results indicated that significant erosion of the surface soil occurred at the bottom of the slope during the experiments. Deeper soil layers were exposed to water in which the solute concentrations were higher than in those washed away. Consequently, the solute concentration of the runoff increased as these solutes were transferred from the soil under the influence of the runoff and the splashing caused by raindrops. Soil erosion thus promoted increased solute concentrations in the runoff.

Figure 18C shows that the measured and simulated solute concentrations for different slope gradients also changed with time. The simulated data were highly correlated with the measured data for solute concentration in the runoff. This degree of correlation demonstrated that the model captured the temporal behavior of the solute transport in the runoff. Increasing the gradient of the slope increased the erosion capacity of rain drops and water flow. Increasing the slope gradient also led to increases in the RMSE (**Table 10**) and R^2. **Figures 20** and **21** show the relationships between potassium concentrations observed in the runoff and predicted using Eq. (26). The graph indicates the model accurately predict the solute transport in the runoff with the solute concentration being at a much lower level.

4. Conclusions

In order to understand the whole process of water-solute-heat transport and nutrient loss, we determined water movement, solute, and heat transport through columns of disturbed soil samples. And we also carried out simulated rainfall experiments on an artificial slope to study nutrient loss.

The results were as follows:

1. Data obtained with experimental infiltration under negative hydraulic heads were employed to analyze the relationship between the Philip model and Kostiakov empirical model, showing as well that they were identical in terms of negative hydraulic heads; Wang's equation could describe the infiltration process very well.

2. The Horton empirical model can be used to describe the variation of soil thermal conductivity; the calculated values of Campbell model and Johansen model have large differences with the measured values. However, the calculated results of Côté-Konrad model and Lu-Ren model are in good agreement with the measured values. The improved Côté-Konrad model and improved Lu-Ren model can use the soil texture to predict soil thermal conductivity. For two improved models, the coefficients of determination R^2 are above 0.92 and the relative errors Re are less than 9.6%. For the soils with high sand content or silt

http://dx.doi.org/10.5772/intechopen.76280

content, the improved Côté-Konrad model is superior to Côté-Konrad model, Lu-Ren model, and the improved Lu-Ren model. For the soils with low sand content and silt content, the Lu-Ren model is obviously better than the other three models. The relationship between the parameters of the model, particle composition, and organic matter content can be predicted by two improved models. These models can describe the relationship between the soil's basic physical parameters and thermal conductivity in detail. Thus, soil thermal conductivity can be predicted more accurately by choosing the appropriate improved model based on the different soil texture.

3. The refined power functions of a model of solute transport were illustrated and tested using simple experiments. The model fit the experimental data very well. Our results also indicated that the constant parameter, ρ, was equal to 1 when the slope gradient was 15° or larger and equal to 2 when the slope gradient was less than 15°. The soil detachability was confirmed to be independent of the rain intensity and was a constant in all treatments. The model, however, could not accurately predict the solute concentrations in the runoff under conditions of severe soil erosion. The initial soil moisture content, rainfall intensity, and slope gradient influenced the solute concentration in the runoff, depth of the exchange layer, infiltration rate, and length of time between the initiation of rainfall and the generation of the runoff.

Acknowledgements

This study was financially supported by the National Natural Science Foundation of China (grant nos. 51239009, 41371239), Science and Technology Planning Project of Shaanxi Province (2013kjxx-38), and Doctoral fund of Xi'an University of Technology (106-211301). We also thank Xiaopeng Chen for his helpful comments.

Author details

Quanjiu Wang, Beibei Zhou*, Lijun Su and Yuyang Shan

*Address all correspondence to: happyangle222@aliyun.com

State Key Laboratory of Eco-Hydraulic Engineering in Arid Area, Xi'an University of Technology, Xi'an, China

References

[1] Darcy H. Less Fontaines Publiques de la Ville de Dijon. Paris, France: Dalmont; 1956

[2] Green WH, Ampt CA. Studies on soil physics. I. Flow of air and water through soils. Journal of Agricultural Science. 1911;**4**:1-24

[3] Richards LA. Capillary conduction of liquids in porous mediums. Physics. 1931;**1**:318-333

[4] Philip JR. Theory of infiltration. Soil Science. 1957;**83**(5):345-357

[5] Kostiakov AN. On the dynamics of the coefficient of water percolation in soils and on the necessity for studying it from a dynamic point of view for purpose of amelioration. In: Transactions of 6th Committee International Society of Soil Science; Russia. 1932. pp. 17-21

[6] Horton RE. An approach toward a physical interpretation of infiltration capacity. Soil Science Society of America Journal. 1940;**5**:399-417. DOI: 10.2136/sssaj1941.036159950005000C0075x

[7] Holton HN. A concept of infiltration estimates in watershed engineering. AICHE Journal. 1961;**150**(1):41-51

[8] Shu ST. Capillary-tube infiltration model. Journal of Irrigation and Drainage Engineering. 1993;**119**(3):514-521. DOI: 10.1061/(ASCE)0733-9437(1993)119:3(514)

[9] Ghosh Kumar R. A note on the infiltration equation. Soil Science. 1983;**136**(6):333-338. DOI: 10.1097/00010694-198312000-00001

[10] Parlange JY, Barry DA, Parlange MB, et al. New approximate analytical technique to solve Richards Equation for arbitrary surface boundary conditions. Water Resources Research. 1997;**33**(4):903-906. DOI: 10.1029/96WR03846

[11] Hogarth WL, Parlange JY. Application and improvement of a recent approximate analytical solution of Richards' equation. Water Resources Research. 2000;**36**(7):1965-1968. DOI: 10.1029/2000WR900042

[12] Lei ZD, Yang SX, Xie SC. Soil Water Dynamics. Beijing: Tsinghua University Press; 1988 (in Chinese)

[13] Yang SX, Lei ZD, Xie SC. General program of one-dimensional flow through unsaturated homogeneous soil. Acta Pedologica Sinica. 1985;**1**:24-34 (in Chinese)

[14] Yi L, Mingan S, Wenyan W, et al. Influence of soil textures on the thermal properties. Transactions of the Chinese Society of Agricultural Engineering. 2003;**19**(4):62-65 (in Chinese)

[15] Chung SO, Horton R. Soil heat and water flow with a partial surface mulch. Water Resources Research. 1987;**12**(11):2175-2186. DOI: 10.1029/WR023i012p02175

[16] Campbell GS. Soil Physics with BASIC. Amsterdam, the Nether-lands: Elsevier; 1985. pp. 221-234

[17] De Vries DA. Thermal properties of soils. In: Physics of Plant Environment. Amsterdam: North-Holland; 1963. pp. 210-235

[18] Johansen O. Thermal Conductivity of Soils. Trondheim: Norwegian University of Science and Technology; 1977

[19] Côté J, Konrad JM. A generalized thermal conductivity model for soils and construction materials. Canadian Geotechnical Journal. 2005;**42**(3):443-458. DOI: 10.1139/t04-106

[20] Lu S, Ren T, Gong YS. An improved model for predicting soil thermal conductivity from water content at room temperature. Soil Science Society of America Journal. 2006;**71**(1):8-14. DOI: 10.2136/sssaj2006.0041

[21] Ting L, Wang Q, Jun F. Modification and comparison of methods for determining soil thermal parameters. Transactions of the Chinese Society of Agricultural Engineering. 2008;**24**(3):59-64. (in Chinese). DOI: 10.3969/j.issn.1002-6819.2008.3.012

[22] Wang S, Wang Q, Jun F, et al. Soil thermal properties determination and prediction model comparison. Transactions of the Chinese Society of Agricultural Engineering. 2012;**28**(5): 78-84. (in Chinese). DOI: 10.3969/j.issn.1002-6819.2012.05.014

[23] Wang W, Jianbo L, Wang S, et al. Spatial variability of soil thermal parameters and its fitting method. Transactions of the Chinese Society for Agricultural Machinery. 2015;**46**(4): 120-125. DOI: 10.6041/j.issn.1000-1298.2015.04.018

[24] Wallach R, Galina G, Rivlin J. A comprehensive mathematical model for transport of soil-dissolved chemicals by overland flow. Journal of Hydrology. 2001;**247**:85-89. DOI: 10.1016/S00221694(01)00365-1

[25] Wang QJ, Wang WY, Shen B, Shao MA. Interacting depth of rainfall–runoff–soil solute. Journal of Soil Erosion and Water Conservation. 1998;**2**(4):41-46 (in Chinese)

[26] Hairsine PB, Rose CW. Rainfall detachment and deposition: Sediment transport in the absence of flow-driven processes. Soil Science Society of America Journal. 1991;**55**(2):320-324. DOI: 10.2136/sssaj1991.03615995005500020003x

[27] Gao B, Walter MT, Steenhuis TS, et al. Rain induced chemical transport from soil to runoff: Theory and experiments. Journal of Hydrology. 2004;**295**:291-304. DOI: 10.1016/S0022-1694(04)00174-X

[28] Wang QJ, Robert H, Shao MA. Horizontal infiltration method for determining Brooks-Corey model parameters. Soil Science Society of America Journal. 2002;**66**:1733-1739. DOI: 10.2136/sssaj2002.1733

[29] Gao B, Walter MT, Steenhuis TS, et al. Investigating ponding depth and soil detachability for a mechanistic erosion model using a simple experiment. Journal of Hydrology. 2003; **277**(1–2):116-124. DOI: 10.1016/S0022-1694(03)00085-4

[30] Moss AJ, Green P. Movement of solids in air and water by raindrop impact—Effects of drop-size and water-depth variations. Australian Journal of Soil Research. 1983;**21**(3):257-269. DOI: 10.1071/SR9830257

[31] Willmott CJ. Some comments on the evaluation of model performance. Bulletin of the American Meteorological Society. 1982;**63**(11):1309-1313. DOI: 10.1175/1520-0477(1982)063<1309:SCOTEO>2.0.CO;2

[32] Meyer LD. Soil-erosion research leading to development of the universal soil erosion loss equation. Science. 1982;**26**:1-16. DOI: 10.2136/sssaj1993.03615995005700030007x

[33] Foster GR. Modeling the erosion process. In: Han CT, editor. Hydrological Modeling of Small Watersheds: Monograph No. 5. St Joseph, MI: ASAE; 1982. pp. 297-379

[34] Liebenow AM, Elliot WJ, Laflen JM, et al. Inter rill erodibility-collection and analysis of data from crop-land soils. Transactions of the Chinese Society of Agricultural Engineering. 1990;**33**(6):1882-1888 (in Chinese)

[35] Laflen JM, Elliot WJ, Simanton JR, et al. WEPP soil erodibility experiments for rangeland and cropland soils. Journal of Soil and Water Conservation. 1991;**49**(1):39-44

[36] Sharma PP, Gupta SC, Foster GR. Predicting soil detachment by raindrops. Soil Science Society of America Journal. 1993;**57**:674-680

[37] Sharma PP, Gupta SC, Foster GR. Raindrop-induced soil detachment and sediment transport from interrill areas. Soil Science Society of America Journal. 1995;**59**:727-734. DOI: 10.2136/sssaj1995.03615995005900030014x

[38] Jayawardena AW, Bhuiyan RR. Evaluation of an inter rill soil erosion model using laboratory catchment data. Hydrological Processes. 2015;**13**(1):89-100. DOI: 10.1002/(sici)1099-1085(199901)13:1<89::aid-hyp677>3.0.co;2-t

[39] Proffitt APB, Rose CW, Hairsine PB. Rainfall detachment and deposition: Experiments with low slopes and significant water depths. Soil Science Society of America Journal. 1991;**55**:325-332. DOI: 10.2136/sssaj1991.03615995005500020004x

[40] Walter MT, Gao B, Parlange JY. Modeling soil solute release into runoff with infiltration. Journal of Hydrology. 2007;**347**:430-437. DOI: 10.1016/j.jhydrol.2007.09.033

[41] Catt JA, Quinton JN, Rickson RJ, et al. Nutrient losses and crop yields in the Woburn erosion reference experiment. In: Rickson RJ, editor. Conserving Soil Resources: European Perspective. Oxford: CAB International; 1994. pp. 94-104

[42] Hansen AC, Nielsen JD. Runoff and loss of soil and nutrients. In: Correll A, editor. Surface Runoff Erosion and Loss of Phosphorus at two Agricultural Soils in Denmark. Tjele: Danish Institute of Plant and Soil Science; 1995. pp. 149-188

[43] Hargrave AP, Shaykewich CF. Rainfall induced nitrogen and phosphorus losses from Manitoba soils. European Journal of Soil Science. 1997;**77**:59-65. DOI: 10.4141/S95-034

[44] Teixeira PC, Misra RK. Measurement and prediction of nitrogen loss by simulated erosion events on cultivated forest soils of contrasting structure. Soil and Tillage Research. 2005; **83**:204-217. DOI: 10.1016/j.still.2004.07.014

[45] Guo TL, Wang QJ, Li DQ, et al. Sediment and solute transport on soil slope under simultaneous influence of rainfall impact and scouring flow. Hydrological Processes. 2010;**24**:1446-1454. DOI: 10.1002/hyp.7605

Chapter 7

Relation between Infiltration Rate, Cover Materials and Hydraulic Conductivity of Forest Soils in Japanese Cedar and Hiba Arborvitae Plantation Forests under Artificial Rainfall in Ishikawa Prefecture, Japan

Yoshitaka Komatsu

Additional information is available at the end of the chapter

http://dx.doi.org/10.5772/intechopen.70575

Abstract

To ascertain the relation between the infiltration rate, cover material and hydraulic conductivity of forest soils in Japanese cedar (*Cryptomeria japonica*) and Hiba arborvitae (*Thujopsis dolabrata*) plantations, we conducted artificial rainfall experiments using an oscillating nozzle rainfall simulator. The maximum infiltration rate (*FIRmax*) was higher than that found in previous studies of cypress plantations. Especially in the conditions where surface cover materials are less than 1000 g/m^2, *FIRmax* tends to become higher than the value of the cypress forest by several magnitudes. *FIRmax* was over 100 mm/h, irrespective of the amount of surface cover materials. The result confirmed little or no correlation between *FIRmax* and surface cover materials in either of the studied tree species. *FIRmax* were lower than hydraulic conductivity. *FIRmax* showed no correlation with hydraulic conductivity and fine fraction content. Therefore, the differences of *FIRmax* between cedar and Hiba plantation were not explainable by surface cover materials, hydraulic conductivity or the fine soil fraction content, which contrasts with results of previous studies of cypress plantations.

Keywords: Japanese cedar, Hiba arborvitae, infiltration rate, cover materials, hydraulic conductivity

1. Introduction

Previous studies have reported that the Hortonian overland flow (HOF) generation seldom occurs in forest soils that are generally rich in macropore-containing organic matter, but in recent years, HOF generation has been observed in part of the unmanaged forests.

This overland flow might occur when the infiltration rate decreases with increasing uncovered forest floor [1–3]. Moreover, the heavy rain events have an effect on increasing the risk of floods [4]. Infiltration has been traditionally studied using a flood-type or mist-type rainfall simulator [5, 6]. Because a flood-type rainfall simulator supplies an excessive amount of water and as the mist-type simulator cannot reproduce raindrop impacts, the conventional method using these simulators does not accurately reproduce natural rainfall. For accurate reproduction of rainfall, Onda et al. [7] presented a series of large-scale sprinkling experiments in which water was sprinkled from the upper canopy. It is presumed in this investigation that infiltration rate observed in previous experiments should be about one order of magnitude more than the actual value. A large-capacity tank truck was installed in this investigation, but a small portable oscillating nozzle rainfall simulator recently developed by Kato et al. [8] was used for the experiments. The experiments showed a positive correlation between surface cover materials and infiltration rate [8–10]. A similar experiment was carried out in the semi-arid region, and obtained results indicated a noticeable positive correlation between these two variables [11, 12]. Previous studies showed that the effect of understories and leaf litters on infiltration rate is significant, which suggests that the surface cover materials reduce raindrop impact on the soil surface, and therefore the formation of surface crusts and HOF are restricted [9, 13–16]. A related finding of these studies is that fragmentation process of *Chamaecyparis obtusa* leaf litter progresses rapidly, and thus the *C. obtusa* forests are likely to become uncovered due to soil loss from steep slopes [17]. Fragmentation of Hiba arborvitae leaves is less likely to occur than those of *C. obtusa*, and the fragmentation of cedar leaves is mostly unlikely to occur. Therefore, the Japanese cedar and Hiba arborvitae forests have larger effects on soil conservation than *C. obtusa* forests [18]. The case studies using the recently developed method does not, however, allow for the sufficient examination of infiltration rate in Japanese cedar and Hiba arborvitae forests. Thus, an additional investigation should be carried out in these forests using a new method. Furthermore, Miyazaki et al. [19] argue that the effect of leaf shape on soil conservation is not clear. Further consideration will be needed on this viewpoint. The results from previously published studies have revealed that the hydraulic conductivity of forest floor and soil surface was greater than several hundred mm/h [3, 20, 21]. A correlation was observed between the hydraulic conductivity and infiltration rate [5, 22], as well as surface cover materials and infiltration rate. However, according to the observation results of plot runoff due to rainfall, there was no obvious correlation between infiltration rate and hydraulic conductivity in *C. obtusa* forests, suggesting that soil detachment by raindrops has a potentially significant impact on the infiltration rate [3]. It is usually assumed that the hydraulic conductivities of surface cover materials and surface soil are the significant factors impacting infiltration rate in forests, but more detailed consideration needs to be given to explain this assumption. A portable oscillating nozzle rainfall simulator has been introduced by Refs. [8, 9] as a new method mainly in *C. obtusa* forests. Therefore, field measurement of infiltration rate using the new method was conducted in Japanese cedar and Hiba arborvitae forests to examine the relation between infiltration rate, surface cover materials and hydraulic conductivity of forest soil through Ishikawa Forest Environmental Tax project [10].

2. Background of Ishikawa Forest Environmental Tax project

2.1. Sediment and flood damage in Japan

Japan is a country with many flood disasters. Muroto Typhoon in 1934, Makurazaki Typhoon in 1945 and Isewan Typhoon in 1959 were the three major typhoons in the Showa era, which caused massive damage. In the Makurazaki Typhoon, sediment-related disasters occurred frequently in the logged mountains area that was weakened by heavy rains, taking a heavy toll of life.

Forestry Agency [23] showed that Japanese government had been promoting the creation of artificial forest for Japanese cypress, Japanese cedar and more species to be restored in the site, which was located in the logged forest after World War II and during the high economic growth period. Changing forest economic conditions in Japan since World War II have caused many plantation forests to be poorly managed. Since the 1980s, wood prices in Japan have been on a declining trend, but the profitability of forestry management has deteriorated drastically since the management costs of employment cost and materials had increased. For this reason, forest owners' willingness to management had declined, and forestry production activities have stagnated. Moreover, mountainous areas where many forestry workers and forest owners live have problems of depopulation and the rapidly aging society.

In 1998, severe natural disasters occurred due to 24-h rainfall of an unprecedented 900 mm in Kochi Prefecture, Japan [24]. The area seemed to have many artificial forests of Japanese cypress and landslides occurred under the heavy rain. With such a background, in 2003, Kochi Prefecture became the first local government to introduce forest environment tax before the whole country in order to improve the forests in the reservoir area. Sediment disasters have been caused by typhoons in recent years, and the importance of countermeasures at basin scale including forest management has been reported.

Further, reports on previous flood disasters are available at the homepage of the Ministry of Land, Infrastructure, Transport and Tourism [25] and Japan Society of Civil Engineers [26].

2.2. What is forest environmental tax?

Recently, 'forest environmental tax' as individualistic tax was considered and introduced in each prefecture of Japan. In the background of the introduction of 'forest environment tax', there is a move towards centralization of national economy that the decentralization law was enacted in April 2000. As a result, the Local Tax Act was revised in an effort to expand local taxation autonomy, and conventional local discretionary taxes were changed from approval system by government to prior consultation system. In addition, the establishment of special taxes for specific purposes by the prior consultation system was also conducted, possibly to implement the political management taking advantage of the characteristics of the region [27].

Yamaguchi Prefecture started to collect taxes for the purpose of forest protection called 'Yamaguchi Forest Management Prefectural Residence Tax', from April 2005 [28]. A detailed report on the introduction and introduction effect of forest environmental tax has been reported [29, 30].

In Ishikawa Prefecture, Japan, there are forests accounting for about 70% of the prefecture land, due to the restoration of the devastated forest land after the World War II, afforestation has been rapidly carried out on about 990 km^2 of plantation. The maintenance area of 290 km^2 of the forest degradation without thinning was concerned an urgent task in 2005. Ishikawa prefectural government stressed that it is necessary to focus on maintenance of degraded forests. The maintenance cannot be performed only by the people involved in forestry or forestry-related works as a living but have to involve professionals. The functions of forests have been maintained through forest management. However, forestry business has been concerned with declining public benefit with degradation, water recharge and conservation of mountain area has grown unprofitable due to the soaring prices of timber and depopulation of mountain area. As a result of this, Ishikawa Prefecture introduced 'Ishikawa Forest Environmental Tax' in fiscal 2007, intensive thinning of degraded plantations mainly in the water source area, and an attempt to save public function of forests. Therefore, since it is necessary to verify the effect of introducing the Ishikawa Forest Environmental Tax, we attended/conducted a water soil conservation function survey in the same way as the verification of Yamaguchi Forest Management Prefectural Residence Tax [28]. The verification method is *in situ* permeability test using an artificial rainfall apparatus. As for the effect of introducing the Ishikawa Forest Environmental Tax was conducted along with the soil conservation function survey and function of biodiversity conservation.

3. Infiltration capacity in Japanese cedar and Hiba arborvitae plantation forests

3.1. Site description and test plot design

All sites were located in the environmental community forests developed through a conservation project in Ishikawa Prefecture, Japan. Plantations cover about 40% of the total forests in this Prefecture. In terms of species composition of the plantations, cedar forests tops with 71% followed by Hiba arborvitae with 12% and pine trees with 9%. Cedar forests are distributed equally throughout the entire areas in Ishikawa Prefecture, while Hiba arborvitae forests show unequal distribution with a large part of the forests located in the Noto Peninsula [31]. We selected 22 Japanese cedar and 16 Hiba arborvitae sites in all available forests before thinning or the forests with different years elapsed after thinning (**Figure 1**). The study sites are dominated by brown forest soils, and have a slope of 40°. Forest leaf litter accumulates at all the sites with dense and sparse understories. According to AMeDAS data [32], annual rainfall averages nearly 2302 mm (1976–2012) with an average temperature of 13°C (55°F) in Ishikawa Prefecture.

3.2. Definition of infiltration rate

Horton equation is based on the observation that infiltration capacity is gradually exponentially reduced with time when, although ground surface is always thin, fresh state is sufficiently supplied. Horton [33] proposed excess rainfall/infiltration processes and introduced that while the equation may not actually represent the law governing the physical processes

http://dx.doi.org/10.5772/intechopen.70575

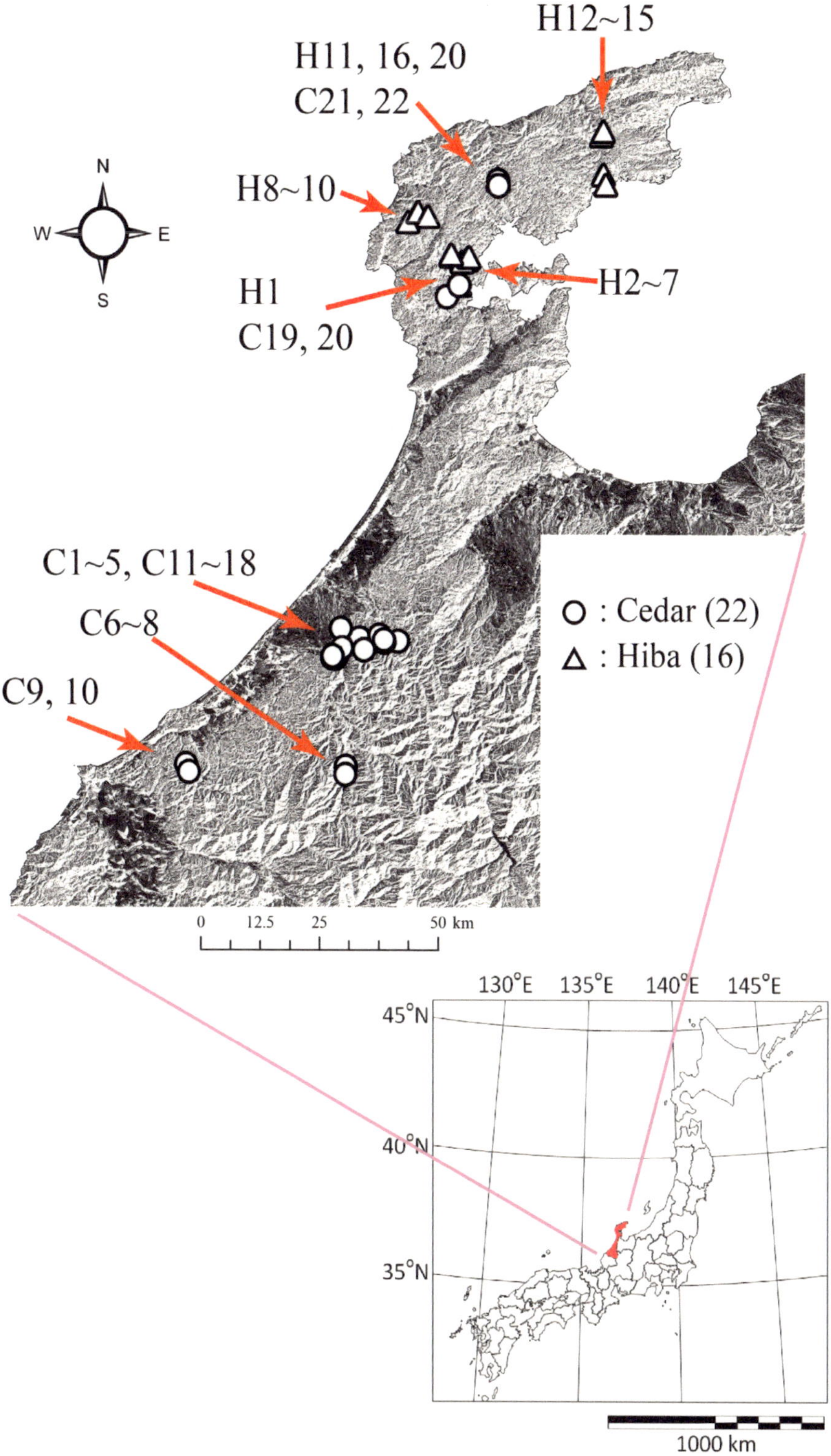

Figure 1. Location of the study site in Ishikawa Prefecture, Japan.

involved, his equation is rationally formed, since it not only represents the observed data within the range of observation but also gives results in agreement with known facts for the limiting or boundary conditions.

Recent studies, however, have shown that infiltration rates can increase with increasing rainfall intensity until it reaches a constant value [5, 34]. Their 'apparent' infiltration rate at steady state (f_s) is defined as area-averaged infiltration rate of which a certain fraction contributes to rainfall excess production. Infiltration capacities can be assumed to have an exponential distribution [34], so f_s can be given by:

$$f_s = f_{max}\left(1 - exp\left(-i/f_{max}\right)\right) \tag{1}$$

where i is rainfall intensity (mm/h), f_{max} is average infiltration rate (mm/h), when the whole plot is contributing to rainfall excess production. With given f and i, the model has only one empirical parameter, f_{max}, which makes it attractive for practical use. Yu et al. [35] and Stone et al. [36] applied the exponential model to their rainfall-runoff data, which yielded much better results than the application of a model with a constant K_e. Langhans et al. [37] adapted his results, so fitting Eq. (1) to the logarithmic data gives a suitable description of the data in terms of fit, without allowing for any physical interpretation. Also, based on experiments in plots with different land-use patterns such as parks and pastures under artificial precipitation system, Tanaka and Tokioka [38] suggested the following hyperbolic function, which gives the relation between rainfall intensity and final infiltration rate:

$$f(i) = FIR_{max} \cdot tanh\left(i/FIR_{max}\right) \tag{2}$$

where i is rainfall intensity (mm/h), f is final infiltration rate and FIR_{max} is the maximum infiltration rate.

3.3. On-site infiltration test using rainfall simulator

Sprinkling experiments corresponding to a certain degree of rainfall intensity were conducted in small size of plots established in each experimental site. We measured the amount of water applied to sprinkling and overland runoff from each plot on a regular basis. The difference between these two amounts means infiltration volume, and the final infiltration rate can be estimated using this infiltration volume. We established a plot in each study site with horizontal projected area of 1 m² (1 m × 1 m). Wave-board panels (about 25 cm height) were placed at the upper end and both sides of the plots. The panels were inserted vertically at a depth of 5 cm under soil surface to prevent the inflow from outside the plot and runoff from the plot. Trays are placed at the cover material-soil interface at the lower end of the plot. This system was adopted to catch and collect overland flow that will contain only the materials above the interface. An oscillating nozzle rainfall simulator was used for sprinkling water. The simulator was set, according to Hiraoka et al. [9] who used this device, with flow rate 12.5 L/min and nozzle height 2 m from the centre of sprinkled plot (**Figure 2**). The angle of nozzle was adjusted to obtain the targeted value for rainfall intensity (180 mm/h). To measure actual

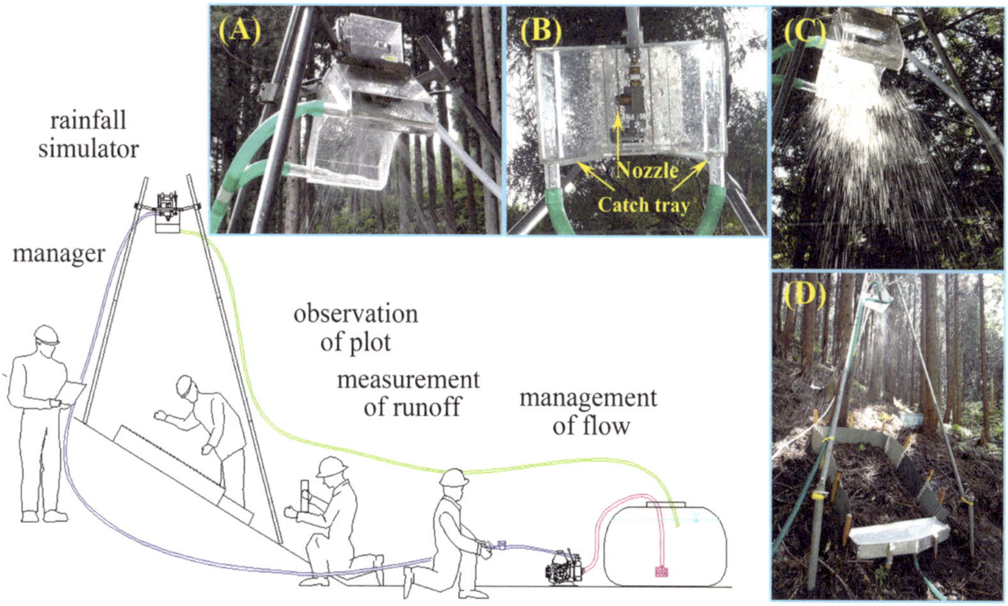

Figure 2. Rainfall simulator and experimental guideline. (A) is machine body (B) is nozzle by under view, (C) is sprinkling condition and (D) is experiment condition.

rainfall intensity, we collected total sprinkled water using plastic sheet. Impact energy of raindrop produced in this experiment was about 16.8 J/m^2/mm, which was similar to the average energy obtained in the experiment in the *C. obtusa* forests [39]. Please also see the work by Kato et al. [8, 12]. The influence of moisture content cannot be ignored, but we sprinkled water with an intensity of 180 mm/h for 2 h to obtain an accurate final infiltration rate. The rainfall intensity of 180 mm/h is approximately equal to the maximum rainfall intensity that has ever been observed in Japan (187 mm/h: observed at Nagayo-cho municipal office during the 1982 flood disaster in Nagasaki Prefecture). The amount of water stored in a tray at the lower end of the plot was measured every hour to determine the discharge of overland flow. Rainfall was artificially produced until the discharge returned to original steady-state. The experiment could be usually completed within approximately 20–30 min, and the final infiltration rate was defined as the average value of the results taken over the last 5 min. Infiltration intensity may, in some cases, begin to slightly increase after the initial decreasing trend, followed by leveling off to the decreased values. If this is the case, the final infiltration rate was defined as the average value obtained from the values over 3 min before and after the infiltration intensity reaches minimum values.

3.4. Measurement of surface cover materials

Surface cover materials are composed of understories and leaf litter. In response to the research by Miura [15], small fractions (<2 mm) were excluded from the litter category because protection of the soil surface cannot be achieved. We collected understories (ground layer) and leaf litter after the experiment. These surface cover materials were air-dried for a week, then re-dried in an oven at 70°C for 48 h to determine the dry weight. Photos were taken directly above the plot, and the floor cover percentage was estimated by calculating the percentage of forest floor that is covered with either litter or understories based on image analysis.

3.5. Measurement of soil properties

Generally, soil properties not only change for each type of soil, vary depending on the location in the same type of soil. It is desirable to make laboratory test (e.g. pF test, permeability test) using a small sample of soil *in situ* to obtain the characteristic value.

We collected soil samples after the experiments to estimate soil properties. To investigate soil properties affecting final infiltration rate, particle size, hydraulic conductivity and bulk density were estimated. Sampling and test were conducted by the following methods. After collecting surface cover materials, collection of undistributed sample soils was made using 400 cc core sampler (cross section area of 100 cm^2 and 4 cm in height) to measure particle size. The reason for examining the physical properties of the surface layer (up to 10 cm from the surface) was because the surface layer was found to be a major influencing factor on infiltration rate. Undistributed soil samples were taken to a depth of 0–5 and 5–10 cm using 100 cc core sampler (cross section area of 19.6 cm^2 and 5.1 cm in height). Three samples were collected at each layer to overcome the difficulties caused by inhomogeneity of soil properties. The average size of these three samples was taken to be the representative value.

Saturated hydraulic conductivity was measured by a permeability test after capillary rise for over 48 h. Determination of permeability was carried out using a constant head permeability test, but a falling head permeability test was used for the lower permeability materials. Then, to find the dry bulk density, the soil was oven dried at 105°C for 24 h and the weight of oven-dried soil was measured.

Particle size distribution was determined by means of the sieving method and by using a particle size analyzer (SALD-3100; Shimadzu Corp., Kyoto, Japan) for fine fractions. We observed the content of particles finer than 0.063 mm, especially clay and silt fractions, in this experiment.

4. Functional assessment of some ground surface parameters

4.1. *FIRmax* in Japanese cedar and Hiba arborvitae plantation forests

Table 1 shows the measurement results of surface cover materials, hydraulic conductivity, dry bulk density and *FIRmax*. In all the cases during the experiment, rainfall intensity is in excess of final infiltration rate ('Rainfall intensity' and '*FIR*' in **Table 1**), which suggests that overland flow may occur in Japanese cedar and Hiba arborvitae forests during the intense rainfall events with around 180 mm/h.

FIRmax ('*FIRmax*' in **Table 1**) was obtained by applying the Eq. (2) to the rainfall intensity and final infiltration rate obtained in this experiment. *FIRmax*, as shown in **Table 1**, distributes in

http://dx.doi.org/10.5772/intechopen.70575

		Cover					Soil					Experiment		
Type	Plot ID	Elapsed time after thinning (year)	Ground cover ratio (%)	Dry weight of understory vegetation (g m^{-2})	Dry weight of litter materials (g m^{-2})	Dry weight of total cover (g m^{-2})	Fine fraction content (%)	Bulk density (g cm^{-3})		Hydraulic conductivity (mm h^{-1})		Rainfall intensity (mm h^{-1})	*FIR* (mm h^{-1})	*FIRmax* (mm h^{-1})
								0–5 cm	5–10 cm	0–5 cm	5–10 cm			
Cedar	C1	1	100	70	220	290	18	0.51	0.69	1420	559	171	147	241
	C2	Un-thinned	80	30	490	520	8	0.72	0.78	1060	429	170	141	212
	C3	Over 5	70	70	500	570	26	0.47	0.58	1120	620	186	171	359
	C4	2	90	190	1170	1360	19	0.60	0.75	1650	593	186	180	562
	C5	3	70	270	1270	1540	30	0.66	0.95	430	202	179	152	209
	C6	1	70	200	580	780	19	0.77	0.89	710	584	178	158	270
	C7	Un-thinned	80	70	320	390	33	0.80	0.79	810	565	190	128	142
	C8	Over 5	90	290	1200	1490	12	0.86	0.81	1170	1239	169	146	227
	C9	Un-thinned	100	190	1710	1900	22	0.80	0.83	700	697	178	171	317
	C10	1	80	20	630	650	38	0.56	0.64	1020	1195	170	162	449
	C11	1	70	370	540	910	34	0.81	0.79	840	1218	194	183	488
	C12	Un-thinned	30	10	460	470	4	0.85	0.95	1300	1799	183	173	324
	C13	Over 5	40	180	1160	1340	27	0.55	0.57	1110	1651	180	174	537
	C14	1	100	0	250	250	14	0.44	0.58	1120	169	149	137	282
	C15	Un-thinned	60	290	680	970	21	0.83	0.90	340	330	153	114	146
	C16	3	80	110	1780	1890	32	0.49	0.73	740	673	189	156	238
	C17	2	90	80	1670	1750	26	0.85	0.995	100	155	177	166	389
	C18	1	90	0	360	360	20	0.59	0.69	560	717	206	187	373
	C19	Un-thinned	90	0	880	880	32	0.77	0.81	1	2	192	169	301
	C20	1	80	50	420	470	4	0.60	0.78	1370	330	179	173	538
	C21	Un-thinned	100	0	1150	1150	30	0.74	0.84	210	296	206	196	508
	C22	1	100	160	950	1110	23	0.38	0.49	680	505	188	142	175

			Cover				Soil					Experiment		
Type	Plot ID	Elapsed time after thinning (year)	Ground cover ratio (%)	Dry weight of understory vegetation (g m^{-2})	Dry weight of litter materials (g m^{-2})	Dry weight of total cover (g m^{-2})	Fine fraction content (%)	Bulk density (g cm^{-3})		Hydraulic conductivity (mm h^{-1})		Rainfall intensity (mm h^{-1})	*FIR* (mm h^{-1})	*FIRmax* (mm h^{-1})
								0–5 cm	5–10 cm	0–5 cm	5–10 cm			
Hiba	H1	3	100	60	580	640	18	0.54	0.71	1340	581	230	164	204
	H2	Un-thinned	60	10	1020	1030	34	0.42	0.61	1360	858	185	90	93
	H3	2	60	0	470	470	19	0.51	0.69	1140	674	193	153	214
	H4	1	80	10	860	870	26	0.50	0.59	1630	782	177	165	372
	H5	1	90	0	570	570	23	0.56	0.90	460	489	179	146	217
	H6	Over 5	90	50	930	980	23	0.53	0.81	610	192	176	151	242
	H7	Un-thinned	100	900	800	1700	30	0.44	0.67	530	310	166	128	150
	H8	1	100	0	420	420	36	0.50	0.73	1290	403	192	187	623
	H9	Un-thinned	80	70	1570	1640	20	0.40	0.45	950	944	193	171	278
	H10	Over 5	100	70	1240	1310	28	0.54	0.85	770	533	192	188	641
	H11	3	80	40	450	490	34	0.51	0.58	1050	684	183	163	270
	H12	Un-thinned	70	400	1910	2310	20	0.67	0.84	1840	284	177	156	194
	H13	1	70	0	1200	1200	24	0.37	0.45	2170	625	146	138	347
	H14	Un-thinned	60	310	470	780	25	0.72	0.91	1030	312	178	135	149
	H15	1	100	0	1180	1180	17	0.55	0.74	1290	485	187	178	411
	H16	2	90	10	580	590	23	0.42	0.53	740	818	177	151	225

Table 1. Result of ***in situ*** artificial rainfall experiments using an oscillating nozzle rainfall simulator.

a range from 141.9 to 562.3 mm/h in the Japanese cedar forest and from 93.3 to 641.0 mm/h in the Hiba arborvitae forests. **Figure 3** shows frequency distribution of *FIRmax* for every 100 mm. A peak is observed in the *FIRmax* distribution in Hiba arborvitae forests around 200–300 mm/h, but tends to distribute equally in the range of 100–700 mm/h in both forests. Low *FIRmax* (<100 mm/h) was observed at only one site in Hiba arborvitae forests.

It is well-known that infiltration rate is affected by the amount of surface cover materials [1, 4, 17]. Also reported is an increasing trend of surface cover materials and understories with increasing time elapsed after thinning [40]. However, there was apparent variability among data, and no correlation was found between the elapsed year after thinning and the amount of surface cover materials for both types of forest. In fact, we could not clarify the correlation, because of only a few numbers of data for surface cover materials corresponding to elapsed year after thinning. We also looked at the separate correlation of understories and litter materials with elapsed year, but no correlation was found among these three groups. The effect of thinning on surface cover materials was also examined in both types of forest. We compared the effects on volume of surface cover materials between un-thinned plots and post-thinned sites with a variety of elapsed years. As a result, there was no noticeable difference in volume of surface cover materials between un-thinned and post-thinned sites in Japanese cedar forests, while a marked decrease in cover materials was observed in post-thinned plots in Hiba arborvitae forests. **Figure 4** shows the box-and-whisker plots of *FIRmax* in un-thinned and post-thinned sites. Despite a decrease in surface cover materials due to thinning, *FIRmax* increases in both types of forest. This result challenges the widely accepted notion that there is correlation between surface cover materials and *FIRmax* in Japanese cedar and Hiba arborvitae forests, and no correlation was observed especially in Hiba arborvitae forests.

Previous studies have shown that the effect of understories and leaf litters on infiltration rate is significant, which suggests that surface cover materials reduce raindrop impact on surface soil, and therefore the formation of surface crusts and HOF is restricted (e.g. [9, 12, 15]).

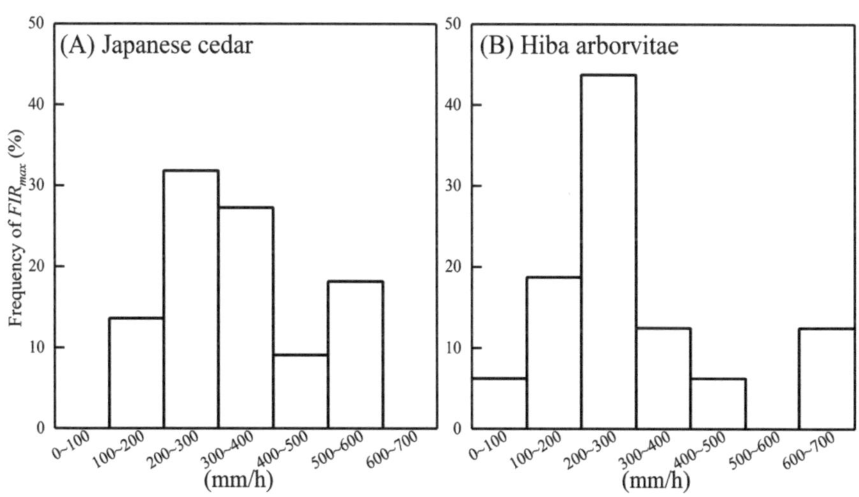

Figure 3. Frequency distribution of FIR_{max}. (A) Japanese cedar and (B) Hiba arborvitae.

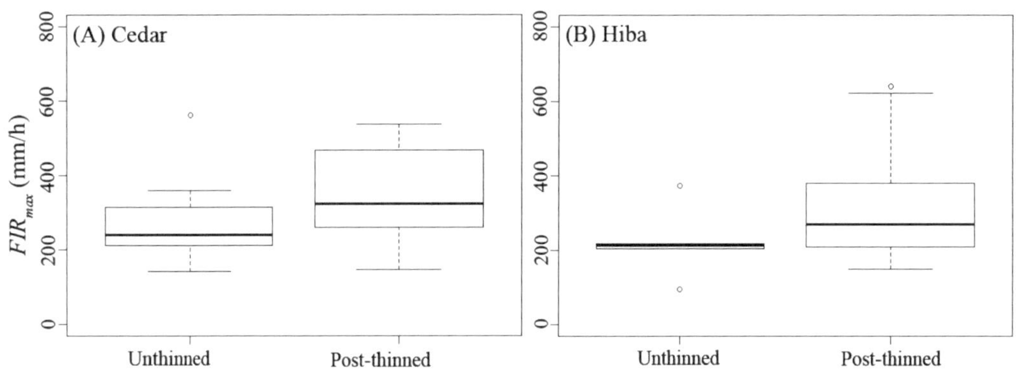

Figure 4. Box-and-whisker plots of *FIRmax* in un-thinned and post-thinned sites. (A) is Japanese cedar and (B) is Hiba arborvitae. The box represents 50% of the data between the 25th and the 75th percentile, the line represents the median and whiskers the minimum and maximum.

Previous study pointed out that relationship surface cover ratio and final infiltration have a positive correlation. **Figure 5** shows the relationship surface cover ratio and final infiltration, but in my case, there is no correlation. Some researchers adapt a linear approximation to the relationship between the two sides. Although statistical analysis was carried out and significance was observed, despite the low correlation coefficient, there was a close relationship between the two, and by using this relationship. It is expected that it will lead to appropriate management guidelines considering maintenance functions. I do not think about denying statistical analysis, but it is unknown whether it is easy to reach such consideration easily.

Surface cover materials were measured using image analysis. A high concentration (>60%) of cover materials was seen both in Japanese cedar and in Hiba arborvitae forests, but no correlation was observed between *FIRmax* and surface cover materials. **Figure 6** shows relation between surface cover materials and *FIRmax* in Japanese cedar and Hiba arborvitae forests. For comparison purposes, also included in the figure are the data of *C. obtusa* plantations [9]. **Figure 6** clearly shows that *FIRmax* in Japanese cedar and Hiba arborvitae forests is generally higher than that in *C. obtusa* plantations. Japanese cedar and Hiba arborvitae forests, especially at sites with a low concentration of surface cover materials (<1000 g/m^2), had a twofold to threefold greater *FIRmax* than that in *C. obtusa* plantations. This means that Japanese cedar and Hiba arborvitae forests may limit the occurrence of overland flow and soil erosion compared to *C. obtusa* plantations. Thus, Japanese cedar and Hiba arborvitae forests may provide a more effective protection of soil surface. The results of our experiment support previous studies and conclusions presented by Ogura and Takahashi [46] and Ogura and Kodani [18]. The regression line in **Figure 8** might show a positive correlation between the amount of surface cover materials and *FIRmax*. A similar finding was also reported by Kato et al. [8]. As shown in **Figure 6** (Japanese cedar: $r = 0.173$, $p = 0.443$; Hiba arborvitae: $r = 0.024$, $p = 0.929$), however, no such correlation was recognized between the amount of surface cover materials and *FIRmax*. Both Japanese cedar and Hiba arborvitae forests gave a relatively high *FIRmax* values (>100 mm/h) for any amount of surface cover materials. We also took the separate correlation of understories and litter materials with *FIRmax*, but no correlation was found among these three groups.

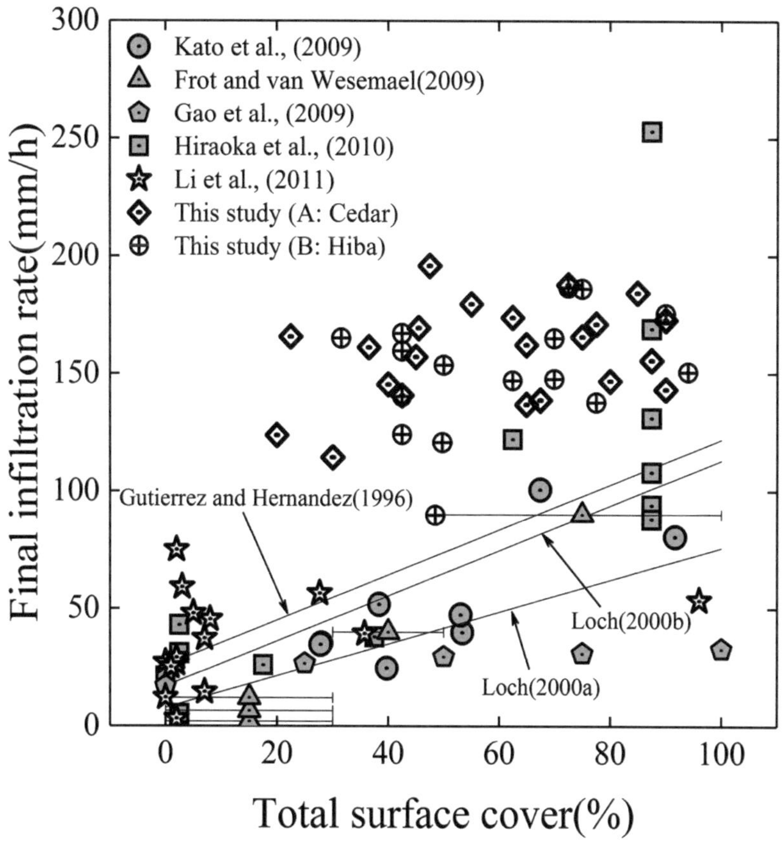

Figure 5. Relationships between the total surface cover and the infiltration rate at the (A) Japanese cedar and the (B) Hiba arborvitae sites (the solid square and open triangle indicate the desert grassland site and shrubland site, respectively. Steppe grassland ($y = 0.92x + 5.74$, $r^2 = 0.59$, $p = 0.029$) and desert grassland ($y = -7.67x + 220.4$, $r^2 = 0.98$, $p = 0.042$) In Mongolia, results by Gutierrez and Hernandez [41] ($y = 0.96x + 25.9$), Loch [11] ($y = 0.96x + 17.0$), Loch [42] ($y = 0.68x + 7.74$), Kato et al. [12], Frot and van Wesemael [43], Gao et al. [44], Hiraoka et al. [15] and Li et al. [45]. Modified from Kato et al. [12]).

4.2. Surface cover materials in Japanese cedar and Hiba arborvitae plantation forests

Measurement of surface cover materials ('Understory vegetation' and 'Litter materials' in **Table 1**) shows that understories comprised approximately 10% of surface cover materials and the remaining 90% of litter materials, which signifies that litter materials are a major component in Japanese cedar and Hiba arborvitae forests. The frequency distribution of surface cover materials per area of 500 m^2 (no figure is presented) shows that the highest peak is at the frequency of 500–1000 g/m^2 in both types of forest, and an increase in weight (1000–1500 g/m^2, 1500–2000 g/m^2) induced the lowering of the frequency.

4.3. Soil properties in Japanese cedar and Hiba arborvitae plantation forests

A high degree of variability in hydraulic conductivity was found, but high values (100–2000 mm/h) were obtained in the 0–5 cm and 5–10 cm layers in both types of forest (except for

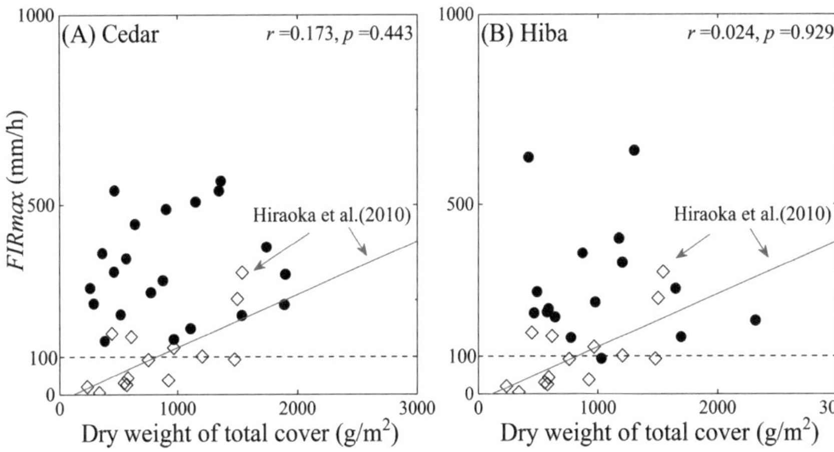

Figure 6. Relation between ground cover materials and *FIRmax*. (A) is Japanese cedar and (B) is Hiba arborvitae and results by Hiraoka et al. [9] (*FIRmax* = 0.14 × (dry weight of total cover) − 15.65).

C19). The same trend was commonly seen in other forests as reported in previous studies [3, 21]. Hiba arborvitae forests had relatively higher hydraulic conductivity than Japanese cedar forests. It was also found that hydraulic conductivity measured at depth of 5–10 cm was comparatively lower than those at depth of 0–5 cm depth.

Dry bulk density ('bulk density' in **Table 1**) was as follows: 0.38–0.86 g/cm^3 in Japanese cedar forests and 0.37–0.72 g/cm^3 in Hiba arborvitae forests at depth of 0–5 cm; 0.49–0.995 g/cm^3 in Japanese cedar forests and 0.49–0.91 g/cm^3 in Hiba arborvitae forests at depth of 5–10 cm. Although there was considerable variability, we found that bulk density increased with depth and was higher in Hiba arborvitae forests compared to Japanese cedar forests. According to Miyata et al. [3], the average bulk density was 0.75 g/cm3 in Hiba arborvitae forests, which was slightly lower than those obtained in our experiment (Japanese cedar forests: 0.67 g/cm3; Hiba arborvitae forests: 0.51 g/cm3). It was presumed that a decrease of porosity induced by an increasing bulk density results in a decrease of hydraulic conductivity with increasing depth. We, however, could not find significant correlation between hydraulic conductivity and bulk density. Therefore, there is another unknown factor for the decrease of hydraulic conductivity.

Fine fraction content ('fine fraction content' in **Table 1**) was 4–38% (average 22%) in Japanese cedar forests and 17–36% (average 25%) in Hiba arborvitae forests, but most of the data for both types of forest were below 35% except maximum values. The maximum values for these forests do not vary significantly as much as the minimum values. The minimum value was 4% for Japanese cedar forests and 17% for Hiba arborvitae forests, respectively. The result obtained in Japanese cedar forests exhibits a larger variability compared to Hiba arborvitae forests.

The high fine fraction (clay + silt) content may increase the effect of clogging under the impacts of raindrops, which can reduce the infiltration rate and hydraulic conductivity [3]. Yokoi et al.

[47] reported that crust was formed when the fine fraction content in the sand pyroclastic flow deposits exceeded 35%. A soil profile at our experimental sites appears to be a brown forest soil. Therefore, the use of a theory developed by Yokoi et al. [47] will be constrained. If the theory is to be applicable, it seems difficult to form crust in Japanese cedar and Hiba arborvitae forests as a whole because the fine fraction content does not generally exceed 35% in these types of forest. In fact, no crust could be visually observed after the experiment, and from the final infiltration rate obtained from the experiment, it also seemed unlikely that the formation of crust occurred. **Figure 7** is the relation between fine fraction content and *FIRmax*, but we could not find significant correlation (Japanese cedar: $r = 0.033$, $p = 0.883$; Hiba arborvitae: $r = 0.013$, $p = 0.154$). **Figure 8** shows the relation between fine fraction content and hydraulic conductivity, in which no correlation could be established.

Figure 9 shows relation between hydraulic conductivity and *FIRmax* in the 0–5 cm surface layer, which revealed no correlation in either type of forests (Japanese cedar: $r = 0.244$, $p = 0.274$; Hiba arborvitae: $r = 0.101$, $p = 0.710$). Similarly, no correlation was detected between hydraulic conductivity and *FIRmax* for soil at 5–10 cm depth. It also became clear that *FIRmax* is definitely lower than hydraulic conductivity at most sites in Japanese cedar forest (19 out of 22 sites) and all sites in Hiba arborvitae forest (16 sites). It is postulated that the hydraulic gradient in the surface soil layer was estimated to be about one and that *FIRmax* had the similar value to hydraulic conductivity, which needs careful consideration. The 2–3 cm layer within 1 h after the sprinkling experiment was moistened, but the dry soil underlay in the deeper soil layers. The presence of macropores in soils might lead to the preferentially infiltration of sprinkled water. With fingering infiltration (unstable infiltration), the preferential water was infiltrated into the soils that are exceptionally conductive. The effect of entrapped air, hyphae respiration and bacteria in unsaturated soil also needs to be examined in future studies. There may be an underlying mechanism to limit infiltration of water into soil profile, which caused lower *FIRmax* compared to the permeability test for fully saturated soil.

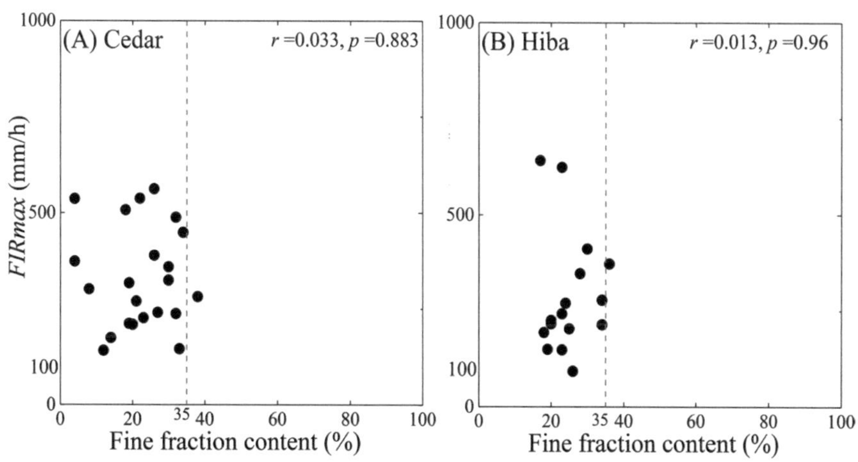

Figure 7. Relation between fine fraction content and *FIRmax*. (A) is Japanese cedar and (B) is Hiba arborvitae.

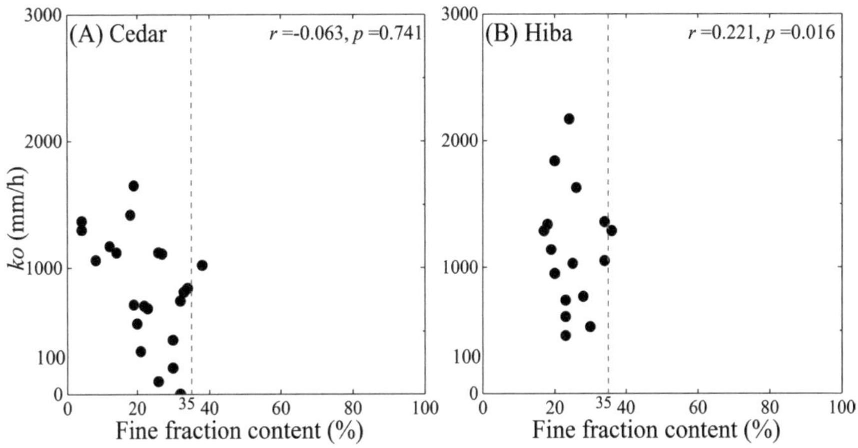

Figure 8. Relation between fine fraction content and hydraulic conductivity (*ko*). (A) is Japanese cedar and (B) is Hiba arborvitae.

In summary, surface cover materials, fine fraction content, and hydraulic conductivity had no correlation with *FIRmax* in either type of the forests examined in this study. Both Japanese cedar and Hiba arborvitae forests gave relatively high *FIRmax* values (>100 mm/h), which is higher than that of the entire *C. obtusa* plantations. These forests, especially at sites with a low concentration of surface cover materials (<1000 g/m^2), had a twofold to threefold greater *FIRmax* than that in C. obtusa plantations. Thus, Japanese cedar and Hiba arborvitae forests may provide a more effective protection of the soil surface. Based on fine fraction content, visual observation, and final infiltration rate, it seemed unlikely that the formation of crust occurs in both types of forest. In both types of forest, *FIRmax* is exceptionally lower than hydraulic conductivity at the soil surface. Fingering might occur during infiltration due to entrapped air and hyphae respiration.

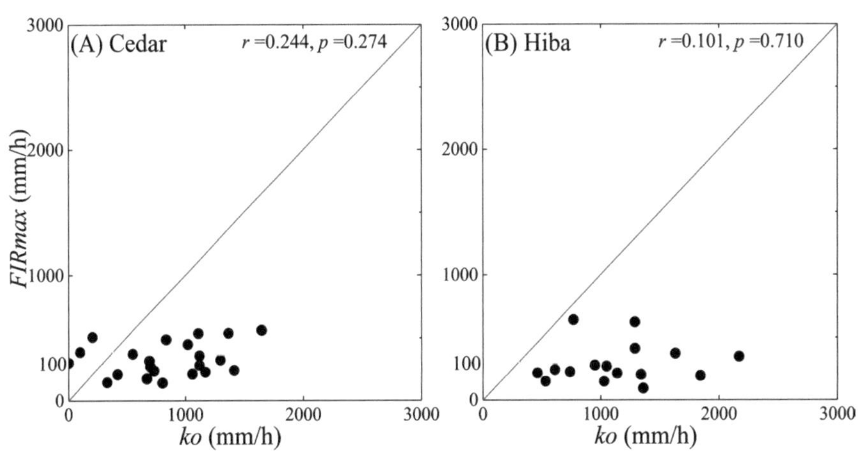

Figure 9. Relation between hydraulic conductivity (*ko*) and *FIRmax*. (A) is Japanese cedar and (B) is Hiba arborvitae.

The above results prove that the change in *FIRmax* in either type of forest cannot be explained by surface cover materials, hydraulic conductivity or fine fraction content. In the present research, we could not clarify the influencing factors to *FIRmax* in Japanese cedar and Hiba arborvitae forests. However, we found that *FIRmax* increased after thinning, and this might be attributed to Ishikawa Forest Environmental Tax effect.

Ishikawa Prefecture is currently collecting 'Ishikawa Forest Environmental Tax', and implementing intensive thinning in forest degradation area throughout the prefecture to improve the public function of forests, such as watershed conservation and preventing landslide disaster. Since fiscal 2017, Ishikawa Prefecture has implemented additional efforts to promote the use of timber.

Author details

Yoshitaka Komatsu

Address all correspondence to: komatsu.yoshitaka.ws@alumni.tsukuba.ac.jp

Graduate School of Life and Environmental Sciences, University of Tsukuba, Tsukuba, Japan

References

[1] Tsujimura M, Onda Y, Harada D. The role of Horton overland flow in rainfall-runoff process in an unchanneled catchment covered by unmanaged Hinoki plantation. Journal of Japan Society of Hydrology & Water Resources. 2006;**19**:17-24 (in Japanese with English summary)

[2] Gomi T, Sidle RC, Miyata S, Kosugi K, Onda Y. Dynamic runoff connectivity of overland flow on steep forested hillslopes: Scale effects and runoff transfer. Water Resources Research. 2008;**44**:W08411. DOI: 10.1029/2007WR005894

[3] Miyata S, Onda Y, Gomi T, Mizugaki S, Asai H, Hirano T, Fukuyama T, Kosugi K, Sidle RC, Terajima T, Hiramatsu S. Factors affecting generation of Hortonian overland flow in forested hillslopes: Analysis of observation results at three sites with different geology and rainfall characteristics. Journal of the Japanese Forest Society. 2009;**91**:398-407 (in Japanese with English summary)

[4] Onda Y. An Overview of the Effects of Forest Devastation on Soil and Water Loss. Tokyo: Japan: Iwanami Shoten; 2008. p. 245 (in Japanese)

[5] Murai H, Iwasaki Y. Studies on function of water and soil conservation based on forest land (I) influence of difference in forest condition upon water run-off, infiltration and soil erosion. Bulletin of the Government Forest Experiment Station. 1975;**274**:23-84 (in Japanese with English summary)

[6] Takeuchi N. A trial manufacture of the simple mountain infiltrometer. Journal of Japanese Forest Research. 1976;**58**:407-409

[7] Onda Y, Tsujimura M, Nonoda T, Takenaka C. Methods for measuring infiltration rate in forest floor in Hinoki plantations. Journal of Japan Society of Hydrology & Water Resources. 2005;**18**(6):688-694 (in Japanese with English Summary)

[8] Kato H, Onda Y, Ito S, Nanko K. Field measurement of infiltration rate using an oscillating nozzle rainfall simulator in devastated Hinoki plantation. Journal of Japan Society of Hydrology & Water Resources. 2008;**21**(6):439-448 (in Japanese with English Summary)

[9] Hiraoka M, Onda Y, Kato H, Mizugaki S, Gomi T, Nanko K. Effects of understory vegetation on infiltration capacity in Japanese cypress plantation. Journal of the Japanese Forest Society. 2010;**92**:145-150 (in Japanese with English Summary)

[10] Ogura A, Onda Y, Komatsu Y. Investigation of water and soil maintenance-field measurement of infiltration rate using an oscillating nozzle rainfall simulator. Bulletin of the Ishikawa Agriculture and Forestry Research Center Forestry Experiment Station. 2012;**4**:1-17 (in Japanese)

[11] Loch RJ. Effects of vegetation cover on runoff and erosion under simulated rain and over and flow on a rehabilitated site on the Meandu mine, Tarong, Queensland. Australian Journal of Soil Research. 2000;**38**:299-312. DOI: 10.1071/SR99030

[12] Kato H, Onda Y, Tanaka Y, Asano M. Field measurement of infiltration rate using an oscillating nozzle rainfall simulator in the cold, semiarid grassland of Mongolia. Catena. 2009;**76**:173-181. DOI: 10.1016/j.catena.2008.11.003

[13] Onda Y, Yukawa N. The influences of understories on infiltration rates in *Chamaecyparis obtusa* plantations (II): Laboratory experiments. Journal of Japanese Forest Research. 1995;**77**(5):399-407 (in Japanese with English Summary)

[14] Tsukamoto J, Kajihara N, Nyuuta S. Prediction of surface soil loss on slopes afforested with Japanese cypress (*Chamaecyparis obtusa* Endl.)—Examination of availability of several features of foresr floor for assessing intensity of soil erosion. Journal of Japanese Forest Research. 1998;**80**:205-213 (in Japanese with English Summary)

[15] Miura S. Proposal for a new definition to evaluate the status of forest floor cover and floor cover percentage (FCP) from the viewpoint of the protection against raindrop splash. Journal of the Japanese Forest Society. 2000;**82**:132-140 (in Japanese with English Summary)

[16] Abe T, Sato H. Influences of forest type, topography, and understory on infiltration rates in eastern Hokkaido, Japan. Journal of the Japanese Forest Society. 2008;**90**:84-90. DOI: 10.4005/jjfs.90.84

[17] Kiyono Y. Analyses of factors affecting the dynamics of coverage and number of species in understories in *Chamaecyparis obtusa* plantations. Journal of the Japanese Forest Society. 1988;**70**(10):455-460 (in Japanese)

[18] Ogura A, Kodani J. Soil erosion in various plantation and skidding road. Annual Meeting of the Chubu Branch of the Japanese Forestry Society. 2008;**56**:57-58 (in Japanese)

[19] Miyazaki H, Furumatsu A, Yamate N, Yoshimura T. A study on the effects of sugi and hinoki leaves on the prevention of soil erosion using an artificial rainfall machine. Journal of the Japan Forest Engineering Society. 1998;**13**(2):67-74 (in Japanese with English Summary)

[20] Tsukamoto Y. Forest Hydrology. Buneido-shuppan; 1992. p. 319

[21] Kosugi K. New index to evaluate water holding capacity of forest soils. Journal of the Japanese Forest Society. 1999;**81**:226-235 (in Japanese with English Summary)

[22] Dunne T, Zhang W, Aubry BF. Effects of rainfall vegetation and microtopography on infiltration and runoff. Water Resources Research. 1991;**29**:2271-2285. DOI: 10.1029/91WR01585

[23] Forestry Agency. Annual Report on Forest and Forestry in Japan, Fiscal Year 2015; 2015. p. 225 (in Japanese)

[24] Hiramatsu S, Ishikawa Y, Miyoshi I, Chiba A. Actual state of sediment-related disaster on September 24-25, 1998 in Kochi prefecture, Japan (prompt report). Journal of the Japan Society of Erosion Control Engineering. 1999;**51**(5):43-47 (in Japanese)

[25] Ministry of Land, Infrastructure, Transport and Tourism [Internet]. Available from: http://www.mlit.go.jp/ [Accessed: 5 June 2017]

[26] Japan Society of Civil Engineers [Internet]. Available from: http://www.jsce.or.jp/ [Accessed: 5 June 2017]

[27] Tachibana S. Introduction aspect and agenda of forest-environmental tax. Wood and Wood Industry. 2005;**7**:4-7 (in Japanese)

[28] Yamaguchi Prefecture. Evaluation Report on the Yamaguchi Forest Management Prefectural Residence Tax Project. Yamaguchi Prefecture; 2009. p. 348 (in Japanese)

[29] Akiyama T. Forest environmental tax and significance of forest environment and forestry. The Norin Kinyu. 2005;**58**(2):32-44 (in Japanese)

[30] Imawaka S, Sato N. Study on new Forest maintenance projects by "forest environmental tax". Bulletin of the Kyushu University Forests. 2008;**89**:75-126 (in Japanese with English summary)

[31] Ishikawa Prefecture Department of Agriculture, Forestry and Fisheries. Hiba arborvitae in Noto Island; 1997. p. 22 (in Japanese)

[32] Japan Meteorological Agency. AMeDAS Data [Internet]. Available from: http://www.data.jma.go.jp/obd/stats/etrn/select/prefecture.php?prec_no=56&block_no=&year=&month=&day=&view= [Accessed: 20 August 2010]

[33] Horton RE. Analysis of runoff-plot experiments with various infiltration-capacity. Transactions of the American Geophysical Union. 1939;**20**(4):697-711

[34] Hawkins RH, Cundy TW. Steady-state analysis of infiltration and overland flow for spatially-varied hillslopes. Journal of the American Water Resources Association. 1987;**23**:251-256

[35] Yu B, Rose CW, Coughlan KJ, Fentie B. Plot-scale rainfall-runoff characteristics and modeling at six sites in Australia and Southeast Asia. Transactions of the ASAE. 1997;**40**(5):1295-1303

[36] Stone JJ, Paige GB, Hawkins RH. Rainfall intensity-dependent infiltration rates on rangeland rainfall simulator plots. Transactions of the ASABE. 2008;**51**(1):45-53

[37] Langhans C, Govers G, Diels J, Leys A, Clymans W, Putte AVD, Valckx J. Experimental rainfall-runoff data: Reconsidering the concept of infiltration capacity. Journal of Hydrology. 2011;**399**(3-4):255-262

[38] Tanaka S, Tokioka T. Evaluating infiltration-runoff by in situ rainfall simulation. In: Japan Society of Civil Engineers, editor. Proceedings of the 62nd Annual Conference of the Japan Society of Civil Engineers; Hiroshima: Japan; 2007. p. CD-ROM 2-003. (In Japanese)

[39] Nanko K, Hotta N, Suzuki M. Assessing raindrop impact energy at the forest floor in a mature Japanese cypress plantation using continuous raindrop—sizing instruments. Journal of Forest Research. 2004;**9**:157-164. DOI: 10.1007/s10310-003-0067-6

[40] Muramoto Y, Nogami K, Takagi M. Effect of line thinning on the changes of undergrowth species composition in matured *Chamaecyparis obtusa* forest-results during 5 years after thinning. Kyushu Journal of Forest Research. 2005;**58**:59-62 (in Japanese)

[41] Gutierrez J, Hernandez II, Runoff and interrill erosion as affected by grass cover in a semi-arid rangeland of northern Mexico. Journal of Arid Environments. 1996;**34**(3):287-295

[42] Loch RJ, Effects of vegetation cover on runoff and erosion under simulated rain and overland flow on a rehabilitated site on the Meandu Mine, Tarong, Queensland. Australian Journal of Soil Research. 2000b;**38**:299-312

[43] Frot E, van Wesemael B. Predicting runoff from semi-arid hillslopes as source areas for water harvesting in the sierra de Gador, southeast Spain. Catena. 2009;**79**(1):83-92

[44] Gao Y, Zhu B, Zhou P, Tang JL, Wang T, Miao CY. Effects of vegetation cover on phosphorus loss from a hillslope cropland of purple soil under simulated rainfall: A case study in China. Nutrient Cycling in Agroecosystems. 2009;**85**(3):263-273

[45] Li XY, Contreras S, Solé-Benet A, Cantón Y, Domingo F, Lázaro R, Lin H, Van Wesemael B, Puigdefábregas J. Controls of infiltration–runoff processes in Mediterranean karst rangelands in SE Spain. Catena. 2011;**86**(2):98-109

[46] Ogura A, Takahashi D. Soil erosion in artificial forest of Hiba arborvitae (*Thujopsis dolabrata*). Bulletin of the Ishikawa-Ken Forest Experiment Station. 2006;**38**:27-32 (in Japanese)

[47] Yokoi Y, Hasegawa S, Sakamoto N. Improvement of hardsetting soils by sandy pryoclastic flow deposit soil dressing in Kamikawa, Hokkaido. Japanese Society of Soil Science and Plant Nutrition. 1998;**69**(6):644-648 (in Japanese with English Summary)

Chapter 8

Field-Controlled Hydrological Experiments in Red Soil-Covered Areas (South China): A Review

Sanyuan Jiang, Qiande Zhu and Seifeddine Jomaa

Additional information is available at the end of the chapter

http://dx.doi.org/10.5772/intechopen.70547

Abstract

Investigation of runoff generation processes and response to changes in catchment characteristics (e.g. land use, soil type, slope, etc.), tillage practice and climate pattern (e.g. rainfall intensity and rainfall duration) is important for understanding of the hydrological cycle and developing land management practices for water and soil conservation. Field and indoor artificial hydrological experiments provide an efficient way for the study of the above processes. This study gave a summary of artificial hydrological experiments using rainfall simulator in China, especially in the red soil-covered region of Jiangxi province. Experiment setting for field and indoor artificial hydrological experiments were introduced; the water balance, runoff components (i.e. surface runoff, subsurface runoff at different depths), runoff amount and relationship to rainfall events were studied and assessment of land coverage and tillage practices on soil and water conservation were conducted. Based on the literature review, it implies that hydrological process at field slope requires more investigation in the following aspects: (1) improvement of monitoring strategies and methodology and isotopic method may be used to improve understanding of hydrological regimes, (2) developing long-term in situ experimental study to analyse soil water movement at different temporal and spatial scales and (3) developing and improving modelling of soil water movement.

Keywords: controlled experiments, runoff processes, water balance, flow pathways, diffuse nutrient losses, soil erosion

1. Introduction

Investigations of rainfall-runoff processes, hydrological flow pathways and water transit time are important for understanding of the hydrological cycle and related nutrient and sediment transport, which may assist developing land management practices with the purpose of water and soil conservation, prevention of nutrient losses and soil erosion. [1]. Field/pilot rainfall-runoff experiments are traditional and sound approaches to uncover runoff generation processes and assess their responses to changes in topography, land use, soil type, underlying geology and climate patterns. In particular, the artificial rainfall-runoff experiments originated from the 1950s in the USA were conducted for investigation of water and soil conservation and later applied worldwide on research of runoff generation processes and nutrient (nitrogen, phosphorus) losses. Up to now, these approaches are still widely used in investigating nutrient export patterns, the impacts of rainfall pattern (amount, duration and intensity), land use, soil type and antecedent soil water content on nutrient losses [2–4].

Original and reliable data can be obtained from rainfall-runoff experiment and hydrometric monitoring in natural rainfall events. However, it takes long time for conducting rainfall-runoff experiment under natural rainfall condition due to high variability, randomness and uncertainty in rainfall occurrence. It is also heavily impacted by environmental and climatic conditions as rainfall condition cannot be controlled. Therefore, it is difficult to obtain ideal results from runoff generation study under natural rainfall condition and therefore is not efficient. Artificial runoff experiment using rainfall simulator can realise simulation of different rainfall patterns (i.e. combination of different rainfall intensities and different rainfall durations). It can be either conducted in the field or in indoor laboratory by simulating physiographic characteristics and rainfall pattern of the concerned study site. Artificial rainfall simulator can increase the efficiency of the experimental study by overcoming the high randomness in occurrence of diffuse nutrient export and soil erosion under natural rainfall condition [5]. Thus, the artificially rainfall-runoff experiments combined with hydrological and geochemical analysis have become effective techniques for investigation of runoff generation, nutrient losses and soil erosion processes and estimation of pollutants' export loads.

In the following sections, we provide a review on artificial hydrological experiments, with particular focus on those conducted in red soil-covered area in South China, from the perspectives of idea and constructions of artificial experimental facilities, experimental methods (such as hydrometrical methods for various parameters, simulated rainfall equipment, hydrogeochemical methods) and findings obtained from relevant studies. Finally, a summary of challenges in artificial hydrological experiments and outlook of future studies are given. The outcome of this work may provide guidelines for the design and constructions of artificial hydrological experiment based on various objectives, improve understanding of hydrological processes and impacts on nutrient and soil erosion, serve for hydrological modelling development by improving parameter setting and support decision-making on watershed management for water and soil conservation.

2. Design and construction of artificial hydrological experiments

Considering the randomness, uncertainty and difficulties of implementing hydrological experiments and relevant studies under natural rainfall condition, the requests of investigating the spatial variability effect of soil characteristics and rainfall characteristics on runoff generation, nutrient fluxes and soil erosion led to the development of rainfall simulators on small plots.

Iserloh et al. have mentioned that using small rainfall simulators, among others, has more advantages due to the low costs, easy to transport and use in inaccessible areas and low water consumption [6]. This technique has been used worldwide by different research groups where it was beneficial for the investigation of runoff generation processes, assessments of soil erosion by surface runoff and decision-making in soil and water conservation. The design and constructions of artificial (indoor and field) hydrological experiments using rainfall simulator for different purposes are described below.

2.1. Indoor artificial hydrological experiment

Figure 1 shows the schematic design of indoor artificial rainfall-runoff experiment in a flume, which is used to investigate soil erosion processes and influential factors, including rainfall patterns (e.g. rainfall intensity, duration) and soil characteristics (e.g. soil gradation, porosity, precedent soil moisture). An example of such experiment is shown in **Figure 1**.

The erosion flume has the size of 6 m long × 2 m wide. The flume and sprinkling system are described in detail in [7–10]. A summary of the key features is given here. The flume sits on a hydraulic piston so that the slope can be adjusted. The flume can be filled with 0.32 m of a natural soil taken from field site and is underlain by 0.10 m of coarse gravel to facilitate the

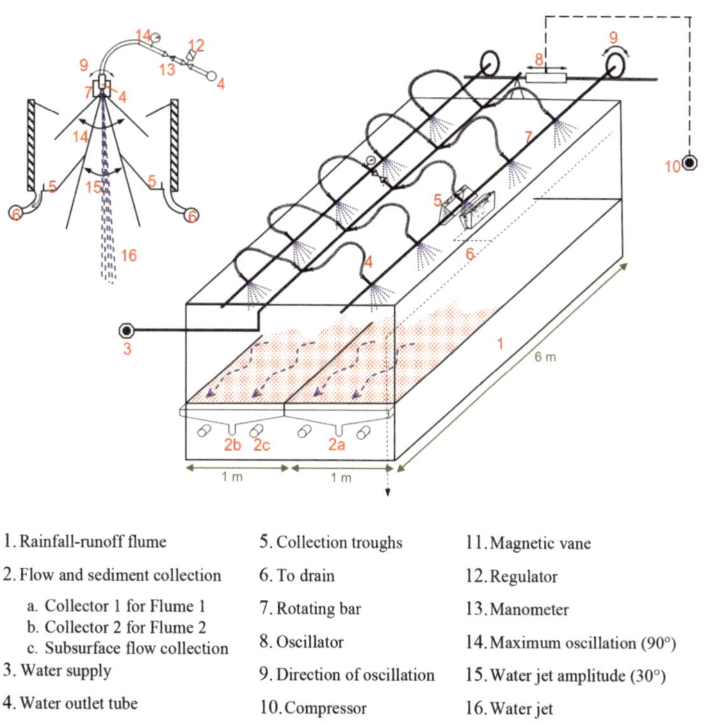

Figure 1. Schematic structure of the designed rainfall-runoff soil erosion flume referring to the Ecole Polytechnique Fédérale de Lausanne (EPFL) erosion flume [7, 8]. The slope of the flume can be adjusted within the intervals 0 and 30%. Precipitation is fallen using 10 oscillating sprinkler systems located at 3 m above the soil surface.

excess rainfall drainage. Also, at the downslope end of the flume, there are eight openings collecting the infiltrated water which drains through the soil and gravel. Other four openings (item 2c, **Figure 1**) at the flume's downslope end capture subsurface flow through the topsoil layer. The deep soil and the openings at the bottom and downslope end of the flume allow us to do experiments on unsaturated soils without fully saturating the soil profile during the experiments.

Water is fallen to the flume by 10 VeeJet 80150 nozzles located on an oscillating bar 3 m above the soil surface ensuring the raindrops reached their terminal velocity. The whole flume is divided into two 1-m width identical flumes allowing replicate experiments. Surface runoff is measured as a function of time over the course of the rainfall event for each collector as shown in **Figure 1**. Water from each flume outlet is sampled in individual bottles. Continuous sampling is conducted for the early experiment stage. Afterwards, the sampling time interval increases towards the steady-state equilibrium.

2.2. Field artificial hydrological experiment

It was recognised that the results obtained under carefully controlled laboratory conditions are rarely directly valid in the field due to the high heterogeneity in terms of the influencing parameters on soil hydrological processes. Thus, additional investigations at the natural field conditions are needed.

The eco-science and technology park of Jiangxi province is an ideal field site for researches on runoff generation, nutrient export and soil erosion [11]. It is located in De'an County of Jiujiang City, Jiangxi province with the total area of 0.8 km^2. For runoff generation study, field soil water leakage experiment plot was constructed with a length of 15 m, width of 5 m (horizontal projection) and slope of 14°. The schematic diagram of the rainfall-runoff experimental plot is shown in **Figure 2**. The four sides and bottom of the plot were constructed using reinforced concrete to isolate from surrounding environment; sandy inverted filter was set on the bottom of the plot. Trapezoidal retaining wall was constructed at the toe of the plot using reinforced concrete. In this way a closed draining soil water infiltration device was built. To prevent water entering the plot, the four sides of the plot were constructed using reinforced concrete, which are 30 cm higher than the land surface. After removal of surficial plant and miscellaneous chips, five layers of undisturbed soil were sampled with thickness of each layer 25 cm, and then each layer was stacked. Mean soil density of each layer was measured at the depth of 10 cm, and then the sampled soil was backfilled in the plot to the depth 105 cm in the order of soil sampling with back-filled depth of 10 cm at each time. The weight of backfilled soil was determined using the formula

$$W = V \cdot \gamma \cdot (1 + S) \tag{1}$$

where V is the volume of backfilled soil at each time (m^3), γ is the dry density of natural soil (kg/m^3) and S is the indoor soil water content (%).

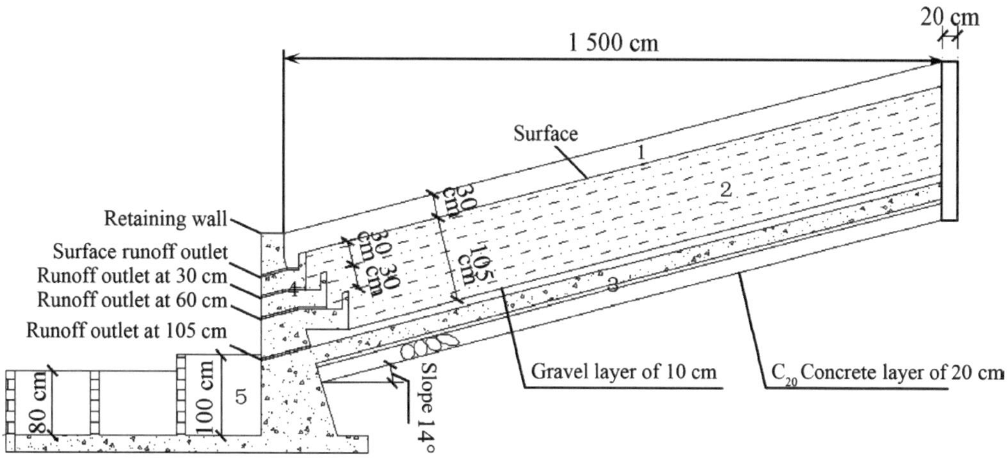

Figure 2. Schematic diagram of the rainfall-runoff experimental plot adapted from [11].

After backfill of soil at each time, soil was compacted to make it achieve the defined depth of 10 cm in order to make the porosity in the backfilled plot similar as that in the natural condition.

In the observation room located at the toe of the plot, four water flow outlets are set from top of the retaining wall to the bottom. The upmost outlet is used for collecting surface runoff and sediment. It connects flow collecting barrel at the bottom of the plot to the surface flow pool via PVC tube. The rest of the three outlets collect underground runoff (i.e. interflow of 30 and 60 cm depth and groundwater runoff at 105 cm below the surface). Horizontal flow collecting barrels were installed in the retaining wall vertically at corresponding depth, which were connected to corresponding runoff pool with bottom area of 1.2 m3. Surface flow pool, interflow pool and groundwater flow pool were equipped with self-record water-level sensor (mode HCJ1) to continuously monitor dynamics of runoff and water leakage generated from different patterns of rainfall events.

3. Artificial hydrological experimental studies

Based on intensive literature review, it is found that artificial hydrological experimental studies can be categorised into the following groups with different objectives: (i) runoff and soil erosion processes and response to rainfall pattern, land use type, slope and tillage approach and (ii) nutrient export and response to land use type, slope and rainfall pattern. Now, a summary of the former studies related to these two perspectives is given.

3.1. Impacts of land use and rainfall on runoff processes

Several parameters are commonly used to describe rainfall-runoff processes, including runoff starting time, share of different runoff components (i.e. surface runoff, interflow at different layers), water residence time, etc. Wang et al. studied the natural rainfall-interflow processes at different soil layers (up layer of 0–40 cm and down layer of 40–110 cm) under natural rainfall condition in hilly region covered with red soil and found that interflow began earlier and had higher value in oiltea camellia than that in resumed field, while the peak fluxes were oppositional; the response time (i.e. lag) of interflow to rainfall and surface runoff increased with increase of soil depth [12]. Based on runoff plot experiments, Yin et al. investigated runoff at different soil layers with respective depth of 30, 60 and 90 cm to the surface with different types of land cover at red soil field slope in Jiangxi province; results showed that grassland cover increased interflow and prolonged interflow duration compared to bare land. Interflow was positively related to rainfall amount, rainfall intensity, rainfall duration and initial infiltrated rainfall, and runoff was mainly affected by rainfall amount at covered field slope while mainly influenced by rainfall intensity at bare field slope [13]. The effects of Bahia grass and its litter on dynamics of soil moisture and water balance were studied using lysimeter in the field slope covered with red soil, and results indicated that surface runoff of bare land was 24.25 times of that with Bahia grass' coverage and 11.78 times of that with Bahia grass' mulching; the effects of ground cover on soil moisture were different between seasons, and different ground covers could increase or decrease soil moisture [14].

Xie et al. (2014) analysed the characteristics of vertical runoff output in different soil strata on a red soil slope plot under three conditions: vegetation coverage, litter mulch and bare land. The results showed that the total runoff of litter mulch treatment was maximum, followed by the bare land and the vegetation coverage treatment under natural rainfall conditions. Surface runoffs of vegetation coverage and litter mulch treatment were far less than that of the bare land treatment, which were 7.9 and 9.8% of the bare land, respectively. The planted grass and the litter mulch can reduce surface runoff significantly. Interflow of the bare land treatment was the least, which was, respectively, 56.4% of the vegetation coverage and 35.6% of the litter mulch treatments. It demonstrated that both vegetation and litter can increase water seepage. Underground runoff was the main way of runoff output on red soil slope under different treatments, and interflow and surface runoff were directly related to the presence or absence of ground coverage. All different runoff components showed seasonal variation characterized by high runoff in spring and summer, whereas lower runoff in autumn and winter due to temporal change in precipitation [11]. In most studies on groundwater runoff at field slope, the sum of interflow and groundwater flow was considered as a whole without division. In forest watershed, runoff was found to be dominated by groundwater runoff (may be above 39% of the total runoff), while surface runoff had small share (around 1–8% of total runoff) [15, 16].

Xie et al. [2010 b] assessed the effect of runoff reduction through different forest vegetation measures for soil and water conservation in hilly-land area of Jiangxi province using sample-plot and runoff-plot methods; results showed that the interception amount by forest canopy followed the descending order of Pinus elliotii, Masson pine, neddle-broad leaf mixed forest and Chinese chestnut, while the effective interception amount for litter and water holding of soil followed the descending order of Chinese chestnut, neddle-broad leaf mixed forest, Masson pine and Pinus elliotii; the rate of runoff reduction through multilayer vegetation measures was above 90% [28]. Based on statistical analysis on runoff plot experiment in red soil area, Xie et al. [17, 18] found that it was an effective approach to raise the level of vegetation structure and vegetation coverage supplemented with necessary engineering measures for soil and water conservation in slope land of red soil in South China.

Xu et al. investigated the characteristics of interflow in the purple soil of field slope under different rainfall intensities and land surface conditions and found that the runoff coefficient and average interflow on undisturbed abandoned lands are 3–15 times and 7–33 times that of bare-cultivated lands, respectively; the difference in runoff generation becomes more evident with increase of rainfall intensity [19]. By the setup through a flow collection system, Liu et al. investigated runoff generation process and found that under condition of small rain, surface runoff and the lag to the flow peak might occur if the soil was dry before the rainfall and surface runoff were primarily controlled by infiltration-excess runoff mechanism [20]. Using rainfall simulation experiments, Zhao et al. assessed different pastures' runoff features under different rainfall intensities, antecedent moisture contents and slope gradients; results showed that ryegrass performs better in delaying time to runoff and reducing runoff coefficient, rainfall intensity and antecedent moisture content as well as gradient that affect runoff features [21].

3.2. Impacts of slope and tillage on runoff processes

Xie et al. studied the effects of tillage measures, such as down-slope tillage, cross-slope tillage and weed clearing in garden on soil and water conservation using field standard runoff plot method and 5 years' monitored data. Results showed that the order of the test plots from superior to inferior in reducing runoff and sediment loss was cross-slope tillage plot (75.33% and 80.57%), down-slope tillage plot (59.56% and 65.11%), weed clearing plot (21.73% and 38.08%) [22]; the runoff from April to September was more than 85% of the annual total runoff, and the sediment loss was more than 90% of the annual sediment loss; interplanting to increase field covering is an effective measure to prevent water loss and soil erosion, and cross-slope tillage is superior to down-slope tillage [27]. In field slope, Huang investigated the impact of different grass-growing methods on soil erosion and found that no matter which method is used, it can effectively decrease surface runoff and soil losses in comparison to bare pure agricultural land [23]. The reduction impact of land coverage and cultivation methods on runoff generation and sediment load as well as flow velocity by grass coverage in orchard were also reported in many other studies through field slope studies and in situ monitoring [24–26]. Xu et al. assessed the impacts of cultivation method, soil thickness, slope and rainfall intensity on interflow at the region covered with purple soil using artificial rainfall-runoff experiment and found that there was a lag between interflow and surface runoff; runoff generation showed single-modal process with slow variation [27]. Based on automatic monitoring of soil moisture, water potentials and runoff on a typical sloping farmland covered with purple soil, Lv [2013] found that the soils on different sections of sloping farmlands had differential water storage ability, following the order: upslope<middle slope<down slope; with the increase of depth, the water content showed less evident change; rainfall characteristics and soil heterogeneity affected soil water content change and soil water movement pattern; deep subsurface flow of a typical rainfall event showed obvious delay between rainfall and runoff; rainfall events of different intensity had different subsurface runoff coefficient (subsurface runoff coefficient was 53.6% to the large

rainfall event while the coefficient was only 1.6% to the small rainfall event); the antecedent soil moisture content also significantly affected the runoff coefficient in the vadose zone [20]. In artificially field slope rainfall-runoff experiment, Ding found that under the same rainfall intensity, the share of interflow in the total precipitation increased with rise of runoff plot's slope, while the percentage of the subsurface flow occupying the total precipitation increased with the decrease of rainfall intensity under the same slope gradient [29]. Wang et al. [2017] investigated the effects of tillage practices and slope on runoff and erosion under simulated rainfall in laboratory plots and found that AD (Artificial Digging), AH (Contour Plow) and CP (Contour Plow) can be adopted as a beneficial summer tillage practice for controlling erosion during summer fallow period because it delayed the time to runoff, decreased runoff and sediment, increased infiltration, which in turn promoted rainfall water and soil conservation [30, 31].

4. Conclusions and outlooks

This study gave a summary of controlled hydrological experiments using rainfall simulator in China, especially in the red soil-covered region of Jiangxi province, including the design and construction of artificial hydrological experiments, studies and outcomes related to the impacts of slope, land use, tillage, rainfall patterns and antecedent soil moisture on runoff generation and sediment export. Results showed that grass covered slope arable land had lower surface runoff and sediment fluxes, higher subsurface runoff, peak flow, and tailing time relative to bare land; the response of subsurface flow was faster to rainfall and surface runoff from lower to upper layer and lag time increased with increasing soil depth; subsurface runoff had close relationship with rainfall amount, rainfall intensity affected peak subsurface runoff substantially, but did not impact starting time and runoff of subsurface flow; rainfall pattern affected subsurface runoff generation considerably including hydrographs and runoff; precedent soil water content had direct influences on runoff generation and runoff of subsurface flow, characterized by decrease of lag time, increase of runoff and peak flow with increase of precedent soil water content. Runoff plot test is a principal method for research of soil and water conservation and also the main approach of runoff sedimentation measurement. The design of runoff plot should obey the principles of improving the accuracy and decreasing the error of tests, saving construction materials and cutting down project costs, reducing the difficulty of construction and enhancing construction quality, as well as being conducive to post observations and lessening operating costs [31]. Although many studies have been conducted worldwide on soil water movement and transformation on field slopes, most of the former researches were focused on surficial soil water movement in indoor artificial rainfall condition with homogeneous initial soil water content and field slope. Some similar studies have been conducted in the field, however, the studies on infiltration of soil water during rainfall period and its redistribution after the infiltration are not sufficient; most studies do not relate rainfall-infiltration, runoff generation at field slopes, and soil water dynamics; spatial heterogeneity of soil property, hysteresis, surface crust, plant interception, and water uptake by roots were not comprehensively considered. In particular, studies on redistribution and export of surface runoff, interflow and groundwater flow are lacking [32].

In the future, hydrological process at field slope requires more investigation from the following perspectives: (1) improvement of monitoring strategies and methodology. Isotopic method provides an efficient way to reveal hydrological process at field slope, which may improve understanding of hydrological regimes; (2) artificial hydrological experiments may contain certain extent of uncertainty during to heterogeneity in rainfall and physiogeographic condition (e.g. soil type, dryness, porosity), design and installation of long-term in situ experimental study to capture variability of soil water movement at different temporal and spatial scales and (3) development and improvement of modelling tool for simulation and prediction of soil water movement under different climatic condition and catchment characteristics, which may assist watershed management for water and soil conservation.

Acknowledgements

This work is also supported by the National Natural Science Foundation of China (41501531), the Natural Science Foundation of Jiangsu Province (BK20151062), the Open Research Fund of State Key Laboratory of Simulation and Regulation of Water Cycle in River Basin (China Institute of Water Resources and Hydropower Research, Grant No: IWHR-SKL-201710) and the Science Foundation of Nanjing Institute of Geography and Limnology, Chinese Academy of Sciences (NIGLAS2014QD07).

Author details

Sanyuan Jiang[1,2]*, Qiande Zhu[2] and Seifeddine Jomaa[3]

*Address all correspondence to: syjiang@niglas.ac.cn

1 Key Laboratory of Watershed Geographic Sciences, Nanjing Institute of Geography and Limnology, Chinese Academy of Sciences, Nanjing, China

2 State Key Laboratory of Hydrology-Water Resources and Hydraulic Engineering, Nanjing Hydraulic Research Institute, Nanjing, China

3 Department of Aquatic Ecosystem Analysis and Management, Helmholtz Centre for Environmental Research – UFZ, Magdeburg, Germany

References

[1] Sidle RC, Hirano T, Gomi T, Terajima T. Hortonian overland flow from Japanese forest plantations – An aberration, the real thing, or something in between? Hydrological Processes. 2007;**21**:3237-3247

[2] Kaufmann V, Pinheiro A, dos Reis Castro NM. Simulating transport of nitrogen and phosphorus in a Cambisol after natural and simulated intensive rainfall. Journal of Contaminant. 2014;**160**:53-64

[3] Luo C-Y, Tu S-H, Pang L-Y, Huang J-J, Lin C-W. Effect of rain intensity on nutrient losses from sloping land of purple soil. Journal of Soil and Water Conservation. 2009;**23**(4):24-27 (in Chinese with English abstract)

[4] Liu R, Wang J, Shi J, Chen Y, Sun C, Zhang P, Shen Z. Runoff characteristics and nutrient loss mechanism from plain farmland under simulated rainfall conditions. Science of the Total Environment. 2014;**468-469**:1069-1077

[5] Ni J-L, He J, Li H-W, Wang S, Lu Z-Y, Qin X-X, et al. Design and calibration of portable rainfall equipment of artificial simulation. Transactions of the Chinese Society of Agricultural Engineering. 2012;**28**(24):78-84 (in Chinese with English abstract)

[6] Iserloh T, Ries JB, Arnáez J, Boix-Fayos C, Butzen V, Cerdà A, et al. European small portable rainfall simulators: A comparison of rainfall characteristics. Catena. 2013;**110**(11): 100-112

[7] Tromp-van Meerveld HJ, Parlange JY, Barry DA, Tromp MF, Sander GC, Walter MT, et al. Influence of sediment settling velocity on mechanistic soil erosion modeling. Water Resources Research. 2008;**44**:W06401. DOI: 10.1029/2007WR006361

[8] Jomaa S, Barry DA, Brovelli A, Sander GC, Parlange JY, Heng BCP, et al. Effect of raindrop splash and transversal width on soil erosion: Laboratory flume experiments and analysis with the Hairsine-rose model. Journal of Hydrology. 2010;**395**:117-132

[9] Viani J.-P. Contribution à l'étude expérimentale de l'érosion hydrique. [Ph.D. thesis]. Lausanne, Switzerland: Ecole Polytechnique Fédérale de Lausanne (EPFL); 1986, 239 p

[10] Baril P. Erodibilité des sols et érodabilité des terres: Application au plateau vaudois. [Ph.D. thesis]. Lausanne, Switzerland: Ecole Polytechnique Fédérale de Lausanne (EPFL); 1991, 218 p

[11] Xie SH, Mo M-H, Tu A-G, Liu Y-Q. Characteristics of vertical runoff output on red-soil slope under natural rainfall condition. Advance in Water Science. 2014;**30**(19):132-138

[12] Wang F, Shen A-L, Chen H-S, Chen J-Z. Experimental study of rainfall-interflow processes on sloping land in red soil hilly region. Journal of Soil and Water Conservation. 2007;**21**(5):15-29 (in Chinese with English abstract)

[13] Yin Z-D, Zuo C-Q, Gao G-X, Peng W-J. Factors affecting interflow of gentle hill in Jiangxi province. Journal of Northwest Forestry University. 2006;**21**(5):1-6 (in Chinese with English abstract)

[14] Liu S-Y, Zuo C-Q, Zhu J-Z. A study on effects of ground cover on dynamics of soil moisture and water balance. Journal of Natural Resources. 2007;**22**(3):424-433 (in Chinese with English abstract)

[15] Cui X-H, Li H-J, Wang B. Water balance of evergreen broad-leaved forest ecosystem in Dagangshan Mountain, Jiangxi province. Scientia Silvae Sinicae. 2006;**42**(2):8-12 (in Chinese with English abstract)

[16] Deng X-W, Kang W-X, Tian D-H, Yan W-D. Runoff changes in Chinese Fir plantations at different age classes, Huitong, Hunan province. Scientia Silvae Sinicae. 2007;**43**(6):1-6 (in Chinese with English abstract)

[17] Xie S-H, Zheng H-J, Yang J, Yu R-G. Effects of runoff reduction through vegetation measures of soil and water conservation in the hilly-land area Southern China. Journal of Soil and Water Conservation. 2010;**24**(3):35-38 (in Chinese with English abstract)

[18] Xie S-H, Zeng J-L, Yang J. Effects of different measures for soil and water conservation in slope land of red soil in South China. Acta Agriculturae Universitatis Jiangxiensis. 2004;**26**(9):624-628 (in Chinese with English abstract)

[19] Xu Q-X, Wang T-W, Li Z-X, Cai C-F, Shi Z-H. Characteristics of interflow in purple soil of hillslope. Advances in Water Science. 2010;**21**(2):229-234 (in Chinese with English abstract)

[20] Liu G-C, Lin S-Y, Liu S-Z. Characteristics of runoff generation and its numerical simulation of surface flow in hilly area with purple soil under conventional tillage systems. Shuili Xuebao. 2002;**12**:101-108 (in Chinese with English abstract)

[21] Zhao X, Huang J, Gao X, Wu P, Wang J. Runoff features of pasture and crop slopes at different rainfall intensities, antecedent moisture contents and gradients on the Chinese loess plateau: A solution of rainfall simulation experiments. Catena. 2014;**119**:90-96

[22] Xie S-H, Zeng J-L, Yang J, Yuan F. Effects of different tillage measures on soil water conservation in slope farmland of red soil in Southern China. Transactions of the Chinese Society of Agricultural Engineering. 2010;**26**(9):81-86 (in Chinese with English abstract)

[23] Huang Y-H, Yang X-Z, Jiang F-S. Effects of different ways of sod in eroded slope orchard on soil and fruit tree growth. Journal of Soil and Water Conservation. 2007;**21**(2):111-114 (in Chinese with English abstract)

[24] Yang J, Mo M-H, Song Y-J, Chen X-A. Hydro-ecological effects of citrus land under vegetation measures of soil and water conservation in red-soil slope. Resources and Environment in the Yangtze Basin. 2012;**21**(8):994-999 (in Chinese with English abstract)

[25] Wang L-L, Yao W-Y, Shen Z-Z, Yang C-X. Effects of grass coverage on shallow flow hydraulic parameters and sediment reduction. Science of Soil and Water Conservation. 2009;**7**(1):80-83 (in Chinese with English abstract)

[26] Li M, Yao W-Y, Chen J-N, Ding W-F. Experimental study on runoff resistance of hilly slope-gullied surface with grass coverage. Shuili Xuebao. 2007;**38**(1):112-119 (in Chinese with English abstract)

[27] Xu P, Wang Y-K, Fu B, Wang D-J, Wang X-T, Wang Y-Q. Interflow occurrence characters and their analysis on slope cropland with purple soil. Bulletin of Soil and Water Conservation. 2006;**26**(6):14-18 (in Chinese with English abstract)

[28] Lv Y-J. Soil hydraulic characteristics and responds of soil water and runoff to rainfall on sloping farmland of purple soil. [dissertation]. 2013. (in Chinese with English abstract)

[29] Ding W-F, Zhang P-C, Wang Y-F. Experimental study on runoff and sediment yield characteristics on purple soil slope. Journal of Yangtze River Scientific Research Institute. 2008;**25**(3):14-17 (in Chinese with English abstract)

[30] Wang L, Dalabay N, Lu P, Wu F. Effects of tillage practices and slope on runoff and erosion of soil from the loess plateau, China, subjected to simulated rainfall. Soil and Tillage Research. 2017;**166**:147-156

[31] Xie S-H, Fang S-W, Wang N. Discussion on design of runoff plot in soil and water conservation experiment. Yangtze River. 2013;**44**(17):83-86 (in Chinese with English abstract)

[32] Xie S-H, Mo M-H, Zhang J. Advances in the research of the law of slope runoff and solute output. Ecology and Environmental Sciences. 2014;**23**(9):1551-1556 (in Chinese with English abstract)

Chapter 9

Fluid Flow, Mass, and Heat Transport Laboratory Experiments in Artificially Fractured Rock

Nicola Pastore

Additional information is available at the end of the chapter

http://dx.doi.org/10.5772/intechopen.74259

Abstract

The knowledge of flow, mass, and heat transport phenomena in fractured rocks represents an important issue in many situations of science and engineering, such as the groundwater resource management, fractured petroleum reservoirs, nuclear waste disposal, as well as geothermal energy development. Experimental data obtained under controlled conditions such as in laboratory increase the knowledge of the fundamental physics that interest the flow and transport in fracture media and allow us to investigate the behavioral differences between fracture and porous medium. An artificially fractured rock sample of parallelepiped shape ($0.60 \times 0.80 \times 0.08$ m^3) has been created, and the flow, mass, and heat transport behavior have been observed. The carried out experiments show the existence of non-Darcian flow regime that cannot be neglected. The latter influences the mass transport behavior giving rise to a delay in mass migration, enhancing the nonequilibrium behavior, whereas the dispersion phenomena seem not be influenced. Heat transport shows a very different behavior compared to mass transport. Convective thermal velocity is lower than mass velocity, whereas thermal dispersion is higher than solute dispersion.

Keywords: laboratory experiments, fluid flow, mass transport, heat transport, fractured media

1. Introduction

In the study of hydrodynamic processes in fractured media, flow is described with a linear relationship between pressure gradient and flow rate [1], valid at low flow regime (Re < 1). For Re > 1, a nonlinear flow can be observed.

In fracture networks, heterogeneity intervenes even in mass and heat transport: due to the variable aperture and heterogeneities of the fracture surfaces, the fluid flow will seek out preferential paths through which mass and heat are transported.

The study of mass and heat transport behavior is based on dual porosity theory. Fractured medium is conceptualized in two domains: connected fractured network where transport processes is governed mainly by advection and rock matrix as well as poorly connected fractures which represent stagnant zones.

The present chapter reports several laboratory experiences at bench scale and their interpretation of flow and transport behavior on artificially fractured rock sample of parallelepiped shape. The experimental results of flow and transport tests in a fractured block at bench scale are interpreted by means of explicit network model (ENM) where the fracture network geometry has been taken into accounts.

2. Artificial fractured rock sample preparation and characterization

A limestone block of parallelepiped shape ($0.6 \times 0.4 \times 0.08\ m^3$) has been recovered from the "Calcare di Altamura" formation located in the Apulian region in Southern Italy. Using a 5-kg mallet blow, a fractured network has been made artificially. The fissured system and their aperture have been recorded with a high-resolution digital camera. The *Perspective Rectifier* (www.rectifiersoft.com) has been used in order to scale and rectify the picture. The fracture traces and the aperture measurement have been extracted from recorded images through the built-in *Scilab Image Processing Toolbox* (www.scilab.org) [2] (**Figure 1**).

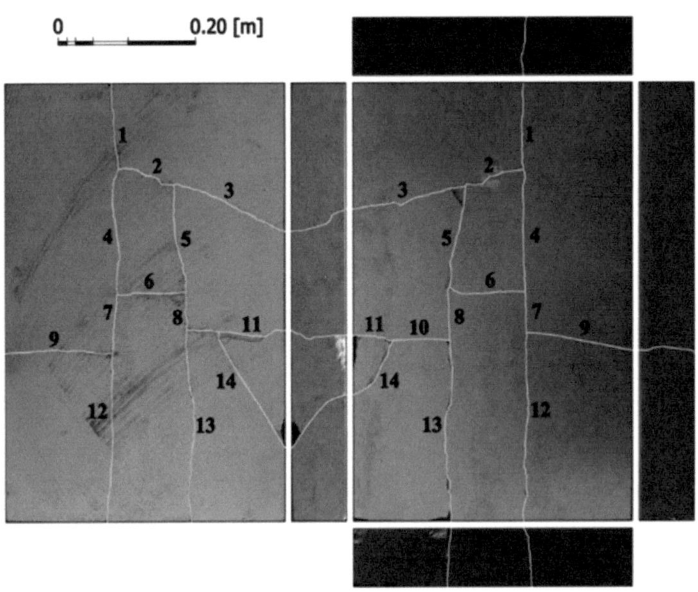

Fracture	Aperture (mm)	Roughness (mm)
1	0.811÷2.907	0.180
2	0.541÷1.977	0.187
3	0.573÷2.784	0.187
4	0.271÷3.900	0.101
5	0.811÷2.194	0.181
6	0.272÷1.794	0.141
7	0.540÷1.510	0.010
8	0.540÷3.097	0.150
9	0.540÷3.577	0.450
10	0.810÷1.300	0.063
11	0.270÷2.901	0.152
12	0.544÷1.912	0.135
13	0.541÷1.9097	0.096
14	0.344÷3.050	0.222

Figure 1. Rectified images with characterization of artificial fractures.

http://dx.doi.org/10.5772/intechopen.74259

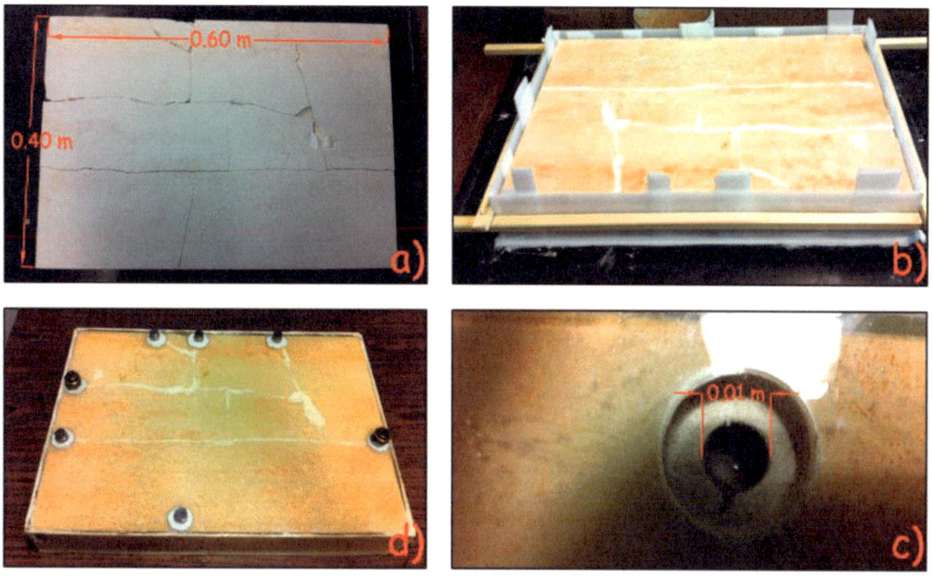

Figure 2. (a) Artificially fractured rock, (b) preparation for epoxy resin casting, (c) particular of the hole at the end of the artificial fracture, and (d) sealed fractured block with the hexagonal bushing for the connection to the hydraulic circuit (from [3]).

Once the artificially fractured block sample has been characterized, the surface of the block sample has been sealed with the epoxy resin, and a hole of 1 cm of diameter has been opened on the boundary of the block sample in correspondence of each discontinuity. Finally, an extruded polystyrene panel with thermal conductivity equal to 0.034 Wm^{-1} K^{-1} and thickness of 0.05 m has been used to thermally insulate the fractured block sample (**Figure 2**).

3. Experimental setup flow, mass, and heat transport test

The sealed and thermally insulated fractured block sample is connected with a hydraulic circuit. The sketch of the experimental apparatus is shown in **Figure 3**. Water moves from the upstream tank to the downstream tank and returns to the upstream by means of a pump. An electric boiler with a volume of 10^{-2} m^3 has been used to heat the water. The instantaneous flow rate that flows across the block is measured by an ultrasonic velocimeter. Beside the inlet port, a syringe for instantaneous injection of a conservative tracer (NaCl) has been placed, while at the outlet port, there is a flow cell in which a multiparametric instrument can be positioned. Two thermocouples have been placed at the inlet and the outlet of a selected fracture path of the limestone block. A TC-08 thermocouple data logger (Pico Technology) with a sampling rate of 1 s has been connected to the thermocouples.

The study of flow, mass and heat transport regarding the active path highlighted in **Figure 3**. **Figure 4a** shows the mechanical aperture distribution obtained from 13,688 measurements and **Figure 3b** shows the reconstructed three-dimensional geometry of the selected path. The

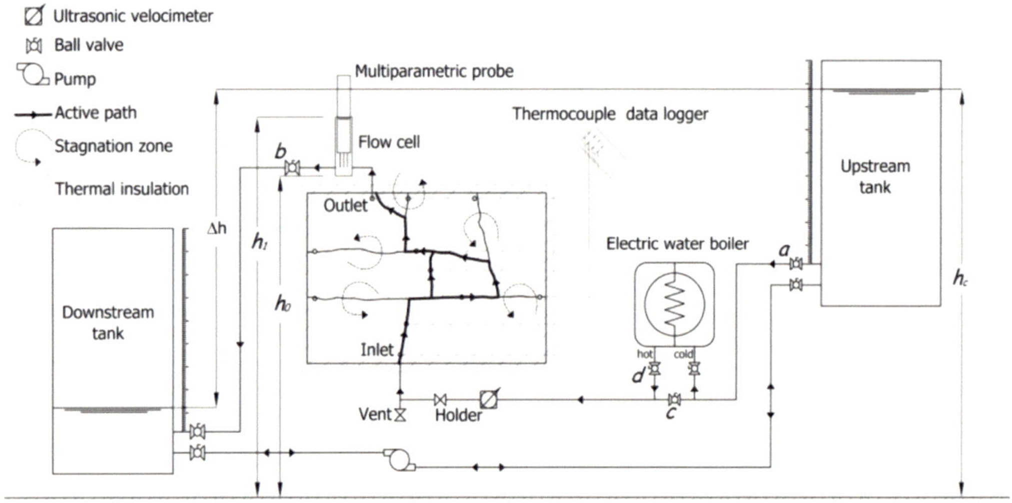

Figure 3. Schematic diagram of experimental setup (Cherubini et al. [9]).

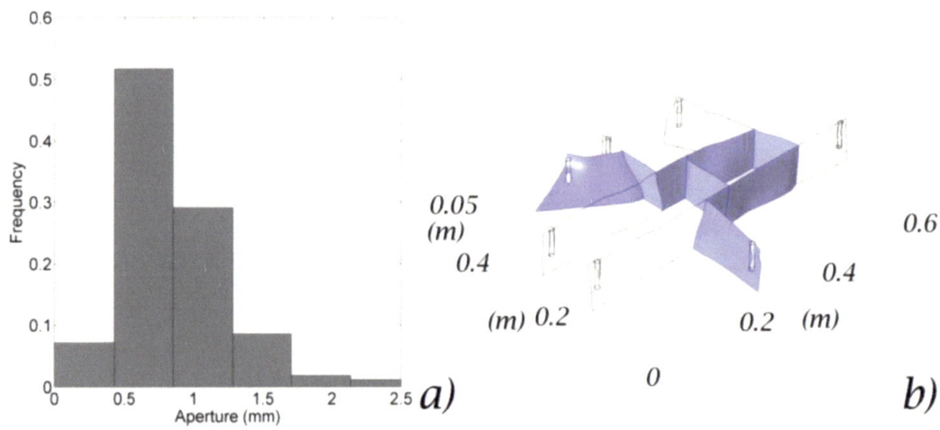

Figure 4. (a) Distribution of mechanical aperture evaluated on the 13,688 samples and (b) three-dimensional reconstruction of the fracture network. The selected path is highlighted (from Cherubini et al. [3]).

average cross-sectional area of the path is equal to 993 μm whereas the average path length is equal to 0.7531 m [3].

4. Flow test

The air in the hydraulic circuit and the fractured media has been removed bringing the system to full saturation with the hydrostatic head equal to h_0. Opening the valve "*a*," water flows in the system, and the snapshot flow rates are measured by the ultrasonic velocimeters. When the hydraulic head in the flow cell reaches h_1, the valve "*a*" is closed, and opening the valve "*b*," the hydraulic head in the flow cell is reported to the value h_0.

The volumetric method can be used in order to estimate the injection flow rate that flows within the fractured block:

$$Q_0 = \frac{S_1}{\Delta t}(h_1 - h_0) \tag{1}$$

where S_1 (L) is the cross-sectional area of the flow cell and Δt is the time required to fill the flow cell from h_0 to h_1. The average flow rate estimated was compared with the snapshot flow rate measured through the ultrasonic velocimeter in order to check the absence of the leaks and losses due to obstruction in the hydraulic system. The experiment is repeated changing the hydraulic head h_c (L) of the upstream tank.

Given that the capacity of the upstream tank S_2 (L) is higher than S_1 and the compressibility of the block and the hydraulic circuit is very low, the flow dynamics can be described through the following expressions:

$$S_1 \frac{dh}{dt} = \kappa(\Delta h)(h_c - h_0) \tag{2}$$

where h (L) is the hydraulic head of the downstream flow cell and $\kappa(\Delta h)$ is the hydraulic conductance term representative of both the hydraulic circuit and the active fracture network configuration.

For the fractured media, a Δh-Q relationship can be represented through the following polynomial expression:

$$\Delta h = AQ + BQ^2 \tag{3}$$

where A (TL^{-2}) and B ($T^2 L^{-5}$) are the linear and nonlinear hydraulic loss coefficients, respectively, and are related to the roughness, aperture, lengths, and shape of fractures.

The hydraulic head differences Δh are given by

$$\Delta h = h_c - \frac{h_0 + h_1}{2} \tag{4}$$

The hydraulic conductance term $\kappa(\Delta h)$ ($L^2 T^{-1}$) representative of the whole hydraulic system assumes the following expression:

$$\kappa(\Delta h) = \frac{2}{A + \sqrt{A^2 + 4B|\Delta h|}} \tag{5}$$

The inverse of $\kappa(\Delta h)$ represents the average resistance to flow $R(Q)$ (TL^{-2}).

Substituting Eq. (5) in Eq. (2) and integrating the latter from $t = t_0$ to $t = t_1$ with the initial condition $h = h_0$, the following equation is obtained:

$$S_1\left(-\sqrt{A^2+4B(h_c-h)}\right)-A\ln\left(\sqrt{A^2+4B(h_c-h)}-A\right)\Bigg|_{h=h_0}^{h=h_1}=t_1-t_0 \quad (6)$$

Then, fitting experimental relationship between the time $\Delta t = t_1 - t_0$, an estimate of parameters A and B can be made.

5. Mass transport test

Tracer tests by means of sodium chloride have been used in order to study the mass transport dynamics through the selected path. First, a steady-state flow condition has been carried out imposing a hydraulic head difference between the upstream tank and downstream tank. Using a syringe, a mass of sodium chloride of 5×10^{-4} kg is injected into the inlet port. The pulse injection assumption can be considered because the source release time is very small (1 s).

The multiparametric probe positioned in the flow cell measures the breakthrough curve (BTC) of the injected sodium chloride and the hydraulic head, in the meanwhile the flow rate is measured using the ultrasonic velocimeter. For different flow rates, a BTC can be registered at the outlet port.

6. Heat transport test

The study of heat transport dynamics in the selected path takes place by injecting a volume of 10 l of hot water into the block sample. At time t = 0 s, the hydrostatic heads in the block sample and in the downstream tank are equal. At time t = 10 s, valve "*a*" is opened and water flows into the block sample. At time t = 60 s, the valve "*d*" is opened, and in the same time the valve "*c*" is closed. In such a way, the hot water enters in the block sample, and the step temperature function at the inlet port is measured using the first thermocouple. The second thermocouple placed inside the outlet port measures the thermal BTC. Analogous manner to the solute transport of the experiment is repeated varying the flow rate.

7. Calibration of the experimental apparatus

A calibration procedure is needed in order to eliminate the effect of the experimental apparatus on the measure of the state variables.

As far as the flow experiment, the hydraulic circuit gives rise to a hydraulic loss that must be estimated. Hydraulic losses due to only the hydraulic circuit can be expressed according the Chezy law:

http://dx.doi.org/10.5772/intechopen.74259

$$Q = C\sqrt{|\Delta h|} \Rightarrow \Delta h = \frac{1}{C^2}Q^2 \tag{7}$$

where C ($L^{5/2}T$) is the characteristic coefficient related to the roughness, section, and length of the tubes of the hydraulic circuit. Then, C^{-2} is added to the parameter B in Eq. 6.

The parameter C is estimated conducting several hydraulic tests only on the hydraulic circuit in the range 0.14–0.93 m, corresponding to the average flow rate in the range of 1.77×10^{-5}–6.80×10^{-5} m^3 s^{-1}. The relationship h_c-Δt has been fitted using Eq. (6) with the parameters A and B equal to 0. The coefficient C result equals to 0.14×10^5 m$^{5/2}$ s^{-2}.

In order to overcome the problem linked with the fact that the residence time of the solute in the probe and in the fracture network are the same order of magnitude, convolution technique has been used.

The BTC recorded by the probe $w(t)$ is represented by the convolution product between the BTC of the fracture network $c(t)$ and the BTC of the probe $s(t)$:

$$w(t) = c(t)*s(t) = \int_0^t c(t-\tau)s(\tau)d\tau \tag{8}$$

The tracer injection device has been directly with the flow cell in which the multiparametric prove is positioned. Several tracer tests have been conducted on this configuration varying the input flow rate in the range 3.53×10^{-6} m3 s^{-1} to 5.32×10^{-6} m^3 s^{-1}. The observed BTCs show an exponential decay function like

$$s(t) = c_0 \exp\left(-\frac{Q}{Vol}t\right) \tag{9}$$

where Vol (L^3) is the volume of flow cell and $c_0 = M_0/Vol$ (ML^{-3}) is the concentration observed at $t = 0$. The observed BTCs have been fitted using Eq. (9), and the volume of the flow cell Vol has been estimated resulting equally to 1.237×10^{-4} m^3 closing the real volume of flow cell equal to 1.417×10^{-4} m^3.

For the heat transport tests, no correction is required; thermocouples at the inlet port and outlet port measure directly the thermal BTCs of the media.

8. Explicit network model

The 2D explicit network model considers the fracture network geometry permitting a more accurate estimation of flow and transport dynamics. The single fractures are represented as one-dimensional pipe elements forming a 2D pipe network [4].

Assuming that a single fracture (SF) j can be represented by a one-dimensional pipe element, the relationship between the head loss Δh_j (L) and the flow rate Q_j (L^3 T^{-1}) can be written in finite terms on the basis of Forchheimer model [5]:

$$\frac{\Delta h_j}{l_j} = aQ_j + bQ_j^2 \Rightarrow \Delta h_j = \left[l_j\left(a + bQ_j\right)\right]Q_j \tag{10}$$

where l_j (L) represents the length of fracture and a (TL^{-3}) and b ($T^2\ L^{-6}$) represent the Forchheimer parameters in finite terms. The term in the square brackets represents the resistance to flow $R_j(Q_j)$ (TL^{-2}) of fracture j.

The flow field can be obtained in a simple way using Kirchhoff's laws. The flow rate Q_j crossing the generic fracture j along a parallel circuit is as follows:

$$Q_j = \Sigma Q\left[\frac{1}{R_j}\left(\sum_{i=1}^{n}\frac{1}{R_i}\right)^{-1}\right] \tag{11}$$

where ΣQ (LT^{-3}) is the sum of the discharge flow evaluated for the fracture intersection located at the inlet bond of j fracture.

Regarding the mass and transport phenomena, once known as the flow field in the fracture network, the probability density function (PDF) of the residence time at the generic node can be obtained as the sum of the PDFs of each elementary path that reach the node. The latter can be expressed as the convolution product of PDF of each individual fracture belonging to the elementary path. Using the convolution theorem, the PDF at generic node $\Gamma\ (t)$ in the network can be represented as follows:

$$\Gamma(t) = L^{-1}\left[\sum_{i=1}^{N_{ep}}\prod_{j=1}^{n_{f,i}} P_j\overline{\Gamma}_j(s)\right] \tag{12}$$

where L is the Laplace transform operator, N_{ep} is the number of elementary paths, $n_{f,j}$ is the number of the fractures in i^{th} elementary path, $\overline{\Gamma}_j(s)$ represents the PDF of the jth SF in the Laplace space, s is the integral variable of the Laplace transform, and P_j is the probability of the particle transition at the inlet bond of each individual SF. The rules for particle transition through the fracture intersections play an important role in mass and heat transport. The simplest rule is represented by the "perfect mixing model" in which the mass sharing is proportional to the relative discharge flow rates. Based on the perfect mixing model, the probability of the particle transition crossing the SF is

$$P_j = \frac{Q_j}{\Sigma Q} \tag{13}$$

The term in the square brackets in Eq. (4) corresponds to P_j.

The PDF represents the density distribution of the time that the mass or heat spends in the fracture network.

In order to know $\overline{\Gamma}_j(s)$, a transport model can be assumed and consequently the transport parameters of each *SF*.

Ref. [6] presented an analytical solution for solute transport in a semi-infinite single fracture embedded in a porous rock matrix. Given that the governing equation of heat and mass transport highlights similarities between the two processes, the analytical solution for solute transport can be used also for heat transport.

According of Tang's solution, the PDF in Laplace space can be expressed as

$$\overline{\Gamma}(s) = \exp(vL)\exp\left[-vL\left\{1+\beta^2\left(\frac{s^{1/2}}{A}+s\right)\right\}^{1/2}\right] \tag{14}$$

In terms of mass transport, the coefficients v, A, and β^2 are expressed as follows:

$$v = \frac{u_f}{2D_f}; A = \frac{\delta}{\sqrt{\theta_m D_e}}; \beta^2 = \frac{4D_f}{u_f^2} \tag{15}$$

where u_f (LT^{-1}) is the convective velocity, D_f ($L^2\ T^{-1}$) is the dispersion, θ_m is the matrix porosity, D_e ($L^2\ T^{-1}$) represents the effective molecular diffusion, and δ (L) is the thickness of the boundary layer. For small fractures, δ became the aperture w_f (L) of the SF.

In terms of heat transport, the coefficient of the PDF is

$$v = \frac{u_f}{2D_f}; A = \frac{\delta}{\sqrt{\theta D_e}}; \beta^2 = \frac{4D_f}{u_f^2} \tag{16}$$

where $\theta = \boldsymbol{\rho}_m c_m / \boldsymbol{\rho}_w c_w$ and $D_e = k_e / (\boldsymbol{\rho}_w c_w)$. ρ_m (ML^{-3}), c_m ($ML^{-3}T^{-3}K^{-1}$) and ρ_w (ML^{-3}), c_w ($ML^{-3}T^{-3}K^{-1}$) represent the density and the specific heat capacity of the matrix and water, respectively.

Definitely, the BTC describing for the mass and heat transport in the fracture network as function of time at the generic node, using the convolution theorem, can be obtained as follows:

$$c(t) = c_0 + c_{inj}(t) * L^{-1}\left[\sum_{i=1}^{N_{ep}} \prod_{j=1}^{n_{f,i}} P_j \overline{\Gamma}_j(s)\right] \tag{17}$$

where c_0 (ML^{-3}) is the initial concentration, c_{inj} (ML^{-3}) is the injection function, and * is the convolution operator. Abate et al. (2006) algorithm is used to perform the inverse Laplace transform.

For heat transport dynamics, the temperature BTC at the generic node of the fracture network is

$$T(t) = T_0 + T_{inj}(t) * L^{-1}\left[\sum_{i=1}^{N_{ep}} \prod_{j=1}^{n_{f,i}} P_j \overline{\Gamma}_j(s)\right] \tag{18}$$

where T_0 (K) is the initial temperature and T_{inj} (K) is the temperature injection function.

Three characteristic time scales can be defined as

$$t_u = \frac{L}{u_f}; t_d = \frac{L^2}{D_f}; t_e = \frac{\delta^2}{D_e} \tag{19}$$

where L (L) represents the characteristic length and t_u (T), t_d (T), and t_e (T) are the characteristic time scales of convective transport, dispersive transport, and loss of the mass or heat into the surrounding matrix [7].

Peclet number (Pe) can be defined as

$$Pe = \frac{t_d}{t_u} = \frac{u_f L}{D_f} \tag{20}$$

Transport processes are convectively dominated when Peclet number is high; on the contrary, when Peclet number is low, they are dominated by dispersion.

In order to investigate the influence of diffusion in the matrix on convective phenomena, the Damkohler number can be used. Da can be defined as

$$Da = \frac{t_u}{t_e} = \frac{\alpha L}{u_f} \tag{21}$$

where α (T^{-1}) is the exchange rate coefficient corresponding to

$$\alpha = \frac{D_e}{\delta^2} \tag{22}$$

The exchange rate coefficient α is equivalent to the inverse of t_e. When t_e reaches t_u (Da = 1), concentration and temperature distribution profiles are subject to dual porosity effect characterized by a long tail.

When $t_e \gg t_u$ $(Da \ll 1)$, the fracture-matrix exchange is very slow, and it does not influence mass or heat propagation. On the contrary, when $t_e \ll t_u$ $(Da \gg 1)$, the fracture-matrix exchange is rapid, there is an instantaneous equilibrium between fracture and matrix, and they have the same concentration or temperature. These two circumstances close the standard advective-dispersive transport equation.

9. Experimental results

9.1. Flow

Different flow tests that have been carried out vary the control head h_c in the range of 0.17–1.37 m observing the average flow rates in the range of 1.85e-6–1.11e-5 m^3/s. The linear and nonlinear terms are equal to A = 4.11e4 and B = 6.61e9, respectively. In **Figure 5** the fitting results are reported.

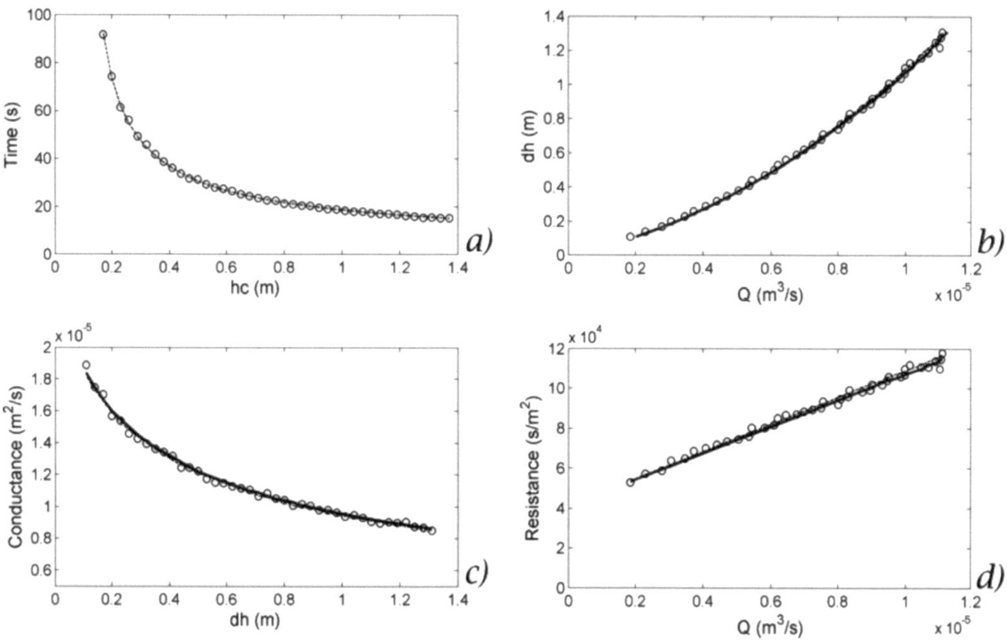

Figure 5. Experimental results obtained for the hydraulic test performed on the selected path. (a) Control head h_c vs. time. (b) Average flow rate Q vs. difference head. (c) Difference head vs. conductance term. (d) Average flow rate Q vs. resistance term evaluated as the inverse of conductance. The circle represents the experimental values, the dashed line represents the fitting of experimental values, and the marked line represents the functions without the effect of circuit (from Cherubini et al. [3]).

Once the linear and nonlinear terms are known, the value of the resistance to flow can be determined as function of the injection flow rate.

The flow field in each single fracture can be estimated in analytical way through the application of Kirchhoff's laws. In **Figure 6** a sketch of the 2D pipe conceptualization of the fracture network is reported.

Constant Forchheimer parameter for the whole fracture network has been assumed. They can be derived matching the experimental resistance to flow evaluated experimentally as the inverse of the conductance Eq. (5) with the resistance to flow evaluated as

$$\begin{aligned}\overline{R}(Q) = R_1(Q_0) + R_2(Q_0) + \left(\frac{1}{R_6(Q_1)} + \frac{1}{R_3(Q_2)+R_4(Q_2)+R_5(Q_2)}\right)^{-1} \\ +R_7(Q_0) + R_8(Q_0) + R_9(Q_0)\end{aligned} \tag{23}$$

with Q_1 evaluated using the following iterative equation

$$Q_1^{k+1} = Q_0\left[\frac{R_3(Q_0 - Q_1^k) + R_4(Q_0 - Q_1^k) + R_5(Q_0 - Q_1^k)}{R_3(Q_0 - Q_1^k) + R_4(Q_0 - Q_1^k) + R_5(Q_0 - Q_1^k) + R_6(Q_1^k)}\right] \tag{24}$$

and Q_2 evaluated as

$$Q_2 = Q_0 - Q_1 \tag{25}$$

The Forchheimer parameters have been estimated, and they are, respectively, equal to $a = 7.345 \times 10^4$ sm^{-3} and $b = 11.65 \times 10^9$ s^2 m^{-6}.

In order to analyze the experimental results, two dimensionless numbers can be evaluated: the Reynolds number and the Forchheimer number.

Reynolds number (Re) is defined as the ratio of inertial forces to viscous forces:

$$\mathrm{Re} = \frac{\rho \overline{v} D_h}{\mu} \tag{26}$$

where $\overline{v}$ represents the average velocity evaluated on the active path and D_h (L) is the hydraulic diameter. For the fracture having small aperture with respect to its height, the hydraulic diameter is equal to $D_h = w_f/2$. Re can be reformulated as

$$\mathrm{Re} = \frac{\rho Q}{2d\mu} \tag{27}$$

where d is the thickness of the fractured block.

Forchheimer number (F_o) represents the ratio of nonlinear to linear hydraulic gradient contribution:

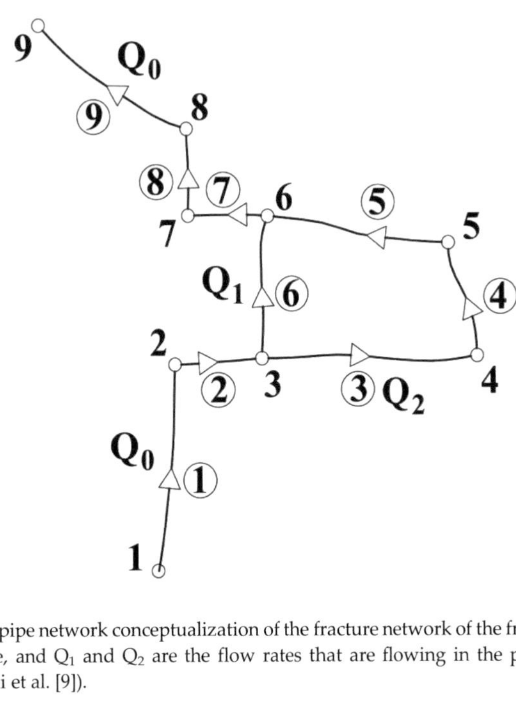

Figure 6. Two-dimensional pipe network conceptualization of the fracture network of the fractured rock block in **Figure 1**. Q_0 is the injection flow rate, and Q_1 and Q_2 are the flow rates that are flowing in the parallel branches 6 and 3–4–5, respectively (from Cherubini et al. [9]).

$$\mathrm{Fo} = \frac{bQ}{a} \tag{28}$$

The Forchheimer number can be used for evaluate non-Darcian flow. Inertial effects dominate over viscous effects at the critical Forchheimer number ($F_o > 1$) [8].

Reynolds number indicates when microscopic inertial effects become important. It is inappropriate on the macroscopic level because microscopic inertial effects do not directly lead to macroscopic inertial effects. High microscopic Reynolds number does not necessarily imply non-Darcian flow. Forchheimer number takes into account both velocity and structure of the medium because the nonlinear term is structure dependent. The linear term inherently contains information on the tortuosity of the flow paths that leads to changes in the microscopic inertial terms. In fact, if the structure of the medium is such that microscopic inertial effects are rare, then the nonlinear term will be small, and the Forchheimer number will remain small until the Reynolds number is large. Instead, both the nonlinear term and the Forchheimer number will be large if the structure of the medium is such that microscopic inertial effects can be expected.

For the range of injected flow rate, investigated Re is in the range 11.56–69.37, and Fo are in the range 0.29–1.76. Inertial forces dominate the viscous ones when Fo = 1 corresponding to a flow rate equals to $Q_{crit} = 6.30 \times 10^{-6}\ \mathrm{m^3\ s^{-1}}$.

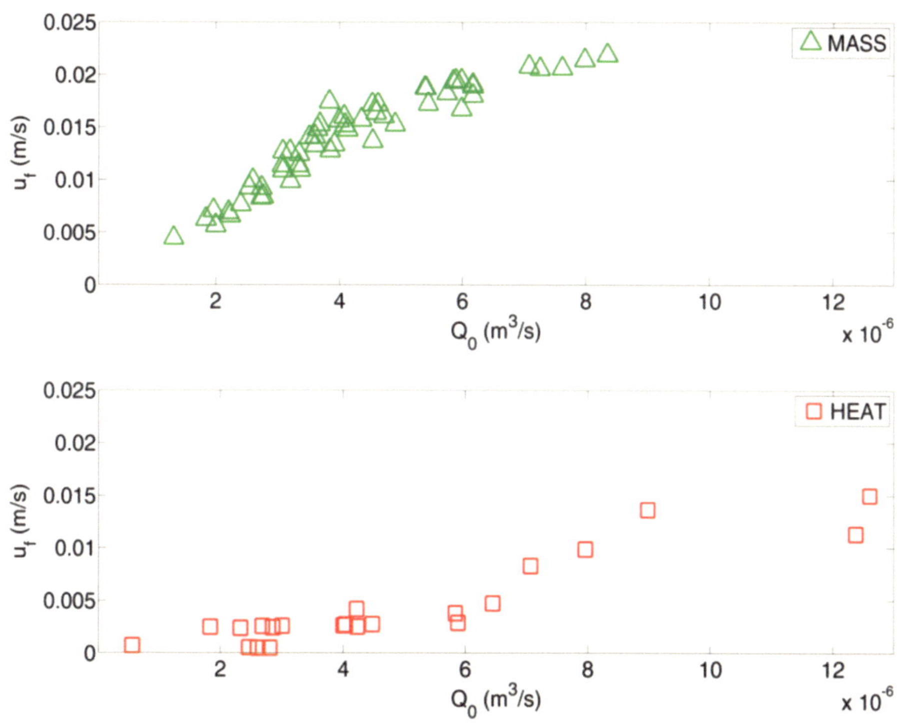

Figure 7. Velocity u_f ($\mathrm{m \cdot s^{-1}}$) as a function of the injection flow rate Q_0 ($\mathrm{m^3 s^{-1}}$) for ENM with Tang's solution for both mass transport and heat transport (from Cherubini [9]).

The term in square bracket in Eq. (26) is equal to the probability of the particle transition evaluated for the branch 6 as function of Q_0. When the flow injection increases, the probability of the particle transition decreases. This means that when the injection flow rate increases, the resistance to flow of branch 6 increases faster than the resistance to flow of the branches 3–5.

9.2. Transport

The behavior of mass and heat transport has been compared varying the injection flow rates. In particular 55 tests in the range 1.32×10^{-6}– 8.34×10^{-6} $m^3\,s^{-1}$ (Re in the range 8.2–52.1) and 21 tests in the range 1.83×10^{-6}– 1.26×10^{-5} $m^3\,s^{-1}$ (Re in the range 17.5–78.71) for heat transport have been conducted [9].

The observed heat and mass of BTCs for different flow rates have been individually fitted using the ENM approach. For simplicity the transport parameters u_f, D_f, and α are assumed equal for all branches of the fracture network.

The experimental BTCs are fitted using Eq. (19) and Eq. (20) for mass and heat transport, respectively.

The determination coefficient (r^2) and the root-mean-square error (*RMSE*) have been used in order to evaluate the goodness of fit.

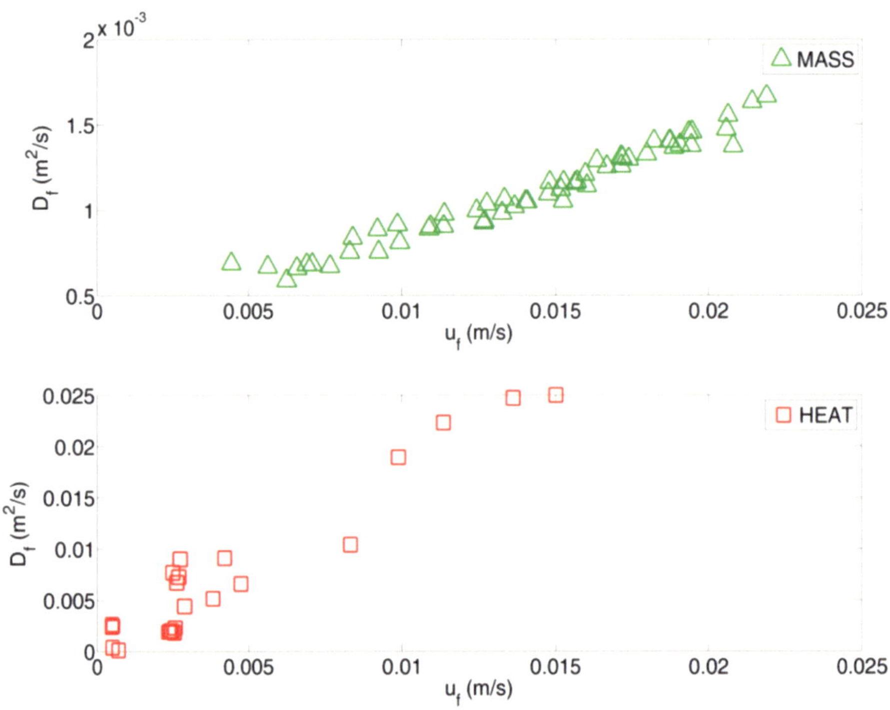

Figure 8. Dispersion D_f ($m{\cdot}s^{-2}$) as a function of velocity u_f ($m{\cdot}s^{-1}$) for ENM with Tang's solution for both mass transport and heat transport (from Cherubini [9]).

The results highlight that the estimated convective velocities u_f for heat transport are lower than for mass transport, whereas the estimated dispersion D_f for heat transport is higher than for mass transport.

A different behavior has been observed for the transfer rate coefficient for mass and heat transport. For mass transport, it can be neglected relatively to the convective velocity, whereas for heat transport, the transfer rate coefficient reaches the convective velocity, and having a characteristic length equal to $L = 0.601$ m, a dual porosity effect is evident.

In **Figure 7** the relationship between u_f and Q_0 is reported. Regarding mass transport, experiments for values of Q_0 higher than 4×10^{-6} m^3 s^{-1} u_f increase less rapidly. This behavior was due to the presence of inertial forces that gave rise to a retardation effect on solute transport.

A very different behavior is observed for heat transport. Heat convective velocity does not seem to be influenced by the presence of the inertial force, whereas u_f is influenced by fracture-matrix exchange phenomena resulting in a significant retardation effect.

Figure 8 reported the dispersion coefficient D_f. Linear relationship between u_f and D_f suggests that inertial forces did not exert any effect on dispersion and that geometrical dispersion dominates the Aris-Taylor dispersion. Even if heat convective velocity is lower than solute advective velocity, the longitudinal thermal dispersivity assumes higher values than the longitudinal solute dispersivity.

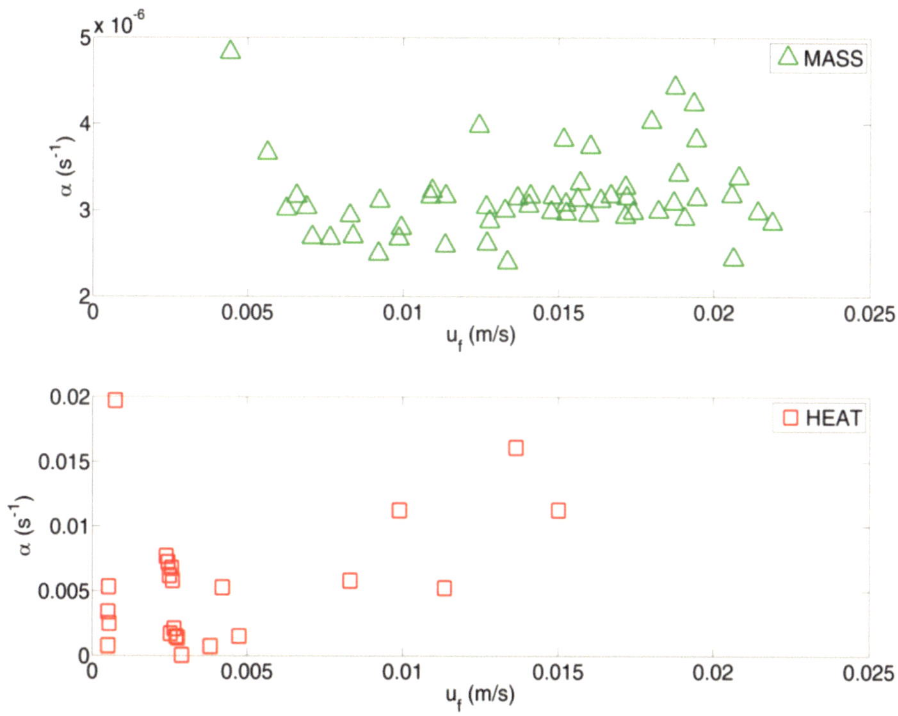

Figure 9. Transfer coefficient α (s^{-1}) as function of velocity u_f (m·s^{-1}) for both mass transport and heat transport (from Cherubini et al. [9]).

Figure 9 shows the exchange rate coefficient α as function of the convective velocity u_f for both mass transport and heat transport. Regarding the mass transport, the fracture-matrix interaction does not exert any influence. The observed non-Fickian behavior is attributable to the presence of the inertial force and secondary path.

Figure 10 shows for mass and heat transport the relationship between Pe and Q_0. For mass transport Pe increases as Q_0 increases, reaching a constant values $Pe = 7.5$. For heat transport Pe has a constant trend lower than the unit.

Figure 11 shows for mass and heat transport the relationship between Da and Q_0. For mass transport Da presents a very low value. The Fickian model can be assumed to represent mass transport in each single fracture.

For heat transport Da reaches the unit. It presents a downward trend as Q_0 increases switching from higher to lower values than unit.

Even if Da presents a downward trend as Q_0 increases, when the latter exceeds Q_{crit}, a weak growth trend for Da is detected that however assumes values lower than the unit.

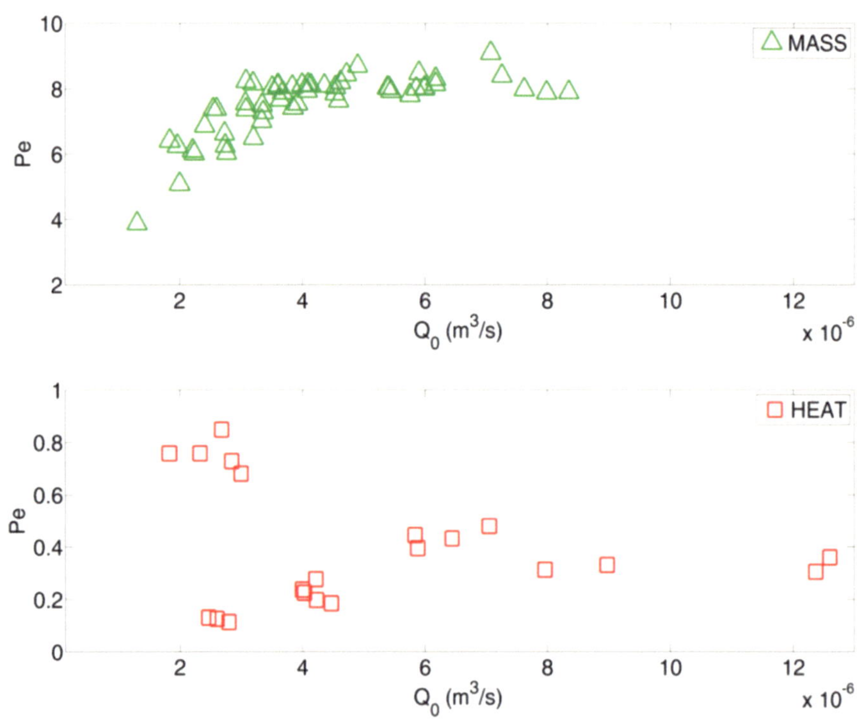

Figure 10. Peclet number as function of injection flow rate Q_0 ($m^3 s^{-1}$) for both mass transport and heat transport (from Cherubini et al. [9]).

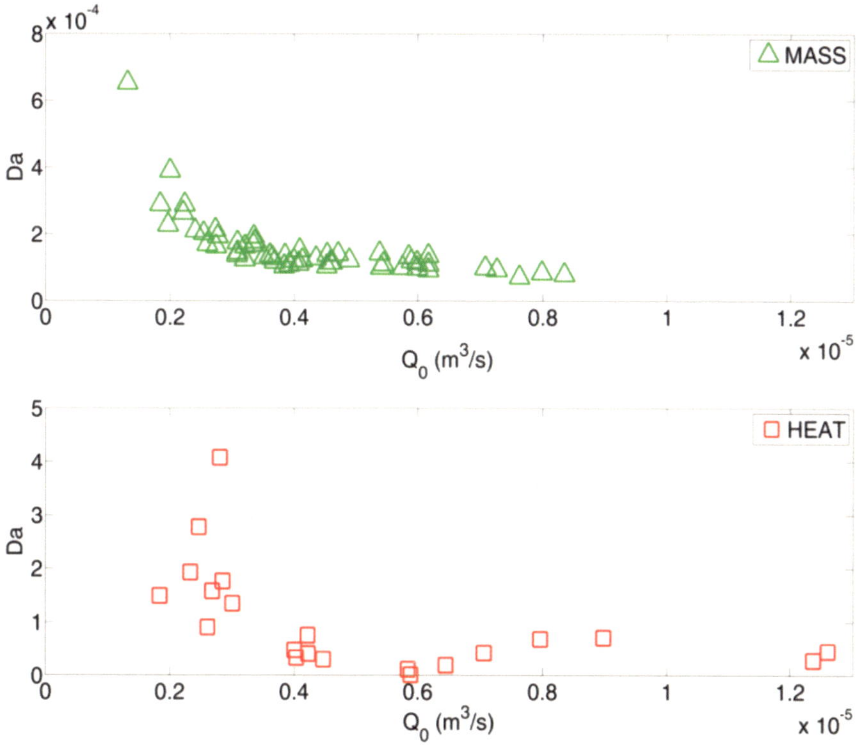

Figure 11. Da number as function of injection flow rate Q_0 ($m^3\ s^{-1}$) for both mass transport and heat transport (from Cherubini et al. [9]).

10. Conclusion

Laboratory experiments on the observation of flow, mass, and heat transport in a fractured rock sample have been carried out. The parameters that control flow, mass, and heat transport have been estimated using the ENM model.

The explicit network model is a useful technique to describe flow, mass, and heat dynamics in fractured media, reducing 2D and/or 3D fracture geometry to a network of 1D pipe elements. However, the real case study can be impracticable because the ENM model approach needs the full knowledge of the fracture network. In order to overcome these difficulties, the ENM can be coupled with equivalent porous media model to represent the bigger fracture which represents the main pathways for flow, mass, and heat dynamics.

Regarding the flow process, the experiments highlighted the dependency between the hydraulic conductivity and the specific discharge in fractured media. Lous and Maini (1970) and Elsworth and Doe (1986) found an underestimation errors of several orders of magnitude in the estimation of transmissivity using non-Darcian constant head data derived by packer tests in fractured rock. The experiments show an underestimated error of 46.59% (average value) of the Darcian flow hydraulic transmissivity.

The flow experiments demonstrate that the flow behavior in the fracture network can be described using the Forchheimer law. Due to the presence of nonlinear flow regime, the probabilities of water distribution between the main and secondary paths are the function of the injection flow rate decreasing as injection flow rate increases.

Several dissimilarities have been detected between heat transport and mass transport. The main discrepancies regarding the transport parameters are as follows: convective solute velocity is higher than heat velocity, whereas solute dispersion is higher than heat dispersion, and the solute exchange rate is negligible compared with the solute velocity, whereas for heat transport, the exchange rate is comparable with the heat velocity giving rise to a very strong dual-porosity effect which delays the heat propagation. On the contrary for the solute transport, the non-Fickian behavior observed in solute BTCs is attributable to the presence of the secondary path combined with the nonlinear flow regime.

Author details

Nicola Pastore

Address all correspondence to: nicola.pastore@poliba.it

DICATECh, Department of Civil, Environmental, Building Engineering and Chemistry, Politecnico di Bari, Bari, Italy

References

[1] Cherubini C, Pastore N. Modeling contaminant propagation in a fractured and karstic aquifer. Fresenius Environmental Bulletin. 2010;**19**(9):1788-1794

[2] Cherubini C, Giasi CI, Pastore N. Bench scale laboratory tests to analyze non-linear flow in fractured media. Hydrology and Earth System Sciences. 2012;**16**:2511-2622

[3] Cherubini C, Giasi CI, Pastore N. Evidence of non-Darcy flow and non-Fickian transport in fractured media at laboratory scale. Hydrology and Earth System Sciences. 2013;**17**:2599-2611

[4] Cherubini C, Giasi CI, Pastore N. On the reliability of analytical models to predict solute transport in a fracture network. Hydrology and Earth System Sciences. 2014;**18**:2359-2374

[5] Forchheimer P. Wasserbewegung durch Boden. Zeitschrift Verein Deutscher Ingenieure. 1901;**45**:1781-1788

[6] Tang DH, Frind EO, Sudicky EA. Contaminant transport in fractured porous media: Analytical solutions for a single fractures. Water Resources Research. 1981;**17**(3):555-564

[7] Pastore N, Cherubini C, Giasi CI, Allegretti NM, Redondo JM, Tarquis AM. Experimental study of heat transport in fractured network. Energy Procedia. 2015;**76**:273-281

[8] Ruth D, Ma H. On the Derivation of the Forchheimer Equation by Means of the Averaging Theorem Transport in Porous Media. Kluwer Academic Publishers. Printed in the Netherlands; 1992

[9] Cherubini C, Pastore N, Giasi CI, Allegretti NM. Laboratory experimental investigation of heat transport of fractured media. Nonlinear Processes Geophysics. 2017;**24**:23-42

Chapter 10

Experimental Variant Slope Soil Tank for Measurements of Runoff and Soil Erosion

Lihu Yang, Simin Qu, Yifan Wang and
Xianfang Song

Additional information is available at the end of the chapter

http://dx.doi.org/10.5772/intechopen.76167

Abstract

Rainfall-runoff processes and the related soil erosion are pivotal research regions in hydrology, soil science, and environment science. Thus, physics model experiments in laboratory scale on the aspect of measuring runoff and soil are one of the best tools in this field. This chapter aims to specify the experimental variant slope soil tank at home and in the USA. The developing of experimental soil tank of variant slopes with artificial simulating rainfall system will assist to understand soil water motivation, runoff yield, and nonpoint source pollution.

Keywords: experimental variant slope soil tank, artificial rainfall simulation system, runoff, soil erosion

1. Introduction

Rainfall-runoff processes and the related soil erosion are highly complex and are affected by many parameters, such as rainfall depth and intensity, soil characteristics, and slope morphology. Early studies showed that general slope parameters influencing runoff generation and soil erosion are slope gradient, length, and shape [1]. Thus, it is critically important to understand hydrological and sedimentological processes in the field or in the laboratory. The major disadvantage in field is their lack of experimental control, and the performance of experiments is wholly dependent upon the vagaries of the weather. The magnitude of such significant variables as the changes of slope gradient, length, shape, or wetness of the catchment cannot be preselected. A laboratory experiment provides control over hydrological variables which is notably absent in the monitoring of experimental and representative basins [2]. Therefore, an experimental variant slope soil tank is used to create artificially the conditions under which hydrological processes occur for quantifying how the inflow rate affects erosion processes in the world.

2. Experimental variant slope soil tank

The study on experimental variant slope soil tank began in the 1950s at home and abroad. The famous labs were Ven Te Chow Hydrosystems Lab at the University of Illinois at Urbana-Champaign in the USA and runoff laboratory of Institute of Geography, Chinese Academy of Sciences, in China. Later, many types of experimental soil tank were developed by Xi'an University of Technology, Institute of Soil and Water Conservation Chinese Academy of Sciences, Beijing Normal University, and Hohai University.

Ven Te Chow from the University of Illinois at Urbana-Champaign in the 1950s developed a kind of hydrological process. These processes include rainfall hydrographs for catchment areas of varying permeability; the abstraction of groundwater by wells, both with and without surface recharge from rainfall; and the formation of river features and effects of sediment transport. This tank measures 24.25 ft. long, 9 ft. wide, and 7.5 ft. deep and has a volume of approximately 12,000 gallons (45,000 L; **Figure 1**). The tank has a fixed elevation platform at its "upstream" end and two movable plate sections that can be used to set a bottom slope. These plates can be set to provide a constant slope along their entire length or can allow for a break in slope at the junction in the two plates. The tank is equipped with an overflow standpipe to maintain a constant water surface in the tank.

An experimental variant slope soil tank was installed in the runoff laboratory of Institute of Geography Chinese Academy of Sciences in 1965 ([3]; **Figures 2** and **3**). It was the first device in China that was used for study the rainfall-runoff and soil water movement. The tank was 8.0 m long, 3.0 m wide, and 1.0 m deep, with drainage holes at the surface and bottom to facilitate water discharge. The slope gradient ranges from 0 to 30°. A runoff collector was installed at the bottom of the soil tank, which was used for collecting runoff samples during the experimental

Figure 1. The margins tank of Ven Te chow Hydrosystems lab.

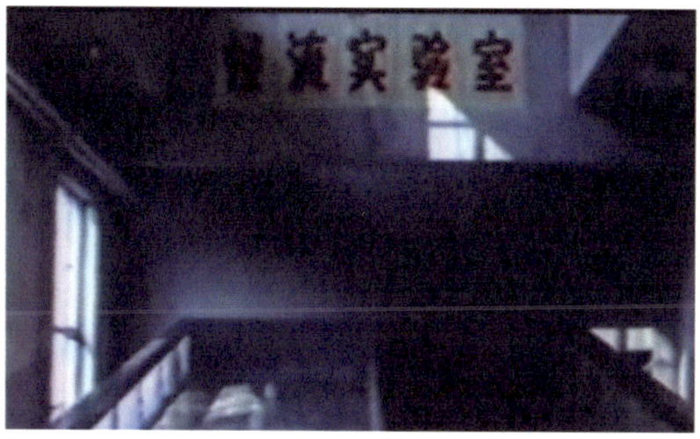

Figure 2. Runoff laboratory of Institute of Geography, Chinese Academy of Sciences.

Figure 3. The experimental variant slope soil tank in the runoff laboratory of Institute of Geography.

process. A rainfall simulator system was used to apply rainfall. This rainfall simulator can be set to any selected rainfall intensity ranging from 0.3 to 3.0 mm/min by adjusting the nozzle size and water pressure. The fall height of raindrops was set at 16 m above the ground, which allows all the raindrops to reach terminal velocity prior to impact with the soil surface.

A new experimental sink of runoff and erosion was reinstalled in the Geographic Sciences Museum of Institute of Geographic Sciences and Natural Resources Research, Chinese Academy of Sciences (IGSNRR, CAS) in 2014 (**Figure 4**). It consists of two metal rectangular boxes,

Figure 4. The new experimental sink of runoff and erosion in IGSNRR.

10 m long, 3 m wide, and 0.8 m high, and each one is located under artificial rainfall system. The slope of the experimental sink could be adjusted automatically from 0 to 35°. One 5 cm hole is cut into the downslope end of each plot. A short metal stub pipe is welded onto the hole to form an outlet. Two water flow monitors are horizontally set up in front of each box for the measurement of the runoff. For simulated rainfalls, runoff volume measurements and sediment sample collection are performed every 5 s and 5 min, respectively.

The experimental variant slope soil tank of State Key Laboratory Base of Eco-hydraulic Engineering in Arid Area in Xi'an University of Technology is manipulated by test water tank of constant waterhead ([4, 5]; **Figure 5**). The size of tank is 5.5 m long and 0.3 m wide and 0.6 deep. The range of slope is from 0 to 30°. The tank is divided into three parts of sub-tank with length of 2.0, 2.0, and 1.5 m, setting a dust-pan-shaped water gathering area in the terminal of tank. From top to bottom, this tank is divided into four parts; every layer of soil has a tight connection without separating and mixing mutually.

Figure 5. The experimental variant slope soil tank in Xi'an University of Technology.

The radioactive sources of ^{137}Cs (34–44 × 10^7 Bq) were used in this equipment to measure the water content. A hole (Φ8 mm) in the lead was made to produce a beam into the detector. Discharge testing system is consisted of a water tank ensuring steady water pressure, water pipes connecting every component, a water meter controlling discharge, and a steady-flow flume to stabilize and homogenize water flow.

The experimental soil tank of Institute of Soil and Water Conservation, Chinese Academy of Sciences, is installed on the ground fixedly using hydraulic system regulating slope (**Figure 6**). The slope varied from 0 to 30° of slope regulating step 5°. The size of tank is 8 m long × 3 m wide × 1 m deep, with many drainage holes (2 cm aperture) at the bottom [6]. The inflow experiment equipment consisted of an overflow tank to produce inflow water and a runoff collector to collect runoff samples. The overflow tank was attached to the upper end of the soil pan; the runoff collector was installed at the bottom of the soil pan [7].

A rainfall simulator system was used to apply rainfall. Computer self-control system can control four independent rainfall zones in different rainfall intensities, which composed of jet rainfall simulator and side rainfall simulator. The jet rainfall simulation system was introduced from Japan, with the rainfall height of 18 m, the rainfall intensity varying from 30 to 350 mm/h, and the biggest duration of rainfall of 12 h. The side rainfall simulation system was developed by Institute of Soil and Water Conservation Chinese Academy of Sciences. Its rainfall height is 16 m meeting the demands of final speed of raindrops. The rainfall intensity variation range is 40–260 mm/h, and the biggest duration of rainfall is 12 h [8].

The experimental variant slope soil tank in State Key Laboratory of Earth Surface Processes and Resource Ecology of Beijing Normal University is approximately 1 m long × 0.2 m wide × 0.05 m deep, using polyvinyl chloride as whole material (**Figure 7**). The projected area on the horizontal plane of soil plate for every slope is 50 × 50 cm. On the side of outflow, a V-type collector with more than 2.5 cm of height on other three sides is set up to prevent soil material in the tank spilling out during rainfall procession. Before packing soil into the tank, a good water permeability cloth is paved at the bottom of tank plate to ensure well drainage and evitable soil particle leakage of soil tank in experiment procession. At the bottom of tank, nine leakage drain holes (diameter is 5 mm) are distributed uniformly to make soil water freely permeate. The slope of tank ranges from 0 to 60°.

Figure 6. Experimental variant slope soil tank in Institute of Soil and Water Conservation Chinese Academy of Sciences.

Figure 7. The experimental variant slope soil tank in Beijing Normal University.

The artificial rainfall simulation testing ground is approximately 300 m^2, concluding 15 trough rainfall simulators, which have various combinations of types of sprayer, water pressure, and rainfall intensity. Every trough rainfall simulator has three swayed sprinklers, spacing at 1.1 m, installed with the height of 5.2 m [9, 10].

The slope-adjustable soil tank is made of welding steel plates splitting into two parts of the main tank and overflow tank in State Key Laboratory of Hydrology-Water Resources and Hydraulic Engineering of Hohai University (**Figure 8**). The two tanks are connected by four holes at the bottom. Effective volume of main tank measures 12 m long × 3 m wide × 1.5 m deep, while that of overflow tank measures 1.5 m long × 10 m wide × 6.2 m deep. The whole tank is divided into two 1.5-m-wide parts, varying from 0 to 30° under the drive of hydraulic pressure. The decision of its size is optimized profoundly after referencing international

Figure 8. The experimental slope-adjustable soil tank in Hohai university.

comparison of similar design products and former research results. The width of half tank is 1.5 m considering that boundary dimension effect can be ignored under this size. What's more, taking into consideration of the most lower than 1.5 m vadose zone in humid region in the south, depth of tank is determined to be 1.5 m, meanwhile soil water and plant root activities are mainly concentrated within this range in the arid and semiarid region.

The total area of rainfall simulation hall is 468 m^2, divided into I(18 × 18 m^2) area of 10 m high and II(18 × 8 m^2) area of 20 m high. The biggest rainfall intensity can achieve to 5 mm/min (300 mm/h). Based on two poles' control, the facility has 77 rainfall sprinklers in the whole distributed 5 water pipes. There are two solenoid valves controlling water inflow in every side of pipes. Among the pipes at least 15–16 rainfall sprinklers are distributed which are controlled by solenoid valves. These solenoid valves are departed into two groups based on test which can combine randomly and controllably, respectively.

3. Water monitoring and sample collection system

Rainfall monitoring system: Rainfall gauges are installed uniformly, recording the average value as rainfall intensity of the whole rainfall. The dataset is collected by computer automatically with 0.1 mm of precision. Laser rainfall intensity monitor is installed in Experimental Hall of Water and Soil Process of IGSNRR. It is composed of an array of laser transmitters and receivers. It achieves the rain non-touch measurement using orthogonally multiplexed laser beams according to the light attenuation law. The measurement error is less than 2%.

Discharge monitoring system: A simple and effective artificial flow measurement is applied for the monitoring system. Because water flows out from pipe, the discharge can be measured using a volume-known container and a stopwatch at the outlet.

Soil water potential monitoring system: The method of automatic collection of soil water potential is to transfer soil water potential (kPa) measured by negative pressure meter into electronical signals (mV) and then transfer analog signals into digital signals through A/D converter inputting the computer. Finally, the signals mentioned above are translated into soil negative pressure value based on mV-negative pressure relationship after calibration. The computer gives commands and instructions getting through specialized microcontroller and makes choice between sections and channels, to realize sample automation.

Water and soil collection system: Surface water collection system has a simple mechanism. Runoff flows into sampling bottles through water pipe collected at regular intervals. Collecting soil water corporates with soil water vacuum extraction device, which distils soil samples in a vacuum and then condensates water using liquid nitrogen.

The set of system consisting of various monitoring devices can develop runoff generation mechanism of different underlying surface, movement law of soil water, pollution law of nonpoint source pollution, migration and transformation law of pollution in soil, soil erosion law, and landslide formation mechanism.

4. Main results

Under the condition of experimental variant slope soil tank, some studies on hydrological physical mechanism experiment can be conducted, for example, rainfall-runoff formation under conditions of different rainfall intensity, surface slope and underlying surface, the law of surface water-biowater-groundwater cycle and transformation, pollution on migration, and transformation in unsaturated soil water. The margins tank has been used in the past to examine the characteristics of two-dimensional mudflows, the gully formation by turbidity currents [11], and the feasibility of using jets to manage sediments in a combined sewer overflow storage reservoir. The experimental soil tank can also be used to analyze the soil infiltration feature during rainfall [12, 13] and hydrodynamic mechanism of soil erosion [14] in Institute of Soil and Water Conservation, lateral downslope unsaturated flow in Hohai University [15], soil water movement using gamma ray in Xi'an University of Technology [5, 16], and rainfall-runoff relationship for different rainfall intensities in Institute of Geography [17]. Solute transport experiments have used tank arrangements for various purposes, including migration of infiltrated NH_4 and NO_3 in a soil and groundwater system [18], bioavailable phosphorus loss in runoff [19], and enrichment mechanisms of phosphorus [20].

5. Conclusion

The successful study of variant slope soil tank with artificial rainfall simulation system provides a basic foundation for hydrological and aquatic environmental experiment research under controlled conditions. After the operation of several years, this system has a reliable performance and stable manifestation. It is of great importance for studying rainfall-runoff relationship, establishment of conceptual hydrological model, migration and transformation of pollution material in soil water environment, and soil erosion and sediment yield study.

Acknowledgements

This work was financially supported by Key Program of National Natural Science Foundation of China (Grant No. 41730749) and CAS Key Technology Talent.

Author details

Lihu Yang[1,2]*, Simin Qu[3], Yifan Wang[3] and Xianfang Song[1,2]

*Address all correspondence to: yanglihu@igsnrr.ac.cn

1 Key Laboratory of Water Cycle and Related Land Surface Processes, Institute of Geographic Sciences and Natural Resources Research, Chinese Academy of Sciences, Beijing, China

2 University of Chinese Academy of Sciences, Beijing, China

3 College of Water Resources and Hydrology, Hohai University, Nanjing, China

References

[1] Sensoy H, Kara O. Slope shape effect on runoff and soil erosion under natural rainfall conditions. iForest–Biogeosciences and Forestry. 2014;**7**(2):110-114

[2] Hall MJ, Johnston PM, Wheater HS. Evaluation of overland flow models using laboratory catchment data I. An apparatus for laboratory catchment studies. Hydrological Sciences Journal. 1989;**34**(3):277-288

[3] Liu C. The Runoff Laboratory of Institute of Geography Chinese Academy of Sciences. Geographical Research. 1989;**8**(3):122 (in Chinese)

[4] Wang W, Shen B, Zhang J. Design of rainfall-infiltration-runoff system in Laboratory. Experimental Technology and Management. 1991;**8**(5):12-16 (in Chinese)

[5] Wang W, Zhang J, Wang Z. Application of gamma-ray in research of soil water movement. Transactions of the CSAE. 2004;**20**(1):95-98 (in Chinese)

[6] Liu J, Lei T. Experiment research on the effect of surface PAM coverage on erosion soil critical shear stresses of slope cultivated land by rainfall. Transactions of the CSAE. 2002;**18**(6):59-62 (in Chinese)

[7] Wen L, Zheng F, Shen H, Bian F, Jiang Y. Rainfall intensity and inflow rate effects on hillslope soil erosion in the Mollisol region of Northeast China. Natural Hazards. 2015;**79**(1):381-395

[8] Zhang X, Zheng F, Wang X, et al. Effects of upslope runoff on loessial hillslope erosion pattern evolution process and erosion sediment. Journal of Northwest A&F University(Natural Sciences Education). 2008;**36**(3):105-110 (in Chinese)

[9] Liang T, Wang LQ, Hu YY, et al. Influence of slope and rainfall intensity on the rare earth elements and phosphorous loss from soil with surface runoff. Journal of Basic Science and Engineering. 2010;**18**(5):741-749 (in Chinese)

[10] Liu H, Fu S, Wang X, et al. Effects of slope gradient on raindrop splash erosion. Acta Pedologica Sinica. 2011;**48**(3):479-486 (in Chinese)

[11] Fedele JJ, García MH. Laboratory experiments on the formation of subaqueous depositional gullies by turbidity currents. Marine Geology. 2009;**258**(1):48-59

[12] Li Y, Shao M. Experimental study on characteristics of water transformation on slope land. Shuilixuebao. 2004;**35**(4):48-53

[13] Mu T, Wang Q, Wang H. Study on physical base model of infiltration and runoff formation on loess slope land for a fixed rainfall intensity during rainfall. Journal of Northwest A&F University(Nat. Sci. Ed.). 2009;**37**(10):199-203

[14] An J, Zheng F, Lu J, Li G. Investigating the role of raindrop impact on hydrodynamic mechanism of soil erosion under simulated rainfall conditions. Soil Science. 2012;**177**(8):517-526

[15] Lv M, Hao Z, Liu Z, Yu Z. Conditions for lateral downslope unsaturated flow and effects of slope angle on soil moisture movement. Journal of Hydrology. 2013;**486**:321-333

[16] Zhang J. Experimental study on infiltration characteristics and finger flow in layer soils of the loess area[doctor]. Xi'an: Xi'an University of Technology; 2004

[17] Zhang S, Liu C, Xia J, Tan G, Li L, Liu C, Zhou C, Guo L. Study on the driving factors of rainfall-runoff process in indoor experiment. Science in China (Series D:Earth Sciences). 2004;**34**(3):280-289

[18] Wang C, Wang P. Migration of infiltrated NH_4 and NO_3 in a soil and groundwater system simulated by a soil tank. Pedosphere. 2008;**18**(5):628-637

[19] Yan W, Zhang S, Wu S, Cai Q, Tang Y. Bioavailable phosphorus loss and its prediction in runoff under simulated rainfall conditions. Environmental Chemistry. 1999;**18**(6):497-506 (in Chinese)

[20] Yan W, Zhang S, Tang Y. Sediment enrichment mechanisms of phosphorus under simulated rainfall conditions. Acta Scientiae Circumstantiae. 2000;**20**(3):332-337 (in Chinese)

Chapter 11

Practice on the Watershed Hydrological Experimental System Reconciling Deterministic and Stochastic Subjects Based on the System Complexity: 1. Theoretical Study

Wei-Zu Gu, Jiu-Fu Liu, Ai-Min Liao, Niu Wang, Jia-Ju Lu, Jin Lin, Hong-Wei Liu, Wen-Zhong Wang, Tao Ma, Zhao Cai, Min-Han Liao, Xue-Gang Li, Peng Zhuo and Na Yang

Additional information is available at the end of the chapter

http://dx.doi.org/10.5772/intechopen.78721

Abstract

This is the first of a two-part series on the watershed hydrological experimental system (WHES). Since the foundational stage and developmental stage of hydrological basin study with a duration of more than ca. one century, facing with the changing environment and, the declined risk of field study while the catchment hydrology is trapped in a theoretical impasse, a third phase of renovation on hydrological experiments seems ready to come out inevitably. Learned from Chinese decades' experiences on the field basin study for the question of what is wrong with the status quo, our exploratory idea is reported in this part. From the viewpoint of general system theory based on the paralleled concepts of the ancient Chinese and the Western, it is considered that the adequate method should face the characters of the complex dynamic system instead of previous static, linear system. From the viewpoint of another philosophical paralleled concept of the Middle Way, it should also face the operation and organizing of the mesoscopic systems for the organized complexity. Then, a framework of WHES is suggested with its organization based on the strategy of constrain complexity and add complexity and on the strategy of manipulation including the artificial-natural and controlled-natural objects. Such a trial framework, the Chuzhou WHES, is reported including the suggested critical zone experimental block (CZEB) instead of the experimental basin (EB) in the last decades.

Keywords: general system, hydrology system, watershed hydrology, hydrological experimentation, experimental basin, critical zone experimental block, eight trigrams, Tai-Chi Tu

1. Introduction

The most difficult is that the hydrology science perhaps cannot take the natural objects to laboratory for experimental studies, but the development of hydrology will depend mostly on the field observations and experiments; even the foundation of scientific hydrology recognized by UNESCO in 1974 is not from the work in laboratory but from the two field basin studies in France [1]. The first modern basin studies commenced in Switzerland at Emmental during 1890s; after that, a multitude of basin studies have developed from many parts of the world since the early twentieth century. Following it, a period of rapid development of hydrological basin studies was resulted from the International Hydrology Decade (IHD) Representative and Experimental Basin Program 1965–1974; during this period, it is estimated that ca. 3000 experimental basins (EBs) have been conducted [2]. If we designate the first phase of basin study until ca. the middle twentieth century as the stage of foundational and the second phase during/after the IHD as the stage of developmental, which has been going on for more than five decades up to now, a third phase of renovation seems ready to come out inevitably. A paper coauthored by 12 scientists pointed out that "Yet, most field experiments and observations in watershed science to date, remain largely descriptive..., have not set out to seek fundamental truth or understanding (nor test any formal theory or hypothesis per se)" [3] while Sivapalan alarmed that "catchment hydrology is trapped in a dead-end track, a theoretical impasse" [4], but the field study also meets its risk of decline, more complicated challenges, as Burt and McDonnell asked: "Field work: A dying art?" [5].

Question was then raised that "What's wrong with the status quo" [3]? Yes, there are many objective causes to look for, e.g., "many times things do not work out, weather does not cooperate," "field studies can be difficult to publish in international journals if they are seen as case studies" [5]. "There has been a movement away from field work and towards an almost complete dependence on modeling," as "computing power has become less expensive and field work more expensive (and risky, compared to model approaches)" [6]; this impacts researchers to flinch from field study rather modeling. However, to remind of especially the inherent objective roots have the significance. Learned from Chinese decades' experiences in their field basin studies with zigzagging process [7] especially those based on the concept of experimental basin, the inherent defects have been summarized mainly as the following: (1) designs of the experimental basins have been limited by the black box concept, as well as by many misconceptions (e.g., the linearity, nonheterogeneity, additivity of hydrologic systems, etc.), (2) operation has been substantially bounded by the hydraulic conception of these watersheds as isolated hydrological systems, (3) the studies of experimental basin are focused just on the surface hydrology, and (4) all of these watershed studies monitor only total runoff at the stream-outlet, and the subsurface responses of the watershed are only estimated by hydrograph separation [7, 8].

After the first stage of foundational and the second stage of developmental, if it reveals that experimental watershed hydrology is "inevitably going into a third phase of transition and innovation [8]"? While the planet moves to a new geologic era of Anthropocene [9], what faced is a changing nature of great transition with anthropogenic perturbation, replumbing of the hydrologic cycle [10], and natural climate oscillations. Aimed at the substantial progress in hydrologic science toward "a new unified hydrologic theory" as Sivapalan suggested [4], the answer will certainly be Yes. For this purpose, it, however, "ultimately depends on supporting new experimental work, new field observations, and new data collection networks" [11]. But, what shape will these new experiments and networks take? Twofold is reported here: our exploratory idea and practical tests.

2. Dialog with the hydrological nature

The hydrological experimentation appears as a dialog with the hydrological nature via designed objects aimed at knocking up its hiding doors of "black boxes" involved in for her true features. However, such a dialog seems not easy, it depends mostly on the adequate *method of questioning*, i.e., on the designed objects and the keys for it. Facing with the inevitable transition phase of the experimental watershed hydrology, it appears worth to have a fundamental rethinking on our previous methods of dialog, which seems challenged, sometimes having a successful beginning but becoming increasingly inadapted with the dynamic nature. In this paragraph, we will report the following: (1) from the viewpoint of *general system theory* based on the paralleled concepts of the ancient Chinese and the Western, it is considered that the adequate *method of questioning* should face with the characters of the complex dynamic system instead of previously the static, linear system, (2) from the viewpoint of *the philosophy of the Middle Way* (*"golden mean"*), based also on the paralleled concepts, it is considered that the adequate *method of questioning* should also face with the operation and organizing of the mesoscopic systems of organized complexity, (3) from these considerations, a suggested framework of *watershed hydrological experimental system* (*WHES*) is reported.

2.1. The quest for the watershed hydrological experimental system

2.1.1. What sort of system the watershed hydrology is?

A system is in general a set arrangement of things so related or connected as to form a unity or organic whole, also in interaction with environment [12]. For watershed hydrology, it is essentially an open system, it is bound up with the continuous exchanges, inflow and outflow, materials and energies to maintain in a so-called steady state. The fundamental characters of it consist of twofold: (a) complexity of various constituents of the system includes mainly the components which are correlated each other in the way of nonlinear and nonsymmetric and (b) complexity of the system itself includes mainly the evolution mechanism, never being in a state of dynamic equilibrium but always in changing with also the self-organization mechanism, etc.

So the watershed hydrological experimental system (WHES) is defined as an experimental system, which is designed to dialog with the complex watershed hydrological nature, to drive opening of various door of their black boxes aimed at revealing mechanisms hidden deep in the system.

2.1.2. *Main components of the watershed hydrological experimental system: the Capra's* parallels-1

In a radical manner, the WHES is a system of material, however refer to operation manner it is manifested itself in three aspects: (a) *Mass*. It is the *material aspect* involving "rock, soil, water, air, and living organisms" according to NRC of USA on 2001 for the components of the critical zone [13]. Magically, it happens to coincide with the ancient Chinese philosophy for nature 1000 years ago, the so-called "*Five-Xing*," which holds the five fundamental elements from the Earth nature ("Jin," Mu," "Shui," "Huo," and "Tu") [8]. (b) *Energy and force*. It is the *driven aspect* including external solar energy from its environment, and various inherent forces, gravity, surface tension, intermolecular forces, capillary force, etc., from material system internally itself. (c) *Information*. It is the *philosophical aspect* including organization, entropy, gene, etc. [8]. This living open system of structural dissipative processes and irreversible evolution tends to increase its entropy spontaneously and go towards disorder, even the exchange of entropy fluxes across its boundaries is continuous [14]. However, the role of feedbacks of this nonlinear system will promote "self-organization" as energy dissipates, providing opportunities for dissipating energy to act again within the system towards the direction of order. "Conservation without evolution is death. Evolution without conservation is madness" [15, 16]. The material system with "*Five-Xing*" can be so arranged following the Chinese general philosophical imagine thinking, the archetypal symbols of "eight trigrams" for all cosmic and human situations [8], every trigram is composed with three lines, solid ("*Yang*") or broken ("*Yin*"). Such "*eight trigrams*" corresponding to the "*Five-Xing*" in different weights (shown later) can also be the image thinking of the WHES, schematically shown in **Figure 1** (left) which is quoted from Capra's book [17]. The "parallels" between modern physics and eastern mysticism has been explored firstly by him, and it is even vaguely similar to the meson octet (**Figure 1**, right). "The eight trigrams are symbols standing for changing transitional states; they are images that are constantly undergoing change. Attention centers not on things in their state of being—as is chiefly the case in the Occident- but upon their movements in change" (Richard Wilhelm, from [17]). "In modern physics, we have come to see the 'things' of the subatomic world in very much the same way, laying stress upon movement, change, and transformation and regarding the particles as transient stage in an ongoing cosmic process [17]."

2.1.3. *Boundary and environment of the watershed hydrological experimental system*

It was suggested that in addition to the three operation manners: mass, energy, and information, manifested by the material system itself, there should have the fourth subjective component [12], i.e., the view of the observer. It is true e.g., in our case, for the boundary of the watershed system to be studied, it depends actually on what is the purpose of studying and largely on the view of point for the watershed system as well. During the first and second stages of basin study aforementioned, the boundary of the WHES is limited mostly by the surficial watershed, and it appears as the *narrow system of interest* (*NSI*) according to Yan et al. [12] as shown schematically in **Figure 2**. It holds relevantly an *around environment*, and a *wider environment* as well. For the new stage, it will be inevitably renovated into a *wider system of*

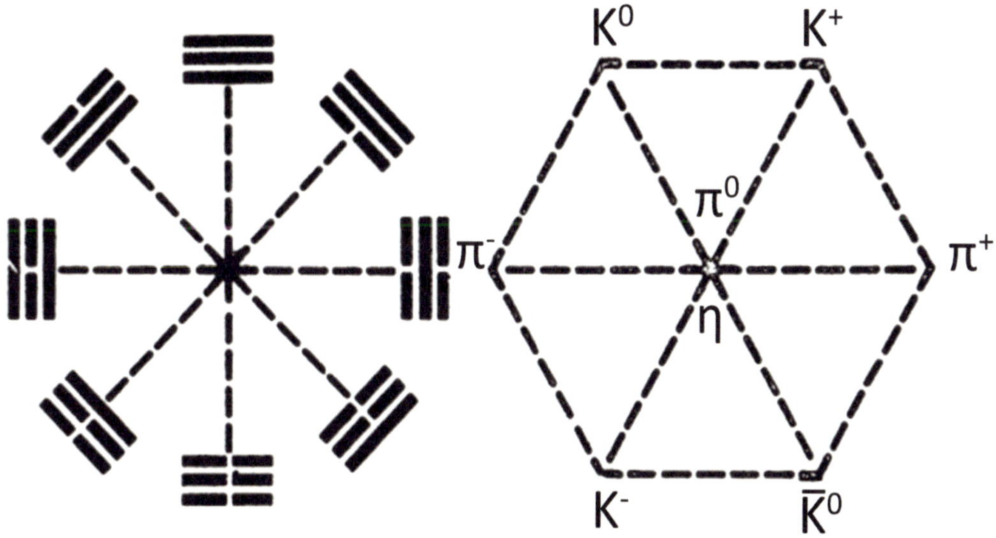

Figure 1. The eight trigrams are symbols with opposite coupled "Yin" and "Yang" lines interchanged, standing for changing transitional states. It appears as an image of the WHES (left). This pattern is even vaguely similar to the meson octet with different charges including π^+, π^o, π^-, κ^+, κ^o, κ^{-o}, κ^-, and η (right) [17].

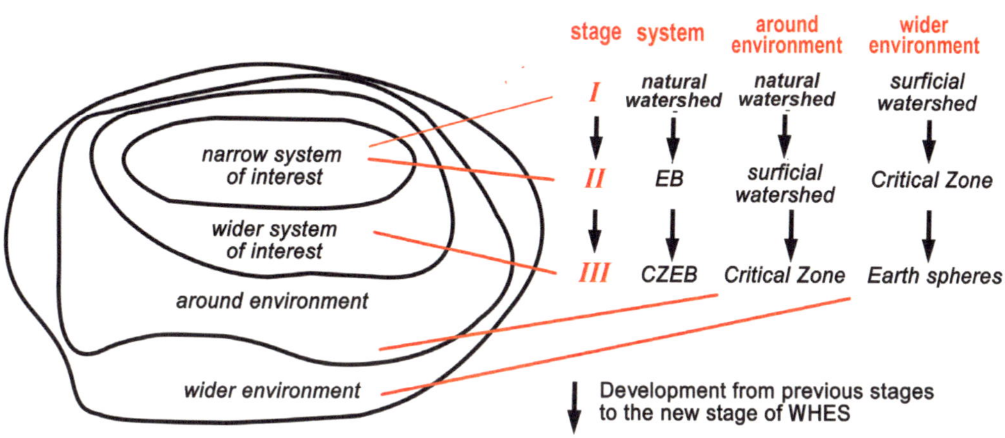

Figure 2. Development of the system boundary, around environment, and wider environment of the watershed hydrological experimental system from its previous stages I and II to the renovation stage III (sketch following Yan et al. [12] with explanations for our systems).

interest (*WSI*) [12], i.e., taking the critical zone as its system boundary with a much wider environment involving the Earth spheres.

2.1.4. *The image thinking for the system evolution: the Capra's* parallels-2

As an open, dynamic, and evolving WHES, all of its system components ("eight trigrams") are never just the aggregate of heaps but are correlated with each other, and are merely in the transitional stages of hydrological process. "The eight trigrams therefore are not representation of things as such but of their tendencies in movement" [17]. In fact, the changing items revealed in the schematic eight trigrams (left of **Figure 1**) are twofold, i.e., internally between different trigrams showing by lines within opposite "*Yin*" and "*Yang*," they are standing in interrelations, and externally between the system and the environment, they are standing in interrelations as well. According to the Chinese saying, it seems that "pulling one hair, the whole body is effected," e.g., a change of soil moisture even partly in the watershed will result in the changes of surface and subsurface runoff generation, the groundwater recharge, the land evapotranspiration, the living organisms, and perhaps the far-sited outlet discharge.

It was illustrated by another symbol called "*Tai-Chi Tu*" or "*Diagram of the Supreme Ultimate*." It is generated by a symmetric arrangement of dark *Yin* and bright *Yang* with cyclic interplay, with fishlike image in opposition, coordination, conjunction, and in blended harmony of *Yin* and *Yang*. Such "archetypal pair of nature is the grand leitmotiv that permeates Chinese culture" [17]. Two dots imply the inherent connections of the primordial pair of opposites *Yin* and *Yang* and symbolize the idea that both contain always the "seed" of the opposite sides. The curved interface can be derived mathematically from periodicity, it also seems to be "the closest analogy to S-matrix Theory in Eastern Thought" [17] while the S-matrix Theory is "a framework which seems to be most appropriate for the description of hadrons and their interactions" [17]. Fritjof Capra used *Tai-Chi Tu* on the cover of his book "The Tao of Physics: an exploration of the parallels between modern physics and Eastern mysticism" with which over 1 million copies was sold worldwide during its third edition (**Figure 3a**). In our case, it can be revealed for different sorts of WHES, or for subsystems, e.g., reconciliations of the WHES and the environment (**Figure 3b**). It is also being used recently as the book cover (**Figure 3c**) on the theory of *irregularities* [18], which is closely related to the rationale of our WHES. Niels Bohr was well aware of the parallel between his concept of complementarity and Chinese thought, when he was knighted, he had to choose a suitable motif for his coat-of-arms, his choice of feel on the Chinese symbol of *Tai-Chi* representing the complementary relationship of the archetypal opposites *Yin* and *Yang*. In choosing this symbol for his coat-of-arms, he also put together the inscription *Contraria sunt complementa* (**Figure 3d**) [17]. In fact "opposites are complementary" just describes a rationale of the WHES as well.

"*Tai-Chi Tu*" (**Figure 3**) and the "*eight trigrams*" (**Figure 1** left) are always coupled, the *eight trigrams* revolved around the *Tai-Chi* including four couples oppositely distributed symmetrically to the circle center of *Tai-Chi* with their trigram lines in contrary (full or broken) with each other as shown in **Figure 3e** (numbers 1 and 8; 2 and 7; 3 and 6; and 4 and 5).The *eight trigrams* are also related to the five materials of system described both by NRC and the *Five-Xing* with different weights (**Figure 3e**). The peripheral arrangement of "*eight trigrams*" is strictly made according to the philosophy of *Five-Xing* i.e., in mutual promotion and restriction to each other to maintain dynamic balance of the system, which in fact are the initiator of the complexity and irregularity of the watershed hydrological system. However, the important parameter *time* of this coupled "*Tai-Chi Tu*" and the "*eight trigrams*," which is the diagram of imagine thinking for the watershed hydrological system and also that of WHES, lies deep hidden, resulted from the philosophical concept of *time*. Traced to the ancient Chinese concepts on time and space since 2000 years ago, OuYang S.C and Lin Y. suggested that " Time dwells in the rotational movements of materials so that it cannot exist independently outside the materials," "time is an attribute of materials so that it does not occupy any material dimension," "time is parasitic on

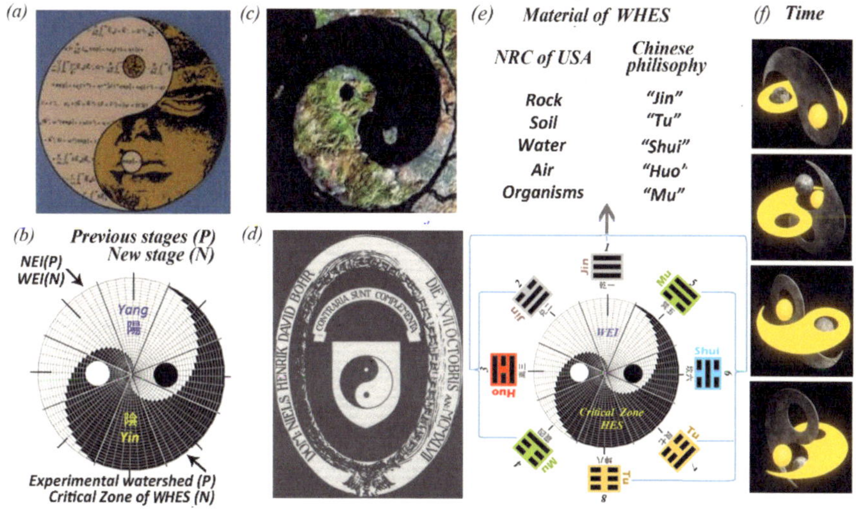

Figure 3. The image thinking for the system evolution based on the ancient Chinese philosophy, the *"Diagram of the Supreme Ultimate"* (*Tai-Chi Tu*) and the *"Eight Trigrams"* (*Bagua*). (a) The *Tai-Chi Tu* Fritjof Capra used on the cover of his book [17], (b) system concept for the experimental catchment of the previous stages and the suggested new stage, (c) the *Tai-Chi Tu*, which OuYang and Lin used on the cover of their book of *irregularities* [18], (d) the *Tai-Chi Tu* Niels Bohr used on his coat-of-arms during he was knighted, put together with the inscription *Contraria sunt complementa* for his theory, (e) the coupled diagram of the *Tai-Chi Tu and Eight Trigrams* (*Bagua*). The *eight trigrams* are related to the five materials of system with different weights, it forms four couples with their peripheral arrangement following the philosophy of *Five-Xing*, in mutual promotion and restriction to each other, and (f) schematically showing the hidden parameter time, it also reveals the unceasing moving of materials and the unceasing evolution of system as well.

the changes of materials," "so that as long as there is material, the material will have to move" [19]. For the coupled "*Tai-Chi Tu*" and the "*eight trigrams*," time is in fact hidden in the continuous movement and evolution as schematically shown in **Figure 3f**.

2.1.5. The mathematical descriptions for the system evolution: the Bertalanffy's differential equations

The concept of changing evolution interrelated between the system components and between the system and its environment as well can be revealed by the combined *eight trigrams* and *Tai-Chi;* they are bound up with the continuing exchanges for maintaining their steady state. Such an image thinking for the system concept of interrelations was illustrated mathematically by Ludwig Von Bertalanffy [19] and discussed by Yan et al. [12].

Suppose there are n elements in a given system, $X_i(i = 1, 2, \cdots n)$, denoting some measure of elements X_i by Q_i, for a finite number of elements, and in the simplest case, while dQ_i/dt is continuous, then the interaction between system elements can be described as follows:

$$\begin{aligned}
\frac{dQ_1}{dt} &= f_1(Q_1, Q_2, \cdots\cdots Q_n) \\
\frac{dQ_2}{dt} &= f_2(Q_1, Q_2, \cdots\cdots Q_n) \\
&\vdots \\
\frac{dQ_n}{dt} &= f_n(Q_1, Q_2, \cdots\cdots Q_n)
\end{aligned} \quad (1)$$

It follows that the change of any measure Q_i, i.e., $\frac{dQ_i}{dt}$, therefore, is a function of all $Q's$, from Q_1 to Q_n, conversely, change of any Q_i in the right side of above equation will introduce the change of other elements $(\frac{dQ_j}{dt})$ and of the system as a whole. It also reveals that not only the $Q_{1,}$ acts on Q_j, i.e., Q_1 is the cause and Q_j is the effect, but also Q_j will act on Q_1 that also existed, they are in *causation*. It means a *feedback*, a *cycle causation*, e.g., the soil and the plant in our system. Furthermore, such an interaction of causation and feedback existed not only for several elements of the system but also for all of the elements, it in fact reveals the systematicness and complexity [12, 19].

The foregoing considerations are a set of *n* simultaneously first-order differential equation called dynamic equations or equations of motion of the system [19]. In order to avoid expressing it in partial differential and integro-differential equations, the above equation abstracts from spatial and temporal conditions, the previous history of the system, etc., even so, it is quite useful for discussing several general system properties [19].

2.2. The middle-ground perspective for the watershed hydrological experimental system

2.2.1. System categories by complexity

In his frontier paper of 1948 on "Science and complexity" [20], Warren Weaver first proposed three terms on complexity: "*Problem of simplicity,*" refers mostly to the physical sciences for study problems, which involved two variables or at most three or four; "*Problems of disorganized complexity,*" referred as "a problem in which the number of variables is very large, and one in which each of the many variables has a behavior, which is individually erratic, or perhaps totally unknown"; and what he emphasized was the *middle region*, the "*Problems of organized complexity,*" with moderate number of variables involved, the really important characteristic of the problems of this middle region "lies in the fact that these problems, as contrasted with the disorganized situations with which statistics can cope, show the essential feature of *organization*" [20]. Later on 1975, Weinberg described these terms as "*organized simplicity* (*mechanisms*)," "*unorganized complexity* (*aggregates*)," and "*organized complexity* (*complex systems*)" [21] as shown in the II, I, and III of **Figure 4a**, respectively. Dooge emphasized this *middle region* as *intermediate systems*, and he pointed out that "most problems arising in catchment hydrology fall in the category of complex systems with some degree of organization," and "a theory of complexity based on the concept of reality as intermediate between determinism and randomness" has emerged [22].

As for the basin studies in China within last decades, it has been conducted mostly by using of the representative basin (RB) and experimental basin (EB) proclaimed by the first IHD more than 50 years ago, and by using of some kinds of artificial physical models e.g., the soil column. It appears that the important *intermediate system* (III in **Figure 4b**) described by Weaver and Dooge is actually a vast gap in between the two extremes, "*unorganized complexity* (*aggregates*)" and "*organized simplicity* (*mechanisms*)" (I and II in **Figure 4**, respectively) for the watershed hydrological experiments [14, 8].

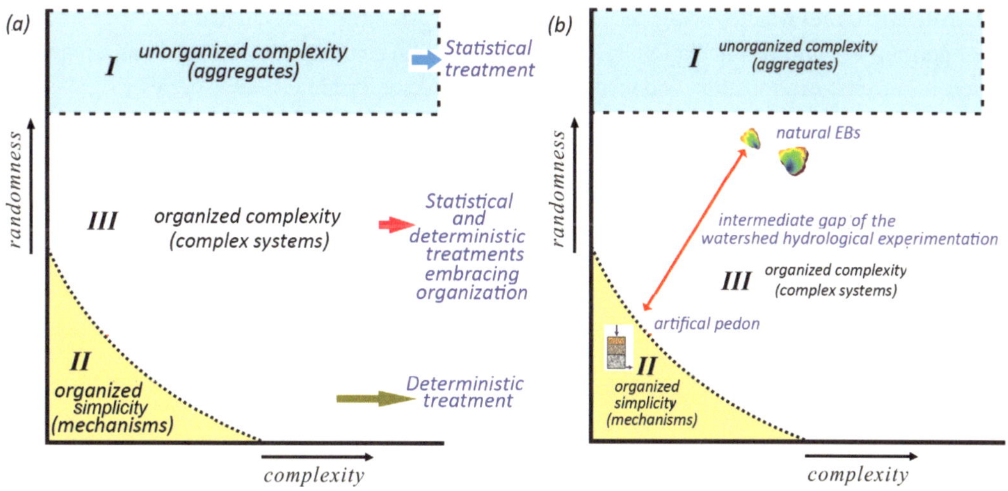

Figure 4. (a) A schematic representation of the complexities adapted from Weinberg [21], Dooge [22], and Sivapalan [4] and (b) there is a gap between the two extremes in watershed hydrological experiments with mostly natural EBs and artificial physical models.

2.2.2. *The middle-ground perspective: one of another* parallels

For the intermediate, it seems hackneyed, e.g., the story of "Boucle d'or les trios ours"; however, for its scientific solution, it seems inevitably to contact with the philosophical concepts about it. In fact as said early by Aristotle, he termed "the golden mean between two extremes of character" in Book IV of his Ethics. Comparing with *Dichotomy*(*dimidiate thinking*) in Western philosophy, the Chinese philosophical mode can be illuminated by the middle-ground perspective, the *mid-view thinking,* with core concept of "holding the two extremes and using the middle impartial," "using the middle impartial for the understanding of the two extremes" [23]. In fact, it happens as a general mode for treating everything both socially and scientifically, "too tight a string will easily be broken while too loose will bring no music" as the saying goes. Parallely, there is also a central concept of "*mid-view thinking*" in Buddhism [24].

Down to watershed hydrological experiments, the intermediate gap (**Figure 4b**) between the deterministic and stochastic extremes is actually a strategic weakness of it; perhaps it may hamper the progress for getting a "unified theory of hydrology at the catchment scale" as Sivapalan has described [4]. It is, thus, very desirable to fill up such a gap, to have new intermediate experimental entities for the watershed music based on the *middle-ground-perspective* (*MGP*) concept. For watershed hydrological experiments, here we refer the entities of deterministic paradigm as the micros (*Micro*), that of stochastic the macros (*Macro*), in between the mesos (*Meso*), from MGP concept; three characters of this intermediate system can be specified as following: (1) character of *Trichotomy*: *MGP* viewed: these are the hierarchies of the whole system, i.e., the *Meso* contains both the information of *Micro* and *Macro, Meso* is composed by *Micro* but also is the subsystem of *Macro*; (2) character of intermediate: *Meso* system acts as the bridging between Micro and Macro, bridging not just means a passageway as current concept revealed, but intermediary and reconciliation independently; and (3) character of *Hypercycle*: as a manifestation of both self-organization and feedback, it composed of four interrelated cycles, i.e., *Micro* ↔ *Meso*, *Micro* ↔ *Macro*, *Meso* ↔ *Macro*, and *Micro* ↔ *Meso* ↔ *Macro* (**Figure 5**) [25].

2.2.3. *Target of the watershed hydrological experimental system (WHES) and its setting*

Learned from the field experiments on watershed hydrology since 1953 in China [7], great efforts have been exerted for "getting right answers [11]" on engineering hydrology, facing with the changing environment since then and lots of new raised water problems coupling with misconceptions, and "in an era of largely model-only research" [5], it needs to have further understanding on scientific hydrology for "the right reasons [11]." The WEHS will specifically target the tests of outrageous hydrological hypotheses, and the generation of theories as Burt and McDonnell [5] and Sivaplan [4] have called upon.

We can take a diagonal in the space of complexity versus randomness of **Figure 4a**, and re-plotting it in **Figure 6a** and **b** with arbitrary points L, A, C, and N (**Figure 6b**) for the sake of discussions later. Aforementioned natural experimental watersheds are falling in the region I (**Figures 4** and **6**), represented by N in **Figure 6b**; it possess main characteristics of an adaptive complex system, and it is defined in **Figure 5** as the "*macro*." In addition, L1 and L2 in **Figure 6b** are in the region II with lowermost randomness and complexity, which is also defined in **Figure 5** as the "*micro*." Evidently, the watershed as a whole cannot be understood directly from individual parts of its own, i.e., from the "micro" leap forward to the "macro." The requirement of the intermediate "*meso*" system is then emerged as have discussed. So the questioning of strategy of setting WHES is then shifted to the "*meso*" system.

2.2.4. *Strategy of "constrain complexity"*

There is a particular difficulty raised to the hydrologists that no possibility for taking the research entity, natural catchment, as a whole to their laboratory for experimental research as many other natural science did, the only way out is to work on the EBs in the field, which in

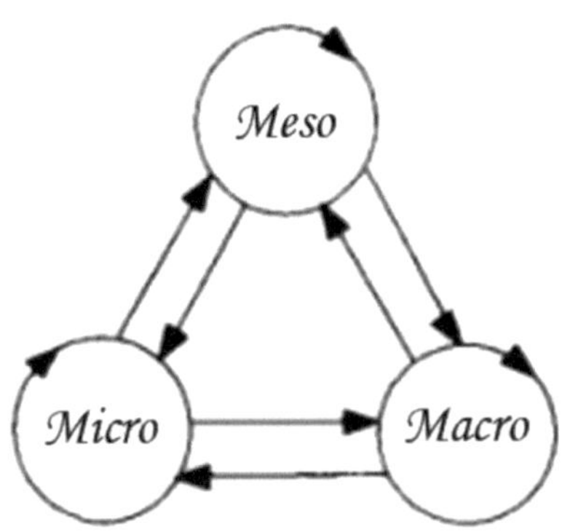

Figure 5. Mechanism of hypercycle among micro, meso, and macro based on the *middle-ground-perspective* of system [25].

http://dx.doi.org/10.5772/intechopen.78721

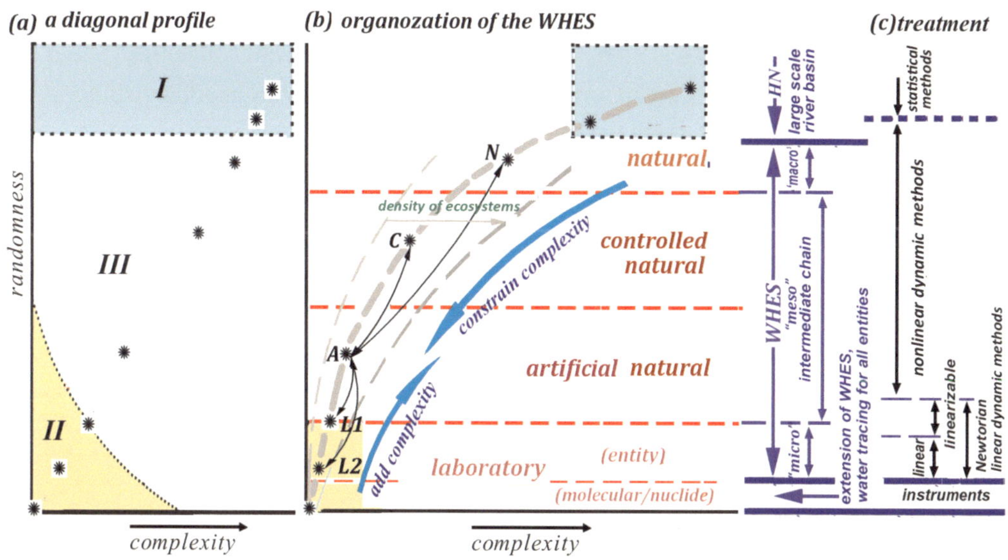

Figure 6. (a) Re-plotting diagonal in the space of complexity versus randomness of **Figure 4a**, (b) the organization of the WHES, and (c) different treatments for the WHES.

fact has been used for decades worldwide. However, as McDonnell coauthored by 11 hydrologists pointed out that "Yet, most field experiments and observations in watershed science to date, remain largely descriptive" [3]. It gives rise to assuming that the impact of the complexities existed inherently within natural watersheds even in different scales? Very likely, yes. The source of complexity very likely is attributed to the materials composing the entity system (five materials provided by NRC, USA, and the *Five Xing* of ancient China) as have argued in a book related to postmodern science that "irregularity just is the basic attribute of materials [26]." By the way, one of our works suggested that the existence of vadose zone appears as the main complexity source to watershed hydrology [27]. It follows that the inherent complexities of natural catchment cannot be avoided except moving away its constituent materials. Nevertheless, making constrains for the complexity seems not impossible, if so, the N in **Figure 6b** could be moved apart from that position with a higher complexity to a lower complexity at e.g., C or even A as shown in **Figure 6b**. For this purpose, several possibilities are suggested under the prerequisites of keeping conformance with the main characteristics of complex system, i.e., taking system theory as the framework. The idea of "constrain complexity" appeared first in the paper of Sivapalan ([4], p. 206; **Figure 6**).

2.2.4.1. Hierarchical subsystems

Classic reductionist framework prefers that the phenomena can be explained completely in terms of relations between other more fundamental phenomena, and to reduce explanations to the smallest entities. It appears as the method of explaining macroscopic properties in terms of microscopic components. General system theory is opposite to such kind of approaches, simply because of the basal principle: "*The whole is more than the sum of its parts*" [19]. It can be shown also from Bertalanffy's differential equations if we take its simplest case with two inputs Q_1 and Q_2, after developing $\frac{dQ_1}{dt}$ and $\frac{dQ_2}{dt}$ in Taylor series, the results will not obey linear additive but a series of higher order terms left, it gives the index to self-organization and feedback mechanisms, which is not explainable from the characteristics of isolated smallest parts. It was also remarked sharply early by Toffler that "One of the most highly developed skills in contemporary Western civilization is dissection: the split-up of problems into their smallest possible components. We are good at it. So good, we often forget to put the pieces back together again…In this way we can ignore the complex interactions between our problem and the rest of the universe" [28].

The framework suggested here is based on two essential features of the organized complex system: (1) the concept of *hierarchical order*. It is obviously "a mainstay of general system theory," "it is intimately connected with those of differentiation, evolution, and the measure of organization," and "the hierarchical order and dynamics may be the very same" [19]. In graph theory, hierarchical order can be expressed by "trees," (2) the existence of complex interactions as described by Toffler. It is also the key characteristic of complex system. "A system or 'organized complexity' may be circumscribed by the existence of 'strong interactions' or interactions, which are 'nontrivial' i.e., nonlinear" [19].

Here, the suggested subsystem framework, the hierarchical order, is similar to a sort of the reductionism framework, i.e., the "hierarchical reductionism" introduced by Richard Dawkins to describe the opinion that complex system can be described with a hierarchy of organizations, each of which is *only* described in terms of objects one level down in the hierarchy [29]. However in our case, not only the level down but feedbacks as Robert Ulanowicz had suggested that "science must develop techniques to study ways in which larger scales of organization influence smaller ones, and also ways in which feedback loops create structure at a given level, independently of details at a lower level of organization" [30]. In order to illustrate the difference between "hierarchical reductionism" and classic reductionism, a schematic diagram of a system is used following Yan [12] as the S in **Figure 7**, it is composed with hierarchically different subsystems S1s, S2s, S3s, etc., while the classic reductionism composes the splitting elements of natural system, i.e., instead of these interrelated subsystems but the different elements in series (Es in **Figure 7**). Hierarchical subsystems will have lower complexities hierarchically compared with the holistic system S as in **Figure 7**. We take this as the way of constrain complexity, it is schematically shown as the A in **Figure 6b**, A is interacted with both C, N of higher levels and L of lower levels, and with feedbacks as well.

2.2.4.2. Reduce the number of system elements

Referring to Bertalanffy's differential equations aforementioned, another way for constrain complexity is trying to reduce the number of elements of the hierarchical subsystems, i.e., the n in his equations. There may be several choices: (1) reduce the number of inflow and/or the number of outflow (I and O in **Figure 7**), (2) reduce the variabilities of some of its elements, e.g., making some or all of the inflow process, e.g., that of water amount and water hydrochemistry of hierarchical subsystems in regular type or in organized patterns, (3) reduce the number of effective parameters from the around environment (**Figure 2**) of hierarchical subsystems, etc.

http://dx.doi.org/10.5772/intechopen.78721

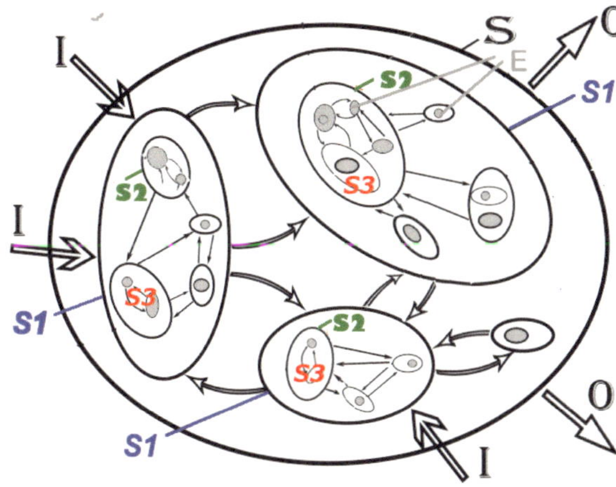

Figure 7. A schematic diagram of a system (adapted from Yan [12], revised).

2.2.5. Strategy of "add complexity"

The artificial physical model e.g., the soil column, is another extreme measure that the current hydrological experimentation studies go for, as shown schematically by L2 in **Figure 6b**. In fact, it is the measure with which many fundamental laws in hydrology; e.g., the Darcy's law and Fick's law were derived from, i.e., in small-scale physics and in simplifications, which is very different from real natural behaviors. Blöschl clearly noted that "in the Darcy example, the geometry of soil pores is far more complex than a bundle of tubes," "but at larger scales where the equations apply the media are considered uniform." [31]. Beven and Germann challenged the Richards equation that "may be predicated on the wrong experimental method for natural conditions" [32]. Sivapalan pointed out that "However, it is often overlooked that the constituent process theories are essentially derived at the laboratory or small scales. They are underpinned by assumptions of homogeneity, uniformity, and time invariance of various flow paths, over the land surface and in channels, and through soils and vegetation" [4]. Despite all of these recognitions, however, modern approaches for modeling hydrologic systems still have to use these problematic equations and "go further and further down the rabbit hole of model uncertainty estimation" [5]. Aimed at tests of outrageous hydrological hypotheses, and the generation of theories [4, 5], it needs to take the upward approach defined by Klemes as "attempts to combine, by mathematical synthesis, the empirical facts and theoretical knowledge available at a lower level of scale into theories capable of predicting events to be expected at a higher, in our case at a hydrological level" [33]. As an example following his idea, for the "10 different formulas for estimating infiltration" [4], the experimentation was extended from the classic one-dimensional scale to two- and three-dimensional scales, i.e., from the lower level schematically as L2 in **Figure 6b** extended to e.g., L1, etc., and it reveals the "add complexity." In our work, the different phenomena got from three-dimensional infiltration experiments were then used to the explanation of one of the generation mechanisms of surface runoff [27].

2.2.6. *Strategy of manipulation*

Remember the warning from Werner Heisenberg that, "what we observe is not nature herself, but nature exposed to our method of questioning," as we have experienced that for achieving our goal of watershed hydrological experimentation, tests of outrageous hydrological hypotheses, and for the generation of theories, it appears very difficult to rely *only* upon the natural EBs even in small scale due to its inherent complexities. However, the natural EBs are indispensable for achieving the goals. A possible solution for this knot maybe taking some actions that are compelling instead to let it be. What follows that the way of constrain complexity and add complexity is never just for the *natural* entities but also for the *manipulated natural entities.* It may have two ways as experienced:

1. ***Artificial-natural***. Starting with the artificial entity, an artificial subsystem of a watershed following different designs, with/without plant systems, maybe a slope, or several combined slopes, or a catchment with slopes. Making artificial boundaries including the bottom and the surface and subsurface dividers, using filling soil or undisturbed soil to form both unsaturated zone and saturated zone. After many years of operation, such an artificial entity approaches gradually to the natural conditions including the promotion of both the physical and chemical features of soils and the ecosystem beneath the ground surface. Here, we term it as the *artificial-natural*, from artificial approaches to natural, e.g., the hydrohill, soil of 1 m in depth was filled up at 1978, equipped and operated at 1982, in the meantime after decades' of exposure under natural rainfalls, it's unsaturated zone can be treated as almost the natural entity. Further step of manipulation is to making around environment e.g., the rainfall input and even the microclimate to be partly controlled.
2. ***Controlled-natural***. Starting with a selected natural EB with natural bedrock and deposit, the artificial controls are then made to it including its bottom and surroundings aimed at reducing its surface and/or subsurface inputs and outputs. Here, we term it as the *controlled-natural*, the natural to be controlled, e.g., the Nandadish, and it will be described in detail later.

It may come down to a point that for tapping their potential, both the strategy of *constrain complexity* and *add complexity* based on the middle-ground perspective still needs to have the wind at their back, the measures of manipulation.

3. Organization of the watershed hydrological experimental system (WHES)

3.1. Requirements of the WHES

It can be summarized based on discussions aforementioned as follows: (1) need all scales including the two extremes of pure natural and pure artificial entities, and various intermediate entities; (2) need both downward or top-down approach and upward or bottom-up approach [33], and it could be realized by strategies of constrain complexity and add complexity, respectively; (3) need the measures of manipulation for both natural and artificial entities; (4) need parallel entities of different coverages especially the plant ecosystems; and (5) need an extension of WHES. As have experienced that for identification, evaluation, and quantification of the interactions between hierarchical subsystems, it needs working on the molecular level and nuclear level for both water and its material mediums, i.e., their isotopic and hydrochemical processes. So the role of it is not merely for getting reference data but for opening another door for the understanding of the hierarchical subsystems.

3.2. Organization and treatment of the WHES

As schematically shown in **Figure 6b**, it includes three levels with both complexity and randomness from high to low, and an extension part:

1. High level: composed by pure natural EBs, the "macros" of WHES represented by N in **Figure 6b**
2. Intermediate level: composed by intermediate chain, the "mesos" of WHES represented by C and A in **Figure 6b**: the controlled-natural entities and the artificial-natural entities, respectively.
3. Low level: composed by pure artificial chain, the "micros" of WHES represented by L1, L2 in **Figure 6b**.
4. Extension of WHES: the laboratory for geochemical processes by water tracing for all entities.

There are different treatments for different organization levels of WHES as shown in **Figure 6c**. Only the "micro" level of WHES is linear, and a small part of the "meso" level, which is very close to the "micro" level is quasilinear or linearizable, these two parts can be treated by Newtonian dynamic methods. For the large-scale river basin, it belongs to the hydrometric monitoring network (HN) (**Figure 6b**), statistics will be the dominate method. In most of the intermediate chain, the "meso" level of WHES (**Figure 6b**) should be treated by nonlinear dynamic methods.

3.3. The WHES of Chuzhou hydrology laboratory

The idea for the necessity of an experimental system for watershed hydrology [7, 34] was summarized from the decades working on field basin studies since the operation of the first hydrological experimental station in China, the Bluebrook, which was established in 1953 [35]. The dream of a WHES finally came to reality with the establishment of the Chuzhou hydrology laboratory (CHL) in 1978 by the Nanjing Hydraulic Research Institutes of the Chinese Ministry of Water Resources on the basis of the Chengxi Runoff Experimental Station established on 1962 with three experimental watersheds. This Chuzhou WHES was designed including both artificial and natural entities of different scales as shown in **Figure 8a**, within which, most of them have established, few of them are in modified and renewed after an interruption of many years. This WHES is settled in a natural Bloomhill (Hua-Shan) watershed (**Figure 8b**), with surficial drainage area of 82.1 km^2 since 1962, 80.0 km^2 since 1998 due to a hydraulic engineering. This basin is situated at the upstream area of a river named as the *Western Stream of Chuzhou,* which was described in an ancient poem of Tang Dynasty by Wei Ying-Wu (737–792) which most Chinese pupil can recite. In the meantime within such a

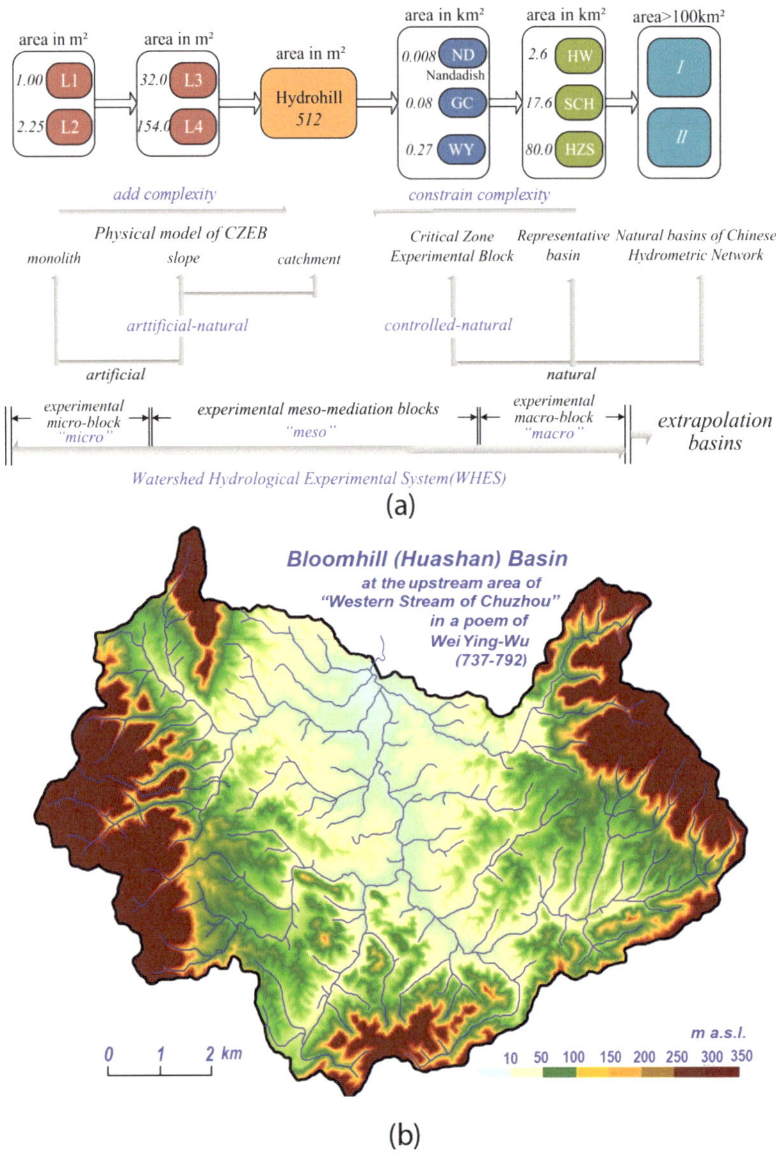

Figure 8. (a) Organization of the Chuzhou WHES of the CHL. ND—Nandadish, GC—Gaocong, WY—Wangying, HW—Huangwa, SCH—Sanchahe, HZS—Huzhuangsan (outlet of the Bloomhill watershed), L1 to L4—lysimeters, area—surficial drainage area and (b) the natural Bloomhill (Hua-Shan) watershed within which the Chuzhou WHES is settled all ad.

traditional cultural atmosphere, the reborn CHL together with WHES is in the process of emotional renovation.

3.4. What is CZEB

The "critical zone" (CZ) is defined by the National Research Council of US in 2001 as "the heterogeneous, near-surface environment in which complex interactions involving rock, soil,

http://dx.doi.org/10.5772/intechopen.78721

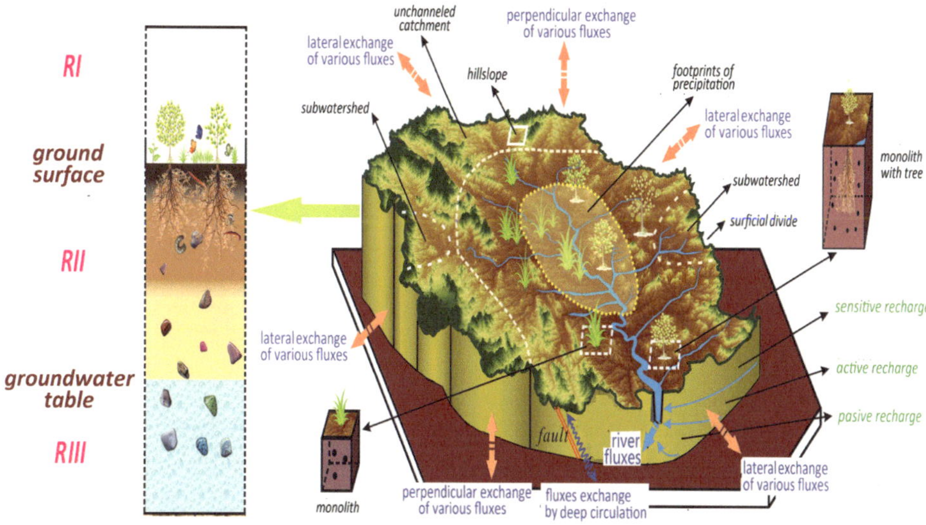

Figure 9. A conceptualized critical zone experimental block (CZEB) [8].

water, air, and living organisms regulate the natural habitat and determine availability of life sustaining resources" [13]. Later in 2005, in a report of a workshop, it was further defined as an extending from "the vegetation canopy to the zone of groundwater" [36], or "the bedrock to the atmosphere boundary layer" [37]. Critical zone becomes one of the most compelling research areas in earth sciences in the twenty-first century [14].

Aforementioned that if we designate the first phase of basin study until ca. the middle twentieth century as the stage of foundational and the second phase during/after the recommendation of EB of IHD since 1965 as the stage of developmental, a third phase of renovation seems ready to come out inevitably. What we suggested is the concept of critical zone experimental block (CZEB) [8], focusing on the fact that the natural watershed hydrological system itself, running on the principle of holism is an extending from watershed surface to its underground deposit layers, never just the dissected watershed surface laterality. The suggested CZEB geologically is a monolith block, with its surface, the watershed, bounded by topographical water divides (**Figure 9**) [8], it is actually an experimental "block" within the critical zone with a surface drainage basin (the watershed). It is a dynamic ecosystem coupled with various supporting systems but using hydrological processes as the unifying theme. It is a living, breathing, evolving boundary layer where rock, soil, water, air, and living organisms interact [38]. The CZEB is the watershed hydrological experimental study in the CZ observatory framework. As "the critical zone observatory network is the only type to integrate biological and geological sciences so tightly" [37]; following this, the CZEB network will perhaps be the network to integrate the hydrological, biological, and geological sciences so tightly as well.

3.4.1. The boundaries of CZEB

Top: if the mean evaporation surface of the canopy is $h(m)$ above the mean ground surface, then the top boundary of a CZEB is defined as $H(m)$, where $H \geq (1.5\text{–}2.0) \times h$ with coefficients of h

varying according to the vegetation, 1.5 for crops and grasses, 2.0 for woods, and forest. This is mainly for the purpose of energy budget and eddy covariance flux observations. *H* is just the lower part of the atmosphere boundary layer.

Bottom: the three cases are discussed below:

Case I: bedrock is situated at a relatively shallow depth from the ground surface while the regolith is shallow, too. The bottom boundary is defined as the geological boundary (**Figure 10a**).

Case II: bedrock is deep. The bottom boundary is defined as the plane where the tritium content of groundwater approaches zero or the detection limit of ±0.7 TU (the "tritium naught line" (TNL) [39]), which is same as that of Case III (**Figure 10b**).

Case III: there are stratified alluvia, potentially with multiple aquifers with aquicludes and aquitards, common in flat plain areas with thick deposit and deep bedrock, up to hundreds of meters or more. Groundwater recharge to the river can be separated into sensitive, active, and passive zones. In this case, the bottom boundary of CZEB is defined as the bottom of the active recharge zone, which is also the plane of TNL (**Figure 10c**).

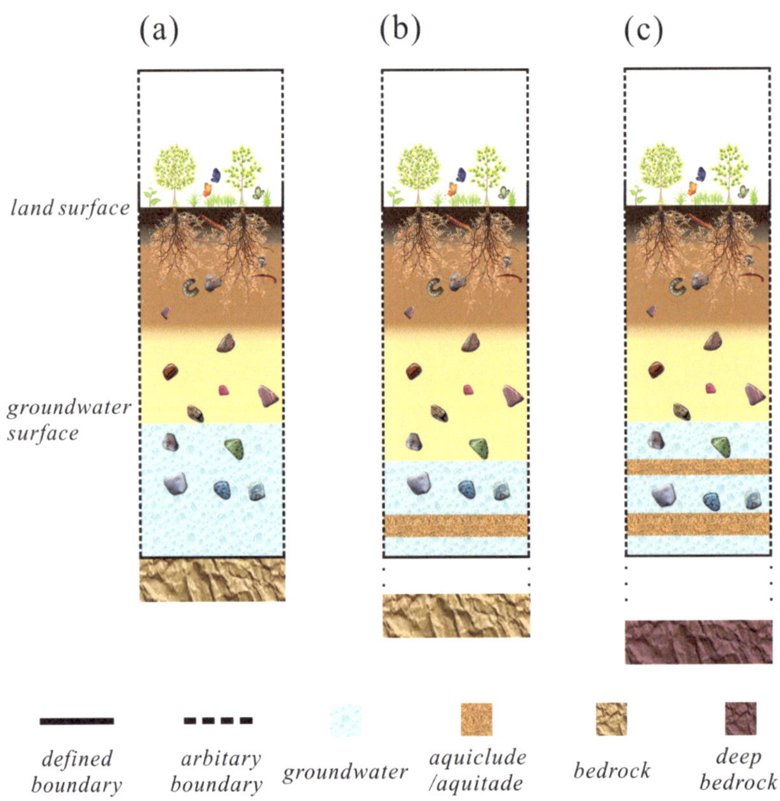

Figure 10. Boundaries of CZEB [8].

http://dx.doi.org/10.5772/intechopen.78721

Lateral sides. There are two parts:

Part I: above the ground surface up to a height *H* as described above, delineated according to the surface topographic watershed boundary (**Figure 11a**). The lateral exchanges of water vapor, heat, CO_2, etc. will be monitored.

Part II: below the ground surface, it is in general defined arbitrarily except in the case of existing geological boundaries. (1) Case I: for a general CZEB, the lateral exchanges as shown schematically in **Figure 9** are subjected to monitoring including water tracing, (2) Case II: for the setting of a *controlled-natural* entity as described above, it needs to close all the underground surroundings until bedrock aimed at *constrain complexity* by using of engineering methods; e.g., in Nandadish, it is ready to use steel sheet piling following its surficial boundary perpendicularly until bedrock.

3.4.2. The functioning of CZEB

3.4.2.1. Interfaces and compartment zones

CZEB encompasses "the near-surface biosphere and atmosphere, the entire pedosphere [8], and the surface and near-surface portion of the hydrosphere and lithosphere" [14]. Within the CZEB, various processes including hydrologic, atmospheric, lithospheric, geomorphic, and geochemical processes are coupled and dynamically interrelated. To simplify the organization of various interfacing processes throughout the CZEB, compartment zones can be broadly separated as follows [14]:

1. The zone aboveground surface, the "aboveground vegetation zone."

2. The unsaturated zone, the "belowground root zone and the deeper vadose zone."

3. The saturated zone, the "saturated aquifer zone."

This layering has a general trend of increasing density with depth, has a dampening effect on state variables with depth, and an increase in distance to energy input at the soil surface [36]. There is also "an overall trend of increasing characteristic response time" [14, 40].

3.4.2.2. The "reactors" of CZEB

Each zone will behave as a "feed-through reactor" according to Anderson et al. [41]. It follows then that there are three feed-through reactors coupled together in the CZEB (**Figure 11**), with probably different rates and residence time of materials.

1. Reactor I. The first zone is the aboveground surface zone, which also contains aboveground vegetation up to its interface with atmosphere (**Figure 11**). Evaporation and plant transpiration may account for 50% or more of the total local precipitation and, use up to ca. 50% of the total solar energy.

2. Reactor II. The second zone is the unsaturated zone, extending from the soil surface to the upper surface of the groundwater table. It consists of the whole soil profile: the O horizon (humus), A horizon (topsoil), B horizon (subsoil), and C and potentially D horizons

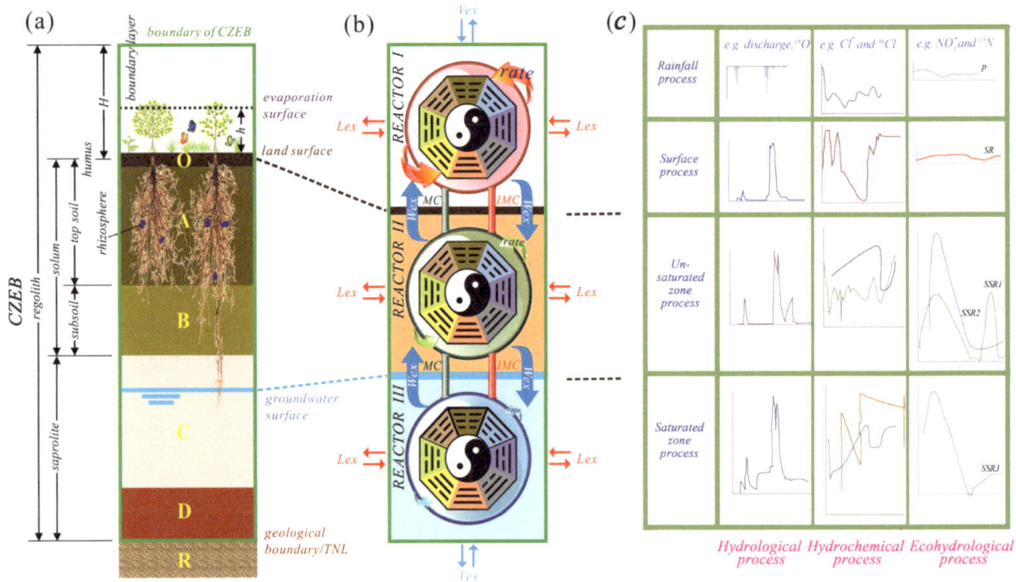

Figure 11. (a) Functioning of the CZEB, (b) the schematic "reactors" corresponding to CZEB, and (c) general monitoring parameters for the CZEB.

(**Figure 11a**). The unsaturated soil zone has been recognized as "the most complicated biomaterials on the planet" [14, 27].

3. Reactor III. The third zone is the saturated zone, extending from the groundwater table down to bedrock, including the capillary fringe (**Figure 11**). There are two general cases for CZEB, phreatic groundwater and confined aquifers.

3.4.2.3. *Functioning of reactors*

1. Reactor I includes hydrometeorological and ecohydrological processes, reactor II includes hydropedological and hydroecological processes, while the reactor III is more exclusive for hydrogeological processes.
2. There is an overall trend of decreasing operation rates from reactors I to III.
3. Each reactor has its own lateral flux exchanges L_{ex} via the CZEB lateral boundaries (**Figure 11b**). Reactor I has vertical fluxes exchange V_{ex} with atmosphere via the upper boundary of CZEB while reactor III has V_{ex} with deep aquifers and/or deep circulation via the bottom boundary of CZEB (**Figure 11b**). Fluxes include all material and immaterial components.
4. Reactors are closely coupled within the boundaries of CZEB. There are flux exchanges W_{ex} (blue arrows in **Figure 11b**) between reactors I and II and, and between II and III within the CZEB. There are flux channels *MC* (**Figure 11b**) through I to III for materials, and flux channels *IMC* (**Figure 11b**) through I to III for nonmaterials.

5. The *MC* appears similar to blood vessel system and *IMC* to meridians and collaterals of the human body from the Chinese art of acupuncture.
6. The general monitoring parameters in reactors are shown schematically in **Figure 11c**.

A trial for CZEB is included in the Chuzhou WHES, i.e., the Nandadish, which will be described in detail later.

4. Conclusions

The hydrological experimentation is actually a dialog with the hydrological nature; however, such a dialog seems not easy, it depends mostly on the adequate *method of questioning*, i.e., on the designed objects and the keys for it. If we designate the first phase of basin study until ca. the middle twentieth century as the foundational stage and the second phase during/after the IHD since 1965 as the developmental stage, which has been going on for more than five decades up to now, a third phase of renovation seems ready to come out inevitably. Facing with this transition of the experimental watershed hydrology, it appears worth to have a fundamental rethinking on our previous methods of dialog, which seems challenged, sometimes having a successful beginning but becoming increasingly inadapted with the dynamic nature. The inherent defects of the field basin studies have been summarized from Chinese decades' experiences in their basin studies with zigzagging process especially those based on the concept of experimental basin system.

Facing with the hydrological complex dynamic system, a framework of *watershed hydrological experimental system* (*WHES*) is suggested, which is raised from both the viewpoints of the *general system theory* based on the paralleled concepts of the ancient Chinese and the Western, and that of the Middle-ground perspective philosophy on the *Middle Way* ("*golden mean*") that is also based on the paralleled concepts. Thus, the WHES is defined as an experimental system that is designed to dialog with the complex watershed hydrological nature, to drive opening of various doors of their black boxes aimed at revealing mechanisms hidden deep in the system.

Three strategies have been developed for the WHES: the strategy of *constrain complexity* for the natural watershed, which is the downward or top-down approach, the strategy of *add complexity* for the physical model, which is the upward or bottom-up approach, and the strategy of *manipulation* for the operation of both ways including the so-called *artificial-natural* and the *controlled-natural*. In fact the suggested WHES is trying to reconcile the deterministic and stochastic extremes, "opposites are complementary" as the basic Chinese philosophy have revealed.

As a trial of it, the Chuzhou WHES is ongoing. Most problems involved in WHES fall in the category of complex systems with some degree of organization. It includes three levels with both complexity and randomness from high to low, and an extension part: (1) high level: composed by pure natural EBs, defined as the "macros" of WHES, (2) intermediate level: composed by intermediate chain, the "mesos" of WHES, it includes both the controlled-natural entities and the artificial-natural entities, (3) low level: composed by pure artificial chain, the "micros" of WHES, and (4) extension of WHES: it is the laboratory for geochemical processes by water tracing for all entities. It is the "nucleus" of WHES, an essential condition of it. Different organization levels of WHES will have different treatments; only the "micro" level of WHES is linear, a small part of the "meso" level close to the "micro" level is quasilinear or linearizable, and these two parts can be treated by Newtonian dynamic methods. Most of the intermediate chain, the "meso" level of WHES, should be treated by nonlinear dynamic methods.

Trying to improve the current idea on the research platform of hydrological experimental basin (EB), which is widely used since IHD of 1965, a CZEB within the Chuzhou WHES was suggested. The boundary and functioning of it are discussed including its interfaces and three compartment zones: aboveground vegetation zone, belowground unsaturated zone including root zone and deeper vadose zone, and the saturated aquifer zone. These zones can be treated as three coupled feed-through reactors. Such CZEB is a watershed hydrological experimental study in the CZ observatory framework.

Acknowledgements

This review is a commemoration of the 65th anniversary of the Chinese basin study and the 37th anniversary of the establishment of the Chuzhou Hydrology Laboratory (CHL). The senior author (W-Z Gu) has been fortunate to render his services to the both from their beginning; with grateful heart, he is indebted to a number of people including colleagues, experts, professors, and local peasants, who have helped, succored, supported to make this possible. We are very grateful for those who have made numerous efforts for the reborn renovation of the CHL after a long abandonment and for its leaping forward from manual to digital: Academician Jian-Yun Zhang, Prof. Jiu-Fu Liu, and many other experts in the Chinese Ministry of Water Resources. We also thank many foreign experts who have encouraged, helped, taught, and guided our work over the years especially during our difficulty, including Jeffrey McDonnell, Keith Beven, Henry Lin, Mebus Geyh, Carol Kendall, Vance Kennedy, Norman Peters, Michael Sklash, А.И. Шикломанов, Joel Gat, Verhagen B, and Klaus Froehlich. We also acknowledge many colleagues in the CHL for their hard work in the field.

This work was supported by the grants from the National Key R&D Program of China (No. 2017YFC0403500) and the National Natural Science Foundation of China (Nos. 91647203, 91647111, and 51609145).

Author details

Wei-Zu Gu*, Jiu-Fu Liu, Ai-Min Liao, Niu Wang, Jia-Ju Lu, Jin Lin, Hong-Wei Liu,
Wen-Zhong Wang, Tao Ma, Zhao Cai, Min-Han Liao, Xue-Gang Li, Peng Zhuo and Na Yang

*Address all correspondence to: wzgu@nhri.cn

Department of Hydrology and Water Resources, Nanjing Hydraulic Research Institute,
Ministry of Water Resources, China

References

[1] UNESCO/WMO/IAHS. Three Centuries of Scientific Hydrology. Paris: UNESCO; 1974

[2] Rodda JC. Basin Studies. In: Rodda JC, editor. Facets of Hydrology. London: John Wiley & Sons; 1976. pp. 257-297

[3] McDonnell JJ, Sivapalan M, Vache K, Dunn S, Grant G, Haggerty R, Hinz C, Hooper R, Kirchner J, Roderick ML, Selker J, Weiler M. Moving beyond heterogeneity and process complexity: A new vision for watershed hydrology. Water Resources Research. 2007;**43**: W07301

[4] Sivapalan M. Pattern, Process and function: Elements of a unified theory of hydrology at the catchment scale. Anderson MG, Encyclopedia of Hydrological Science. London: John Wiley & Sons; 2005. 193-219

[5] Burt TP, Mcdonnell JJ. Whither field hydrology? The need for discovery science and outrageous hydrological hypotheses. Water Resources Research. 2015;**51**:5919-5928. DOI: 10.1002/2014WR016839

[6] Kirkby MJ, editor. Geomorphology: Critical Concepts in Geography. Vol. II. London: Routledge; 2004. p. 16

[7] Gu W-Z, Liu C-M, Song X-F, Yu J-J, Xia J. Hydrological experimental system and environmental isotope tracing: A review on the occasion of the 50th Anniversary of Chinese basin studies and the 20th Anniversary of Chuzhou Hydrology Laboratory. In: Xi R-Z, Gu W-Z, Seiler K-P, editors. Research Basins and Hydrological Planning: Proceedings of the International Conference on Research Basins and Hydrological Planning; Hefei, China; March 2004. London: A.A. Balkema Publishers; 2013. pp. 22-31

[8] Gu W-Z, Liu JF, Lu JJ, Frentress J. Current challenges in experimental watershed hydrology. In: Bradley P, editor. Contaminat Hydrology and Water Resources Sustainability. Rijeka, Croatia: InTech Publishing House; 2013. pp. 299-333. DOI: 10.5772/55087

[9] Vince G. An Epoch Debate. Science. 2011;**334**:32-37

[10] National Research Council. Challenges and Opportunities in the Hydrologic Sciences. Washington D.C.: The National Academies Press; 2012

[11] Kirchner JW. Getting the right answers for the right reasons: Linking measurements, analyses, and models to advance the science of hydrology. Water Resources Research. 2006;**42**:W03S04

[12] Yan ZX, Fan DP, Zhang HX. An Introduction to Systems Science—Exploration of Complexity (in Chinese). Beijing: Peoples' Publishing House; 2006

[13] National Research Council. Basic Research Opportunities in Earth Science. Washington D.C.: The National Academies Press; 2001

[14] Lin H. Earth's critical zone and hydropedology: Concepts, characteristics, and advances. Hydrology and Earth System Sciences. 2010;**14**:25-45

[15] Bateson G. Mind and Nature: A Necessary Unity. New York: Dutton; 1979

[16] Weinberg S. Dreams of a Final Theory: The Scientist's Search for the Ultimate Laws of Nature. London: Vintage Books; 1994

[17] Capra F. The Tao of Physics—An Exploration of the Parallels between Modern Physics and Eastern Mysticism. London: Flamingo; 1982. p. 417

[18] Lin Y, OuYang SC. Irregularities and Prediction of Major Disasters. New York: Taylor and Francis Group, CRC Press; 2010. p. 575

[19] Von Bertalanffy L. General System Theory: Foundations, Development, Application. New York: George Braziller, Inc; 1973

[20] Weaver W. Science and complexity. American Scientist 36, 536 (1948). In: Klir GJ, editor. Facets System Science. New York: Kluwer Academic/Plenum Publishers; 2001. pp. 533-540

[21] Weinberg GM. An Introduction to General Systems Thinking. New York: Wiley-Interscience; 1975

[22] Dooge JCI. Looking for hydrologic laws. Water Resources Research. 1986;**22**(9):46S-58S

[23] Hu WX. The Philosophy of the Middle View (in Chinese). Beijing: Beijing University Press; 2016

[24] Murti TRV. The Central Philosophy of Buddhism (Translated into Chinese). Guiyang: Guizhou University Press; 2013

[25] Zhao GQ, Yang XM. On the "revolution of meso-herarcgy" of the transnational corporation from the view of middle-ground perspective with self organization. Scientology and Management of Science and Technology. 2004;**25**(3):79-82

[26] McNeil DH. Entering the era of irregularity (forward). In: OuYang SC, McNeil DH, Lin Y, editors. Entering the Era of Irregularity (in Chinese). Beijing: Meteorological Press; 2002. p. 3

[27] Gu W-Z, Liu J-F, Lin H, Lin J, Liu H-W, Liao A-M, Wang N, Wang W-Z, Ma T, Yang N, Li X-G, Zhuo P, Cai Z. Why hydrological maze: The hydropedological trigger? Review of experiments of Chuzhou hydrology laboratory. Vadose Zone Journal; **17**:170174. DOI: 10.2136/vzj2017.09.0174

[28] Toffler A. Forward: Science and change. In: Prigogine I, Stengers I, editors. Order out of Chaos. Toronto: Bantam Books; 1984

[29] Dawkins R. The Blind Watchmaker. London: Norton & Company, Inc.; 1986

[30] Ulanowicz RE. Ecology: The Ascendant Perspective. New York: The Columbia University Press; 1997

[31] Bloschl G. On the fundamentals of hydrological sciences. In: Anderson MG, editor. Encyclopedia of Hydrlogical Sciences. John Wiley & Sons; 2005. pp. 1-12

[32] Beven KJ, Germann P. Macropores and water flow in soils revisited. Water Resources Research. 2013;**49**:3071-3092. DOI: 10.1002/wrcr.20156

[33] Klemeš V. Conceptualization and scale in hydrology. Journal of Hydrology. 1983;**65**:1-23

[34] Gu W-Z. On the domain and approach of the experimental hydrology (in Chinese). In: Nanjing Hydrology Institute, editor. Treatise on Hydrology and Water Resources. Beijing: Water Resources and Electric Power Press; 1987. pp. 507-515

[35] Wei-Zu G, Liang S-L. The Bluebrook drainage experimental watersheds. Water Resources and Hydropower Engineering. 1965;**N6**:43

[36] Brantley SL, White TS, White AF, Sparks D, Richter D, Pregitzer K, Derry L, Chorover J, Chadwick O, April R, Anderson S, Amundson R. Frontiers in exploration of the critical zone. Report of a workshop sponsored by the National Science Foundation; 2005

[37] Report prepared by the CZO Community. Future Directions for Critical Zone Observatory (CZO) Science; 2010

[38] National Critical Zone Observatory Program. Available from: http://criticalzone.org/index.html

[39] Seiler K-P, Lindner W. Near surface and deep groundwater. Journal of Hydrology. 1995;**165**:33-44

[40] Arnold RW, Szabolcs I, Targulian VO. Global Soil Change. Laxenburg: International Institute for Applied Systems Analysis; 1990

[41] Anderson SP, von Blanckenburg F, White AF. Physical and chemical controls on the critical zone. Elements. 2007;**3**:315-319

Chapter 12

Practice on the Watershed Hydrological Experimental System Reconciling Deterministic and Stochastic Subjects Based on the System Complexity: 2. Practice and Test

Jiu-Fu Liu, Ai-Min Liao, Niu Wang, Jin Lin, Hong-Wei Liu, Wen-Zhong Wang, Tao Ma, Zhao Cai, Min-Han Liao, Xue-Gang Li, Peng Zhuo, Na Yang, Jia-Ju Lu and Wei-Zu Gu

Additional information is available at the end of the chapter

http://dx.doi.org/10.5772/intechopen.79357

Abstract

This is the second of a two-part series on the watershed hydrological experimental system (WHES) aimed at practice and test of it at Chuzhou Hydrology Laboratory. It constitutes both natural and artificial entities of different scales, within which two typical main subjects are reviewed here. First is a natural watershed Nandadish, which is subjected to be a Critical Zone Experimental Block, under manipulation strategy of constrain complexity to compare with the pure natural watersheds, it is the controlled-natural as we termed. Second is an artificial catchment Hydrohill, under the strategy of add complexity to compare with the simple artificial lysimeters, it is the artificial-natural as we termed. The constructions and instrumentations of these experimental catchments are reviewed, especially their renovation version during recent years after a long abandonment. Some results get during the operation of Chuzhou WHES are outlined here as well.

Keywords: watershed hydrology, hydrological experimentation, critical zone experimental block, artificial catchment, controlled catchment, Hydrohill

1. Introduction

A framework of watershed hydrological experimental system (WHES) is suggested raised from the theoretical study on the complex hydrological system. It is defined as an experimental system to dialog with the complex watershed hydrological nature, to drive the opening of various doors of their black boxes aimed at revealing mechanisms hidden deep in the system with some degree of organization. In fact, the organization of the WHES is trying to reconciling the deterministic and stochastic extremes for the watershed hydrological complex system, "opposites are complementary" as the basic Chinese philosophy have revealed.

As a trial of the suggested WHES, the Chuzhou WHES is ongoing for tests. The constituent parts of the WHES can be resolved into four categories as the "macros", "mesos", "micros," and the "nucleus". "Macros" are composed by pure natural EBs at high level of complexity and randomness; "micros" by pure artificial chain at low level; they are two extreme levels in WHES, while the "mesos" are the chain of intermediate phase; actually, it will play the critical role, the "golden mean," in WHES as described in Chapter 11. The "nucleus" are water isotopes and solute isotopes from all chains of macros, mesos, and micros by overall sampling; it characterizes the internal linkages between them, and reveals their interrelated processes; actually, it is an essential condition of in WHES.

In Chuzhou WHES, there are two extremes in the intermediate "meso" blocks (Figure 8a in Chapter 11): One is the controlled-natural Nandadish based on the strategy of constrain complexity, another one is the artificial-natural Hydrohill on that of add complexity. Nandadish is designed to meet with the idea on Critical Zone Experimental Block with an intention trying to replace the current Experimental basins suggested by the Representative and Experimental Basin Programme of the first International Hydrology Decade (IHD) since 1965. These two typical experimental meso-mediation blocks of Chuzhou WHES are reviewed here.

2. Practice on the WHES: I–Nandadish, a natural CZEB

2.1. The constructions of Nandadish

Nandadish catchment is a forest watershed with a surficial drainage area of 7897 m^2. Before its setting, a geological exploring was made by using of 69 drillings distributed in an area covering not only the surficial watershed drainage area from its surficial divides but more area outside. The bedrock elevation of every drill was then measured, so the hypsographic map of the bedrock can be made, together with that of ground surface, the isopachous map of its Quaternary deposit can then be get as well (**Figure 1**). Also from eight drillings with depth penetrating through the bedrock with core sampling for the formation and lithology explorations, no fault, no fracture, and obvious fissures were found from this igneous stratum of andesitic and tuffaceous facies with a thin-weathered layer. It is good for our idea of *controlled-nature*. The depths of its quaternary regolith resting on the bedrock have a range of 1–7 m with an average of 2.46 m (**Figure 1**). It is deeper near the upper divide but only ca 1 m in thickness near the outlet, making the catchment easy to close via a concrete wall installed to the bedrock at the outlet. The vadose zone consists of brunisolic soil of heavy loam, medium

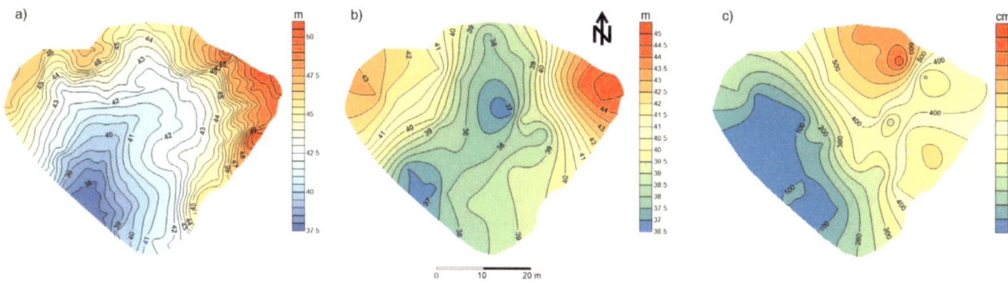

Figure 1. (a) Hypsographic map of the ground surface of Nandadish; (b) hypsographic map of the bedrock surface; and (c) Isopachous map of the deposit thickness (including surficial soil).

and clay loams; saprolite with prismatic and block structures, horizontal and vertical fissures and cracks developed in the upper regolith. The altitude difference of watershed approaches 12.9 m with a surface slope ranging from 6.7 to 17.1%.

Aimed at a CZ hydrological experimental block aforementioned, the main construction tasks as sketchily shown in **Figure 2a** are threefold: (1) To change its original trench into the layered

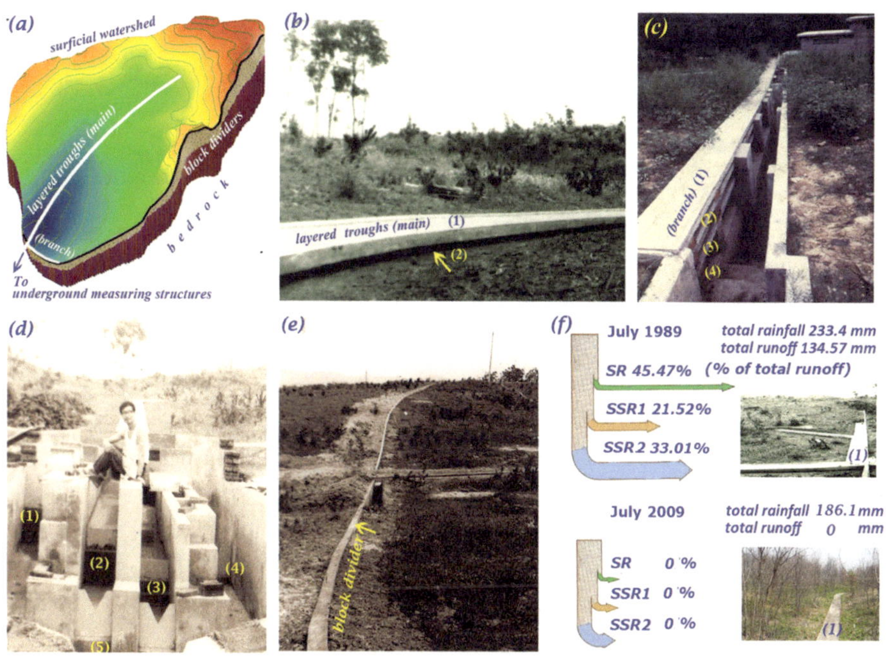

Figure 2. (a) Schematic diagram of Nandadish showing the locations of main construction tasks; (b) original view of the main troughs and the catchment coverages during 1980; 1-trough for rainfall, which is served for separation of "channel rainfall" from surface runoff collected from trough 2, 2-for surface runoff (SR); (c) original view of the branch troughs with a watching gallery for the students practice to seeing the real processes of different runoff components during rainfall event; 1-trough for rainfall, 2-for surface runoff (SR), 3, 4-for subsurface runoff (SSR) from troughs at different depths; (d) measuring structures under construction for different troughs, 1-for rainfall (V-notch sharp crested weir), 2-for SR (V weir and rectangular sharp crested weir), 3-for SSR (V weir and rectangular weir), 4-for SSR (V weir), 5-for total runoff (V weir and rectangular weir, not shown); (e) a part of the underground "block divider" with 0.3 m above the ground surface; (f) the change of runoff compositions within time span of 20 years compared with July 1989 and July 2009 using same measuring structures with different catchment coverages.

troughs aimed at collecting different runoff components for direct monitoring and sampling as well. From the natural surficial topography, a main trench and a branch trench are set up both with four layered troughs with locations shown in **Figure 2a**, the general view is shown in **Figure 2b** and **c**; (2) These troughs are led to discharge measuring structures separately within an underground building, the original view of four measuring structures under construction corresponding to troughs, respectively is shown in **Figure 2d**; (3) For setting of a *controlled-natural* entity, it needs to close all the underground surroundings until bedrock, the "block divider", aimed at *constrain complexity* aforementioned as shown in **Figure 2a** and **e**, it is 367 m in total with average depth of 2.94 m from bed rock to 0.3 m above the ground surface. It is only completed partly because of seeking for a better engineering method for the limited working space.

The coverage during the watershed's construction in 1979 was natural grasses with small shrubs and a few Masson pines aged 5–6 years (**Figure 2b** and **f**). Since then, coverage has shifted to a dense forest with canopy height ca 12 m. There are two dominant tree species (*Q. acutissima Carruth* and *B. papyrifera*) accounting for ~90 and 10%, respectively. The new version of that forest watershed is shown in **Figure 3**.

It follows that a hydrological change is happened, which is summarized in **Figure 2f**, compared with the total runoff 134.57 mm of July 1989 with monthly rainfall of 233.4 mm, the total runoff and all runoff components including surface runoff and subsurface runoff of July 2009 are zero all and even having less monthly rainfall of 186.1 mm. So, the setting of troughs during the renovation of CHL is considered and will be described in the following paragraphs.

2.2. The instrumentation of Nandadish

2.2.1. Precipitation

The redistribution of rainfall intensity by the canopy is one of the main research topics in Chuzhou Hydrology Laboratory. Rainfall observation system (**Figure 4a**) in Nandadish was

Figure 3. The bird's eye view of the forested Nandadish watershed taken at May 2018. 1-trench with layered troughs; 2-block divider; 3-house for discharge measuring structures.

built up to observe rainfall over trees, rainfall under trees (i.e., throughfall), and stem flow to determine the temporal and spatial redistribution of rainfall, and to estimate the canopy interception. To observe rainfall over trees, four tipping bucket rain gauges were mounted on towers located on the four directions and center of the catchment (**Figure 4b**). To observe throughfall, 8 tipping bucket rain gauges, 80 micro rain gauges under trees, and 94 standard rain gauges under trees were installed under trees (**Figure 4a** and **c**). Stem flow was collected in 14 trees from two dominant tree species (*Q. acutissima* Carruth and *B. papyrifera*) using stem flow collection collars and tipping bucket flow meters (**Figure 4d** and **e**).

To accurately determine the ratios of throughfall, stem flow and canopy interception to gross rainfall, a large quadrat with an area of 25 × 25 m (**Figure 4f**) and a rainfall station under trees with an area of 8.45 × 4.05 m (**Figure 4g**) were constructed in the Nandadish. In this rainfall

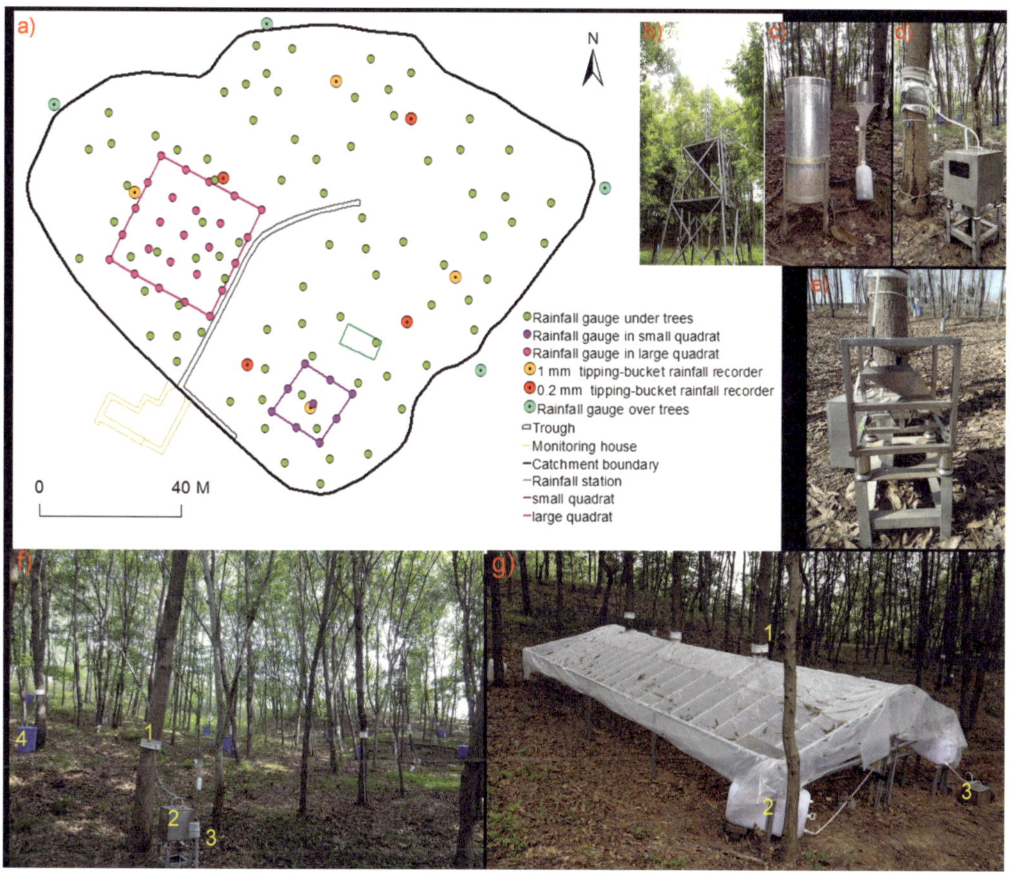

Figure 4. Instrumentation for precipitation measurement in the Nandadish: (a) locations of stainless troughs and rain gauges; (b) four tipping-bucket rain gauges were mounted on towers; (c) micro rain gauges and standard rain gauges under trees; (d) stem flow measurement using collection collars and tipping-bucket flow meters; (e) the inner construe of the tipping bucket flow meter; (f) photograph of a large quadrat with an area of 20 × 20 m (1-the collection collar collecting stem flow, 2-a tipping-bucket flow meter with 500-mL buckets used to determine the stem flow process; 3-a 100-L water container used to measure the total amount of stem flow); (g) a rainfall station under trees with an area of 8.45 × 4.05 m (1-the collection collar collecting stem flow, 2-a 150-L water container used to buffer the throughfall flow so that the following tipping-bucket flow meter is capable of measuring the flow when strong rain occurs, 3-a tipping-bucket flow meter with 2.0-L buckets).

Figure 5. Instrumentation for runoff measurement in Nandadish: (a) Main trench with troughs (not shown), (b) the branch trench with four troughs (previously **Figure 13c**), (c) schematic figure illustrating various troughs, (d) discharge measurement structures for different runoff components from troughs 1 for rainfall; 2 and 3 for surface runoff; 4 and 5 for interflow (50 cm below the soil surface); 6 for interflow and groundwater flow (down to bedrock); 7 and 8 for total runoff (weirs are shown (f)and (g)). 1, 3, 5, 6, and 8 are 90° V-notch weir; 2, 4 and 7 are the full width rectangular weirs; 9 is the probe-type water level gauge; 10 is a video for monitoring the runoff processed, (e) combination of a 90° V-notch weir and a full width rectangular weir for SR, SSR50, and the total runoff, (f) the full width rectangular weir for total runoff, (g) the 90° V-notch weir for total runoff.

station, three trees of about 20 years were included, and three tipping bucket flow meters (0.4 L per bucket) were installed to measure the stem flow of each tree. A closing apparatus like a roof was explored to collect the total through rain. This apparatus is named as "collecting roof". The collecting roof with a large area will result in a total amount of through rain when strong rainfalls occur, and thus large flow rates appear in the two outlets of the collecting roof. Two larger tipping bucket flow meters (2.0 L per bucket) were installed in the two outlets to measure flow rates. In addition to measure stem flow and through rain, a tipping bucket rain gauge was laid over the trees to record the rainfall inputting the rainfall station. According to water balance, the rainfall over the trees equals to the sum of stem flows, through rain and interception of trees.

2.2.2. *Runoff*

The surface and subsurface runoff are monitored directly via four layers of troughs fixed in a trench with a gradient of 6.7% (**Figure 5a**). These troughs are stacked on top of each other to capture rainfall, surface, and subsurface flows (**Figure 5b** and **c**): the uppermost trough captures rain; the next lower trough captures surface runoff (SR); and the two lower troughs capture subsurface flow from soil layers spanning the depths of 0–50, and 50–100 cm, inferred as SSR50 and SSR100 troughs. SSR50 and SSR100 troughs have 20-cm stainless lips that extend horizontally into the soil layer to prevent leakage between layers (**Figure 5c**). Waters captured in troughs are routed into a gauging room and measured by 90° V-notch and rectangle weirs (**Figure 5d** and **e**). For SR and SSR50, 90° V-notch and rectangle weirs are combined to measure discharge: when the large discharge occurs (correspondingly the water head above the rectangle weir is higher than 5.0 cm), the V-notch weir fails to measure discharge and the rectangle weir performs better; when the water head above the rectangle weir is lower than 5.0 cm, the discharge is measured more accurately by the V-notch weir than the rectangle weir. For rainfall and SSR100, only V-notch weir is used due to their less discharge compared with that of SR and SSR50. The trough SSR50 previously is located at 30 cm below the ground surface, this depth was extended to 50 cm during renovation due to the big changes happened to the growing of plants together with the deeper extension of their root system as described in **Figure 5f**.

2.2.3. *Soil moisture*

A network of 34 profile soil moisture sensors (SM-1, ADCON, same as those installed in Hydrohill, see latter) were installed in the different depths of soil. The number of profile soil moisture sensors with a depth at 90, 120, and 150 cm below the ground surface are 9, 10, and 15 (**Figure 6a**). The previous network of 34 access tubes (1982–1994) for neutron moisture gauge from UK Institute of Hydrology is shown in **Figure 6c**, the construction of access tube is shown in **Figure 6e**, while the standards of soil moisture for the calibration of neutron moisture gauge in a special laboratory are shown in **Figure 6f**.

2.2.4. *Groundwater*

For groundwater monitoring, there are 34 galvanized tube wells intersecting through the soil till the bedrock (**Figure 6b**). Water table measurement is performed with 30 level sensors (LEV1, ADCON, see later in Hydrohill). The previous network of wells for groundwater monitoring and sampling (1982–1994) is shown in **Figure 6c**, while the construction of well is shown in **Figure 6d**.

2.2.5. *Sap flow*

Sap flow is measured by Granier-type thermal dissipation probes (TDP) (Yugen, Beijing, China) that were installed in the sapwood of sample trees. A set of TDP includes a heated needle above and a reference needle below (**Figure 7a**). The sap flow velocity is calculated based on the temperature difference between the two needles. After the corky bark within a rectangular with a wide of 4 cm and tall of 10 cm was shaved off, the two probe needles

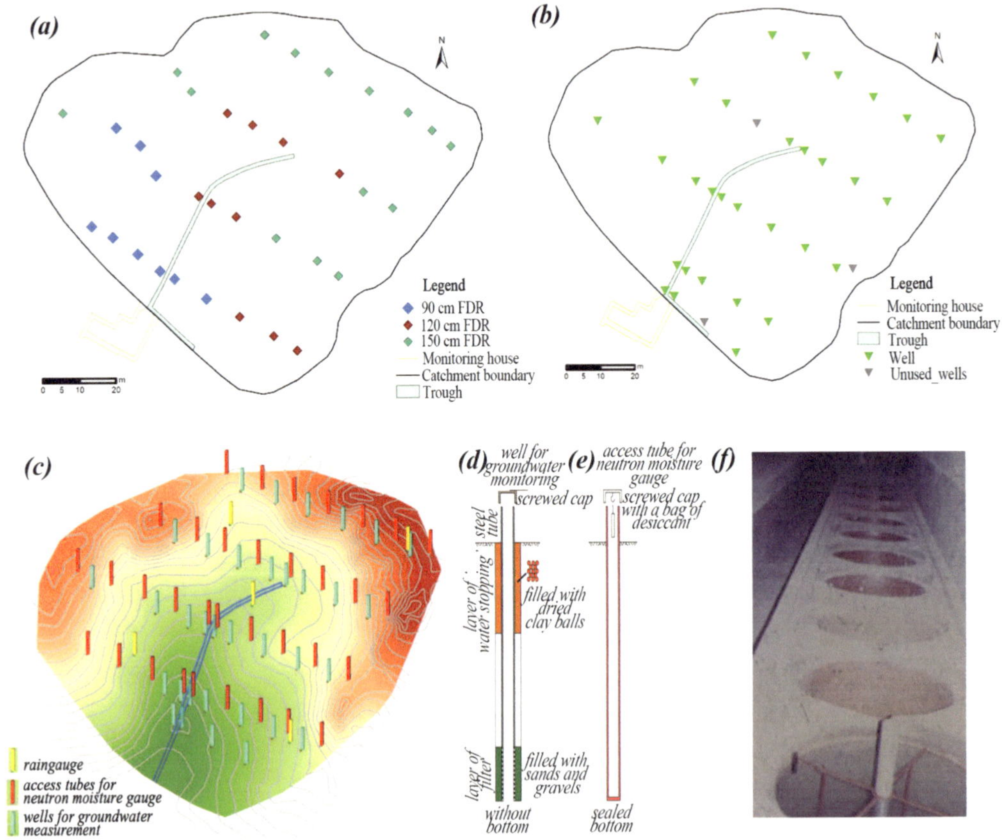

Figure 6. (a) Locations of profile soil moisture sensors with different depths: 90, 120, and 150 cm (since 2012); (b) network of wells for groundwater monitoring and sampling (since 2012); (c) previous network of access tubes for soil moisture monitoring and that of wells for groundwater monitoring (1982–1994); (d) construction of groundwater monitoring well; (e) construction of access tube for neutron moisture gauge (1982–1994); (f) standards of soil moisture with volumetric contents from 100 to 3% in a special laboratory for the calibration of neutron moisture gauge.

was inserted into the sapwood approximately 5 cm apart vertically. The reflective bubble shield was wrapped around the TDP probe to avoid monitoring errors caused by direct sunlight and rainfall leaching (**Figure 7b**). In the Nandadish catchment, totally 24 sets of TDPs were installed for 12 trees from 5 species, including three *Quercus acutissima Carruth*, three *Broussonetia papyrifera, two Populus L.*, two *Celtis L*, and two *Melia azedarach L.*, to match the sapwood width of different tree species with different DBH, three different lengths of the probe (TDP10, TDP20, and TDP30) were adopted. All of the TDPs were installed about 145 cm above ground, and on both of the south and north sides of each sample tree. The temperature difference between the two needles was scanned at 1 min intervals, and the 10 min average value was recorded by a data logger (CR1000, Campbell, USA Scientific Inc., Logan, UT, USA).

http://dx.doi.org/10.5772/intechopen.79357

Figure 7. Installation of thermal dissipation probes: (a) a set of TDP includes a heated needle (1) and a reference needle (2); (b) the reflective bubble shield (3) was wrapped around the TDP probe to avoid monitoring errors caused by direct sunlight and rainfall leaching, and a collection collar (4) for stem flow was installed above the TDP to stop stem flow entering.

2.3. Water sampling

Water samples from precipitation, runoffs, and plants were collected. Rain water samples were collected via a specially designed rain gauge and a standard rain gauge, which were installed on the roof of the gauging room (**Figure 8a**). The specially designed rain gauge is capable of collecting rain samples at 1-hour interval, while the standard rain gauge collects the mixed sample of each rain event. A batch sampling system is designed and constructed based on the negative pressure to easily and fast collect the water samples of runoff components of SR, SSR50, and SSR100 (**Figure 8b**). Water samples for runoff components are collected also via a stainless steel tube head fixed at the connection trough before the runoff reaches the ponding of the weir.

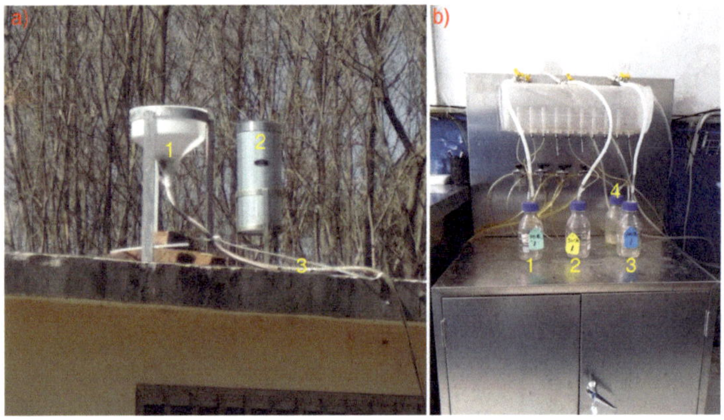

Figure 8. Water sampling in Nandadish: (a) sampling for precipitation, 1-a specially designed rain gauge capable of collecting rain samples at one-hour interval, 2-a standard rain gauge collecting the mixed sample of each rain event, 3-a tube directing rain water into a sampling bottle at the gauging room; (b) sampling for runoff components, 1-a sample bottle for SR, 2-a sample bottle for SSR50, 3-a sample bottle for SSR100, 4-a safeguard bottle.

3. Practice on the WHES: II–Hydrohill, an artificial catchment

3.1. The construction of Hydrohill

The artificial catchment Hydrohill of CHL was designed by Wei-Zu Gu during 1975 while he came back from his peasant life, kindly accepted and supported by Chuzhou administrator Mr. Wu-Min Cao for the laying down of both the Hydrohill and Nandadish on 1978. NHRI completed it and start running with data collection since July, 1982. The technological process can be sketched out summarily as follows.

(1) *Site selection and clearance*: A southeastward hillslope on a small hill, which is protruding integrally outwardly from its main area was selected for our Hydrohill due to its main geologic setting of andesitic tuff with altered volcanic rock. The space of this slope is enough for our initial design including in total three sister catchments with different characters each other. Moving away all the deposits including the weathered rock until the exposure of fresh bedrock, the site of clearance with area of ca 4700 m^2 was then prepared, however, only one catchment was constructed; (2) *Artificial aquiclude and surrounding wall*. A concrete aquiclude with two intersecting slopes dipping toward each other at 10°, an overall downslope gradients of 14°, and the longitudinally extending rectangular drainage trench etc. are constructed following the integral design of Hydrohill (**Figure 9a**). After that, an impermeable wall was set up on this aquiclude across the catchment boundaries (**Figure 9a**) aimed at enclosing the catchment to prevent any lateral exchanges of underground flow. These were completed during 1978 with its bird's eye view shown in **Figure 9b**. (3) *Soil filling*. An agricultural land was selected for the soil source of Hydrohill, a soil profile at the agricultural site with depth ca 1.5 m below the ground surface was dug for general observations, and undistributed soil of different layers were sampled for bulk density using current method. After that, the soil of the agricultural site was started to remove from its top horizon of about 10 cm in depth; it was piled up at a place close to Hydrohill, covered and marked. Then, the deeper parts of the soil were taken layer-by-layer every 20 cm and piled up, covered, and marked again. During soil filling

Figure 9. Constructions of Hydrohill. (a) the artificial aquiclude and surrounding wall; (b) view of the completed concrete aquiclude and wall; (c) view of the artificial catchment while soil was filled up, it was in idle for 3 years waiting for the physical improving of filling soil; (d) setting up monitoring networks for both saturated and unsaturated zone; (e) set drainage trenches; (f) the splicing fiberglass troughs and some steel supports; (g) the rectangular bottom trough for SSR100 and steel supports for the setting of SSR60 trough above it; (h) the gap for inverted filter and the nylon net in preparing; (i) the stainless steel screen in preparation for setting; (j) a curved connection trough; (k) connection troughs in installing; (l) discharge measuring structures using V-notch based logarithm sharp crested weirs (1982–1995).

in the concrete framework of Hydrohill, it was started from the soil pile of the deepest layer, that is, 90–100 cm of the original soil layer, it entered first into the artificial framework, after this layer was filling up, it was sampled for bulk density check. Then, the next piles of soil and so on. The workers are happy to follow our lazy filling and the time-consuming checking on depths and bulk density of whole area by paying hourly wages. After whole area was completed, it was allowed to settle for 3 years before equipped (**Figure 9c**), while the natural grasses were revitalized. We then have opportunity to remeasure the established bulk densities during first exaction for the main trench later during 1981, it is 1.44 g/cm^3 at the top soil of 0–30 cm, 1.42 g/cm^3 of the layer at the depth of 30–50 cm, 1.40 g/cm^3 of 50–75 cm, and 1.60 g/cm^3 of 75–100 cm. (4) *Setting up monitoring networks for both saturated and unsaturated zone.* Three networks including 22 wells for saturated water (groundwater) table measurements and groundwater sampling, 21 aluminum alloy access tubes for neutron moisture gauge were installed, all of them were drilled to the aquiclude (**Figure 9d**). The tubes for groundwater monitoring were slotted along the lowermost 20 cm and wrapped up by plastic net (**Figure 10a** and **b**). After installation into the drilling hole, the space around the slotted lengths were packed with sands to allow movement of groundwater to the well, however, the space above the slotted lengths should be carefully stuffed up by small dried clay balls to prevent any water other than groundwater intruding along the pipes (**Figure 10c** and **d**). Different from the tube for groundwater, the neutron access tubes should have an intact wall with sealed bottom, using threaded cap with a hanging pack of desiccant for sustaining a state of dryness in the tube (**Figure 6e**). A part of resulted networks is shown in **Figure 10d**. (5) *Drainage trenches.* After the installation of groundwater wells and access tubes, the drainage trenches including the main longitudinal trench and the side trench perpendicular to the main trench at the watershed outlet were dug up (**Figure 9e**). (6) *The layered troughs.* It was worked first from the bottom one. The splicing fiberglass troughs each 40 cm wide and some steel supports are shown in **Figure 9f**, **Figure 9g** shows the rectangular trough for SSR100 and steel supports for the setting of trough above it, that is, the SSR60. These troughs were constructed as shown in **Figure 11**, they were stacked on top of each other to create a set of long zero-tension lysimeters, each trough has a 20 cm aluminum lip that extends horizontally into the soil layer to prevent leakage between layers (**Figure 11**).(7) *The inverted filter.* There is a gap between the trough and the soil (**Figure 9h**) for the inverted filter (**Figure 11**) including the filling materials and the nylon net at the soil side and the stainless steel screen at the trough side (**Figure 9h** and **i**, **Figure 11**). (8) *The connection troughs.* With different curvatures, these troughs link the runoff troughs with the approaching part of measuring structures individually (**Figure 9j** and **k**). (9) *Discharge measuring structures.* We combined the V-notch sharp crested weir and the logarithm-notch sharp crested weir together, and defined it as the V-notch based logarithm sharp crested weir (**Figure 9l**) for discharge measurement of SR and SSRs. (10) At last, 30 hole sites for tensiometers at different depths together with the connection tubes from tensiometer to the scanning recorder were set up (**Figure 10e**).

After the finalization of construction, the artificial Hydrohill catchment has a drainage area of 490 m^2 (horizontal projection), 512 m^2 (inclined surface) as shown in **Figure 12** including its previous version since 1982 (**Figure 12a–d**), and the renovation version (**Figure 12e–h**). The lengths of the longitudinal trough (5 in **Figure 12b**, **c**, **e**, **g**) and the transverse trough (4 in **Figure 12b** and **e**) are 29.4 and 6.8 m, respectively. The width of the trough is 0.4 m. The horizontal projected area of the trough is 13.8 m^2, and thus horizontal projected area of soil surface is 487.2 m^2.

Recently, the average thickness of the soil is lower than that in 1982, especially in the upstream (**Figure 13**). To date, the average thickness of the soil in the upstream (~85 cm) has been less than that in the downstream (~105 cm) and that in the midstream (~103 cm).

Figure 10. Construction of groundwater monitoring well. (a) the slotted part of steel tube; (b) the tube was wrapped up by plastic net; (c) making clay balls for using; (d) during installing, the space between the tube and drilling hole above the slotted lengths (shown by a yellow arrow) should be carefully stuffed up by small dried clay balls using a special designed tool. This picture is shown in commemoration of the late prof. M-Q Lu (left) of Nanjing University for his efforts and contributions. (e) a part of resulted networks showing well for groundwater, access tube for neutron moisture gauge and the tensiometer with connection plastic tubes to the scanner.

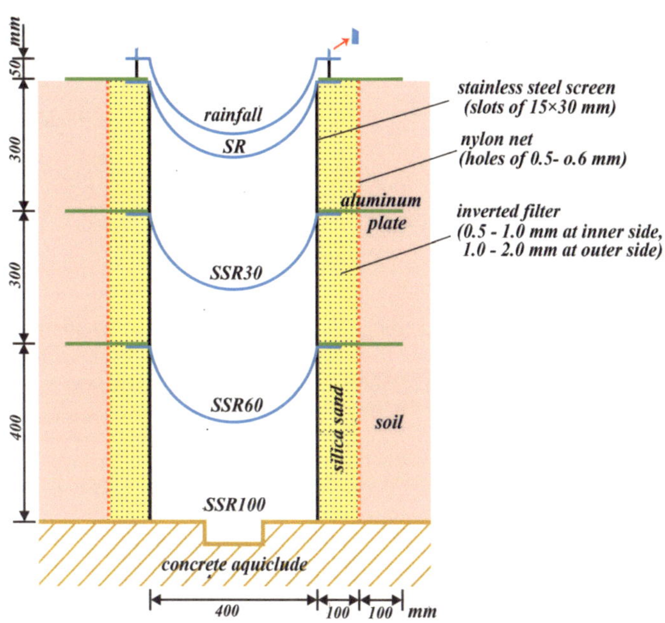

Figure 11. Construction of troughs. SR-trough for surface runoff; SSR30-trough for subsurface runoff at the depth of 30 cm below the ground surface; SSR60-that at the depth of 60 cm; SSR100-that at the depth of 100 cm.

3.2. Instrumentation in Hydrohill catchment

3.2.1. Precipitation

To monitor the distribution of rainfall and test the spatial uniformity of the rainfall generated by a rainfall simulator, 12 standard rain gauges (**Figure 14a** and **b**) and 5 tipping bucket

Figure 12. Finalization of the artificial catchment Hydrohill, its previous version (1982–1995) and renovation version (since 2012). (a) a satellite image showing Hydrohill and the whole hillslope area to be excavated of ca 4700 m^2; (b) and (c) the previous version of equipped Hydrohill; (d) the environment of Hydrohill. 1-land surface of Hydrohill; 2-an uncompleted artificial catchment; 3-the outlet of the Bloomhill Basin showing a discharge measuring structure of the gauging station; 4-the side trench perpendicular to the main trench at the watershed outlet; 5-the main longitudinal trench with layered troughs; 6-energy budget monitoring apparatus; 7-a scanning recorder for soil water potential from distributed sensors; 8-the neutron moisture gauge is in operation; 9-underground house for the discharge measuring structures; 10-laboratory for the calibration of neutron moisture gauge (1982–1988); 11-water supply tower; 12-two lysimeters with layered runoff troughs, 4 × 8 m^2 each, two different soils; 13-movable rainfall simulator at rest; 14-a reservoir for water supply of rainfall simulator; 15-a dual polarization Doppler precipitation radar for the areal precipitation monitoring of the Bloomhill Basin with drainage area of 80 km^2; 16-tracks for the moving of rainfall simulator; 17-laboratory for instrument measuring of isotopes; 18-various monitoring wells, sensors, samplers.

rain gauges (**Figure 14a** and **c**) were installed within the catchment. To test the measurement accuracy of the tipping bucket rain gauge, a 10-L plastic pot was connected with the drainage holes in the bottom of the tipping-bucket rain gauge (**Figure 14c**).

3.2.2. *Runoff components*

The uppermost trough (**Figures 11** and **15**) collects rain; the next lower trough collects surface runoff (SR); the three lower troughs collect subsurface flow from soil layers of the depths of 0–30, 30–60, and 60–100 cm (inferred as SSR30, SSR60, and SSR100 troughs).

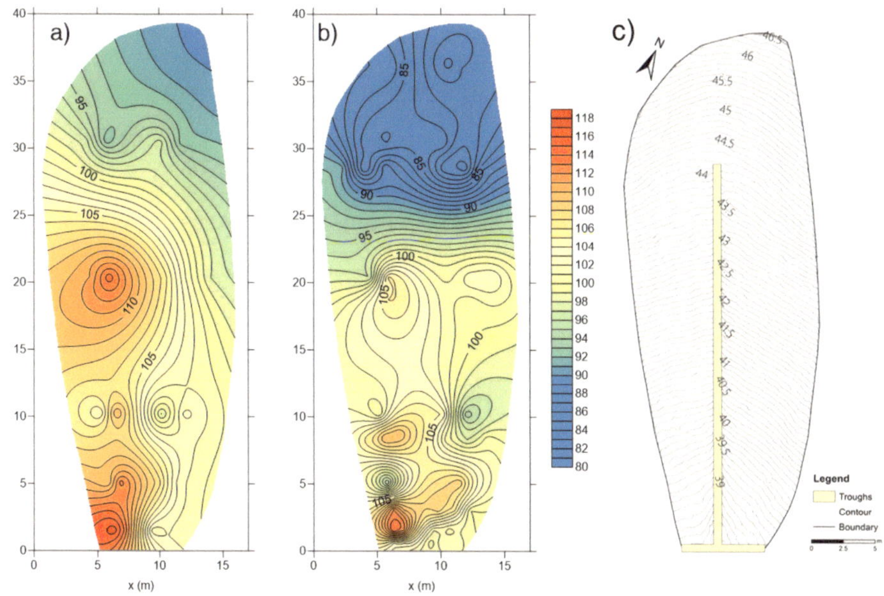

Figure 13. Changes in the filling soil thickness of the Hydrohill catchment: (a) Isoline of catchment soil depth in 1982; (b) that in 2017; (c) hypsographic map of soil surface obtained from a three-dimensional laser scanner (2017).

Figure 14. Instrumentation for precipitation measurement in Hydrohill: (a) locations of 12 standard rain gauges and 5 tipping bucket rain gauges; (b) a standard rain gauge; (c) a tipping-bucket rain gauge. 1-external structure, 2-internal structure, 3-a plastic pot used to collect total precipitation of each event.

Runoff collected from trough SSR30 is the interflow from the unsaturated zone, but that from trough SSR60 will depend on the depth of the saturated zone due to the fluctuation of groundwater table. In case when the saturated zone table (plus its capillary fringe) is lower than the trough, then the runoff measured in SSR60 is the interflow from the unsaturated zone; otherwise, it will be the groundwater flow from the saturated zone. Water collected in troughs is routed by measuring structures as shown in **Figure 15** (the renovation version). For each weir, a pressure-type water level gauge (LEV1, ADCON) and a probe-type water level gauge (NKY08-2, NHRI) are adopted to simultaneously measure the water head above the weir.

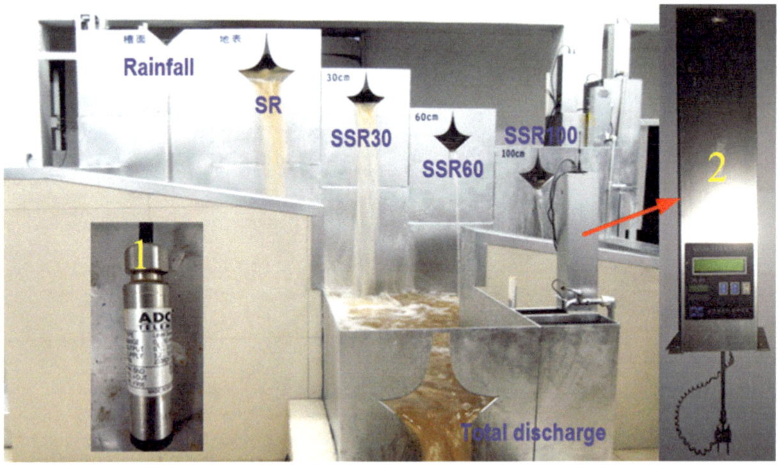

Figure 15. The discharge measuring structures of renovation version and the instrumentation for its water heads. 1-probe-type water level gauge; 2-tracking gauge for the real-time water head process.

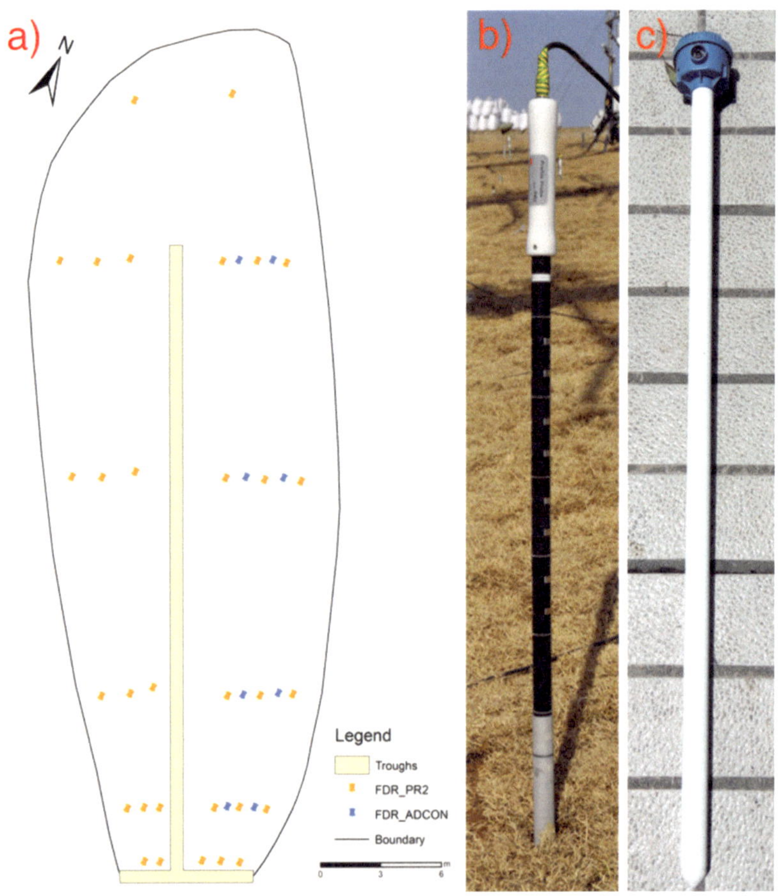

Figure 16. Instrumentation for soil water measurement in Hydrohill catchment: (a) the locations of the soil moisture sensors and soil water sampling points; (b) the UK soil moisture sensor (PR2, Delta-T, UK); (c) the German soil moisture sensor (SM-1, ADCON, Germany).

3.2.3. Soil moisture

A network of 21 aluminum alloy access tubes for neutron moisture gauges were constructed previously (1982–1995). Since then, all aluminum alloy access tubes have been displaced with 31 profile soil moisture sensors (PR2, Delta-T, UK), with their locations shown in **Figure 16a**. Each sensor has six sensor points located at the 10, 20, 30, 40, 60, and 100 cm (**Figure 16b**). Another six profiles with another kind of soil moisture sensors (SM-1, ADCON, Germany) were installed as well. This type of SM-1 has nine sensor points located at the 10, 20, 30, 40, 50, 60, 70, 80, and 90 cm (**Figure 16c**).

3.2.4. Groundwater

An array of 22 galvanized tube wells intersect through the soil till the concrete aquiclude (**Figure 17a**). Water table measurement is performed with level sensors (LEV1, ADCON, Germany, see **Figure 17c**).

3.2.5. Evaporation from land surface

An energy budget system (**Figure 18a**) and an eddy covariance system (**Figure 18b**) were mounted for accurate monitoring evaporation. The systems were equipped with the following sensors: one 3-D ultrasonic anemometer (C150, Campbell, USA), one CO_2/H_2O infrared gas analyzer (Campbell, USA), three air temperature and moisture sensors (HMP155A, Vaisala, Finland), four-way net rasiometers (CNR4, Kipp and Zone, Netherlands), one ground surface infrared temperature sensor (SI-111, Campbell, USA), five soil temperature sensors (109, Campbell, USA), five soil moisture sensors (CS616, Campbell, USA), and four soil heat flux sensors (HFP01SC, Huksflux, USA). In addition, a small aperture scintillometer (SLS-40A, SCINTEC, Germany, see **Figure 18c**) is to be installed with its transmitter unit and receiver unit outside the Hydrohill catchment.

3.2.6. Movable rainfall simulator

A movable rainfall simulator system (**Figure 19a**) was designed and constructed over the Hydrohill catchment in 2012. This system consists of five sub-systems, which can be controlled independently by the control platform (**Figure 19d**) to generate different rainfall intensities (10–200 mm/h) via the combination of different sizes of sprinkle nozzles (**Figure 19e**) with regulations of water pressure. **Figure 20** shows the schematic diagram of the rainfall simulator system. The rainfall area of this modeling system is 656 m^2, which can effectively cover the Hydrohill catchment (**Figure 19b**), and the spatial uniformity of the simulated rainfall is larger than 0.8. After the simulating rainfall event is finished, the rainfall simulator system can be moved to the outside of Hydrohill catchment (**Figure 19c**).

3.2.7. Water sampling

3.2.7.1. Water sample types and analysis indexes

Water samples from precipitation, runoffs, soil, groundwater, and plants were collected. Analysis indexes for water samples are included in three categories:

① General parameters: electrical conductivity (EC), pH, dissolved oxygen (DO), and water temperature;

② Hydrochemistry: K^+, Na^+, Ca^{2+}, Mg^{2+}, HCO_3^-, CO_3^{2-}, Cl^-, NO_3^-, SO_4^{2-}, and SiO_2;

③ Isotopes: ^{18}O, ^{2}H, and ^{15}N.

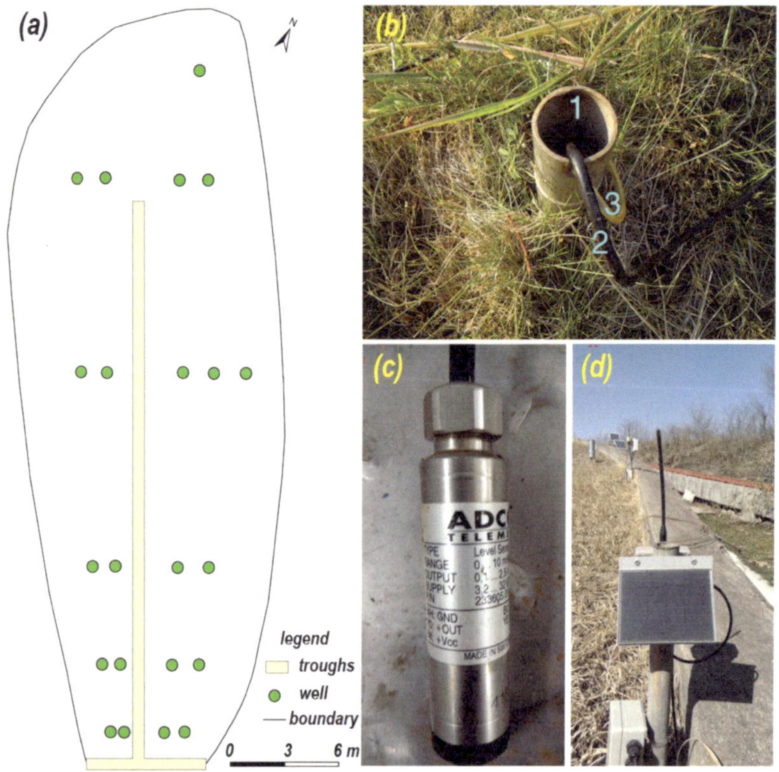

Figure 17. Instrumentation for groundwater measurement in Hydrohill catchment: (a) locations of the monitoring wells; (b) the monitoring well (1-the galvanized tube, 2-the connection cable of the level sensor, 3-a plastic tube for sampling groundwater); (c) the level sensor which is a pressure transducer with atmospheric pressure compensation; (d) a remote terminal unit (RTU) receive the data from the groundwater level sensors (LEV1, ADCON, Germany) and send the obtained data to the gateway via radio together with that from soil moisture sensor as shown in **Figure 16** (SM-1, ADCON, Germany).

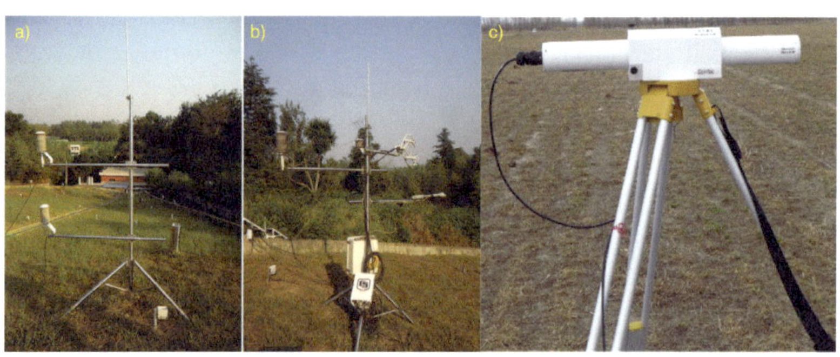

Figure 18. Instrumentation for the evaporation from ground surface in Hydrohill. (a) An energy budget system; (b) an eddy covariance system; (c) a small aperture scintillometer.

Figure 19. Rainfall simulator system in Hydrohill: (a) before the rainfall simulating; (b) during the rain simulating; (c) after the rainfall simulating event (1-an impounding reservoir supplying the rainfall simulator system, 2-a water supplying pipe); (d) the rainfall simulator control platform located in the gauging room; (e) combination of different sizes of sprinkle nozzles.

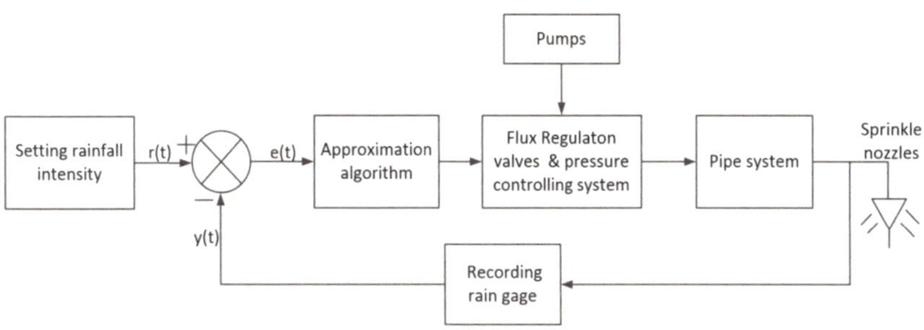

Figure 20. Schematic diagram of the rainfall simulator system.

3.2.7.2. Sampling for precipitation

Rain water samples were collected using two methods: first is via the rainfall trough, which serves as a rain gauge distributed longitudinally with total area of 13.8 m^2 (**Figure 21a**), the second is via a specially designed rain gauge with a diameter of 40 cm, which was installed on the roof of the gauging room (**Figure 21b**). The result from these two methods shows that they are similar to each other with only few exceptions (**Figure 22**).

3.2.7.3. Sampling for runoff components

To easily and to fast collect water samples of different runoff components in Hydrohill, a batch sampling system is designed and constructed based on negative pressure (**Figure 23**). The schematic of the batch sampling system is shown in **Figure 23a**. Water samples for runoff components are collected via a stainless steel tube head fixed at the connection trough before the runoff reaches the ponding of the weir (**Figure 23c**).

Figure 21. Two methods for rainfall sampling: (a) method-1 using the rainfall trough as a longitudinal rain gauge; (b) method-2 using a specially designed rain gauge.

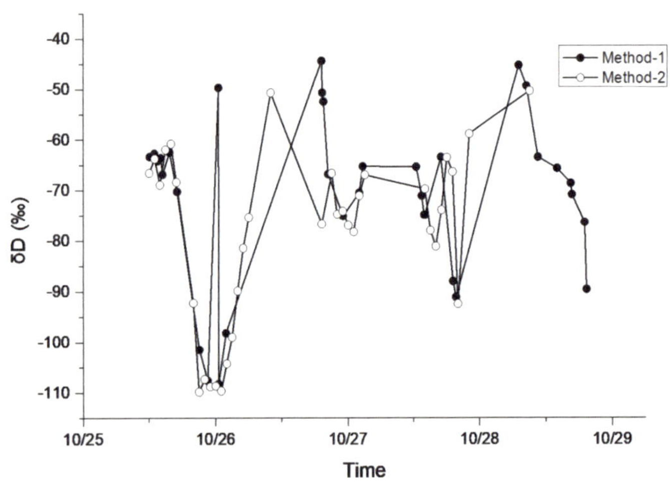

Figure 22. Comparison of results from two methods for rainfall sampling. Method-1: Using the rainfall trough; Method-2: Using a specially designed rain gauge. Data is resulted from a rainfall event in 2016.

3.2.7.4. Sampling for soil water and groundwater

To sample the soil water, 31 suction lysimeters (**Figure 24a** and **d**) were installed at three depths: 9 at 15 cm, 12 at 45 cm, and 10 at 80 cm. To keep the synchronism of sampling, a batch sampling system is designed and constructed based on negative pressure (**Figure 24b** and **e**). Groundwater samples of 22 points were collected with the tubes previously fixed on the level sensors (**Figure 24a** and **f**). Similarly, a batch sampling system is designed and constructed based on negative pressure to keep the synchronism of sampling groundwater (**Figure 24c** and **d**).

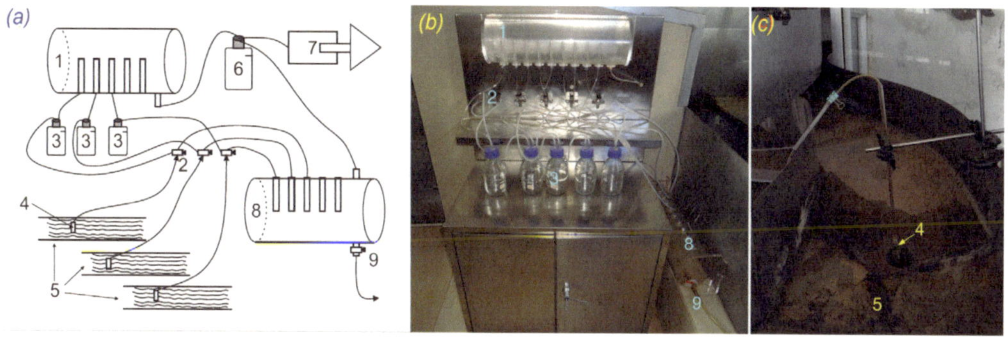

Figure 23. Instrumentation for sampling runoff components in Hydrohill: (a) the schematic of the batch sampling system; 1-the negative-pressure extender, 2-three-way valve, 3-sample bottles, 4-a stainless steel tube head enclosed by a yarn to stop sand and litter into the sampling tube, 5-troughs, 6-a safeguard bottle, 7-a vacuum pump, 8-a container used to drain the old water in the tubes, 9-a valve to drain the water in the container; (b) photo of a batch sampling system based on negative pressure; (c) a stainless steel tube head fixed at the connection trough before the runoff reaches the ponding of the weir.

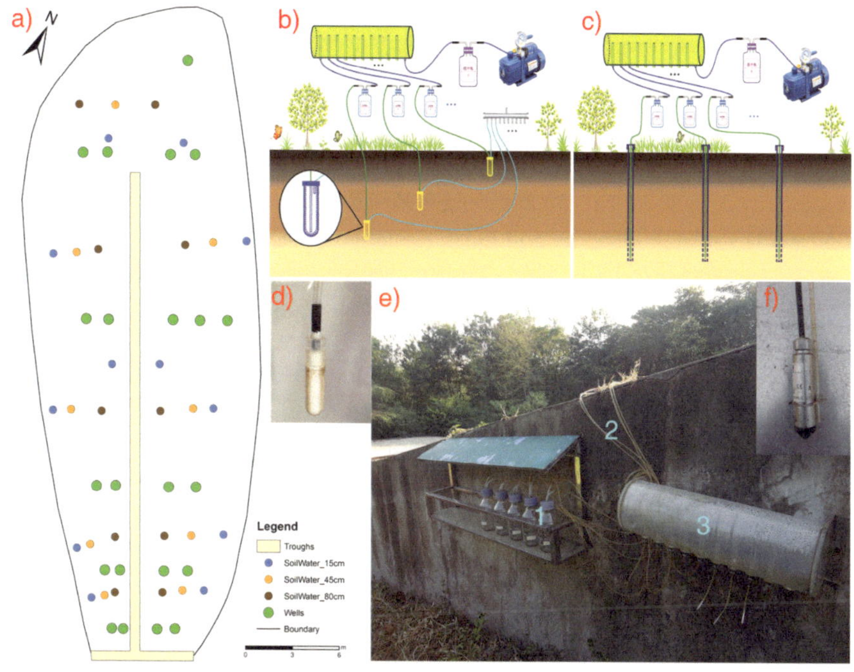

Figure 24. Instrumentation for sampling soil water and groundwater in Hydrohill: (a) locations of the sampling points of the soil water and groundwater; (b) schematic of a batch sampling system designed and constructed based on negative pressure for sampling soil water; (c) that for sampling groundwater; (d) a suction lysimeter; (e) photo of a batch sampling system designed and constructed based on negative pressure for sampling soil water and also for groundwater; (f) the tube previously fixed on the level sensor to sample groundwater. 1-sample bottles, 2-tubes connecting with suction lysimeters or groundwater wells, 3-the negative-pressure extender.

3.3. Analyses of water samples

3.3.1. General parameters

Electrical conductivity (EC) was measured in the field using a portable EC digital analyzer (HQ14d, Hach, USA). Dissolved oxygen (DO), pH, and water temperature were measured in the field using a multi-parameter digital analyzer (HQ40d, Hach, USA). These parameters were determined immediately after water samples had been collected.

3.3.2. Hydrochemistry

Most chemical analyses were undertaken at the Tiexinqiao experiment base of Nanjing Hydraulic Research Institute, China. All water samples were analyzed within 2 weeks of the date of collection. Concentrations of the major cations (Na^+, K^+, Ca^{2+}, and Mg^{2+}) were analyzed by inductively coupled plasma optical emission spectrometry (ICP-OEC, see **Figure 25a**), and the anions (SO_4^{2-}, Cl^-, NO_3^-, and F^-) were analyzed by DIONEX ICS-2100 ion chromatography (**Figure 25b**). All samples were filtered through a 0.45 μm filter before laboratory analysis. The analytical precision of the measurement of ions was determined by calculating the absolute error in ionic balance, and the analytical error was less than ±2% for the anions and between ±1.5 and ±4% for the cations. HCO_3^- concentrations were determined by a titration assay on site or within 24 h of sample collection.

3.3.3. Isotopes

Soil and plant waters were previously obtained via a vacuum extraction system (LI-2000, LICA, China, **Figure 26a**). The $\delta^{18}O$ and δD of water samples determined using a liquid water isotope analyzer (908-0008, LGR, USA, **Figure 26b**) or a liquid–gas water isotope analyzer (L2120–i, Picarro, USA, **Figure 26c**). The dual isotopes of nitrate were prepared by quantitative bacterial reduction of nitrate to nitrous oxide (N_2O) using the denitrifier method followed by automated extraction and purification using Trace Gas Pre-concentrator unit (IsoPrime Ltd., Cheadle Hulme, Cheadle, UK) and analysis of the N_2O product using an isotope ratio mass spectrometer (GV, IsoPrime, **Figure 26d**). Four international nitrate (USGS-32, USGS-34, USGS-35, and IAEA-N3) and experimental reference materials that were treated identically with the water samples were used to calibrate the measured sample data. Each sample was measured in duplicate and the standard error was 0.3‰ for $\delta^{15}N\text{-}NO_3^-$ and 0.5‰ for $\delta^{18}O\text{-}NO_3^-$.

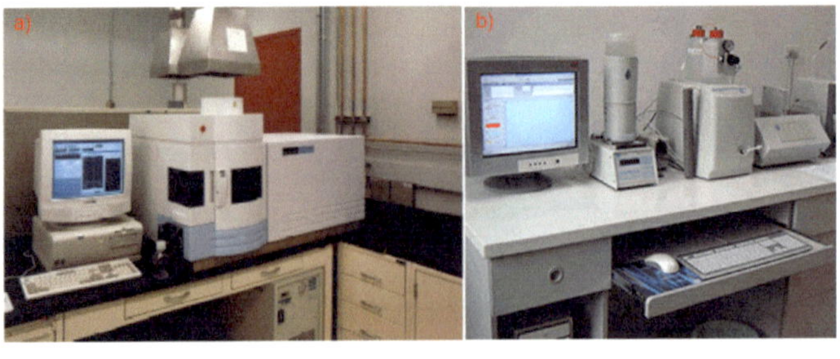

Figure 25. Instruments adopted in hydrochemistry analysis of water samples: (a) inductively coupled plasma optical emission spectrometry (ICP-OEC); (b) ion chromatography (ICS-2100, DIONEX, USA).

Figure 26. Instruments adopted in isotope analysis of water samples: (a) a vacuum extraction system (LI-2000, LICA, China); (b) a liquid water isotope analyzer (908–0008, LGR, USA); (c) a liquid–gas water isotope analyzer (L2120–i, Picarro, USA); (d) IsoPrime100.

4. Some results

4.1. Explore the possible paths

Aimed at ending the scientific stalemate on our watershed experimental studies. Since 1982, the origin of CHL, from classic natural experimental watershed, current pedon lysimeter, and the uncompleted experimental system until the Chuzhou WHES, various possible paths are tried for the emerging of some possible paths ([1–6]), to achieve hopefully the sustainable development of the watershed hydrological experimentation. It is found that the intermediate "mesos" including those of controlled-nature and artificial-nature with constrain and add complexity respectively, show its crucial importance for revealing the individual mechanisms hidden deep. Philosophically, it is "the golden mean between two extremes of character" in Book IV of his Ethics of Aristotle, and the idea of "holding the two extremes and using the middle impartial" in China for the "music" of our watershed experimental studies.

4.2. Explore the subsurface runoff components

- Direct measurement: After progressively improving, the method of longitudinal zero-tension lysimeter (layered trough) is used in catchment scale for the direct measurement of surface and subsurface runoff components ([1, 7]).

- Runoff components: Three components are identified including surface runoff (SR), interflow (IF) from unsaturated zone, and groundwater flow (GF) from saturated zone ([5–8]).
- Amount proportion: From 375 runoff generation-events (1982–1995), the total subsurface contribution accounted for 43% of total runoff, 27% of total runoff was contributed from the *direct interflow* from the unsaturated zone [9].
- Patterns of rainfall-runoff process: Four patterns are identified according to the dominated runoff components, surface flow or subsurface flow [9–11].
- Rainfall-runoff correlation diagram: scattering of data points including that of surface runoff very likely is caused by different runoff compositions of different sources of water rather than simply the rainfall characters or, the curve numbers [9, 12].

4.3. Explore the generation mechanisms of runoff components

Eleven mechanisms types have been identified including 4 of SR, 4 of IF, and 3 of GF [13, 14].

4.4. Explore the composition of pre-event water

- Occurrence of pre-event water: It is identified that the pre-event ("old") water is frequent occurred even in the SR [11, 15].
- Process of pre-event water: A 4-year case studies show that the pre-event (old) water within 4 different runoff patterns accounted for 0–36% in surface dominated pattern and up to 60% in subsurface pattern, 47–77% and 21–75% the other patterns [9, 11, 15, 16].

4.5. Explore the hydrological puzzles

It emerges that the unsaturated zone is the gremlin, the key to revealing the hydrologic maze because it is closely related to the runoff composition, hydrological heterogeneity and the double paradox in catchment hydrology and hydrogeochemistry [6, 9, 17].

4.6. On some parameters

- ^{131}I tracing for infiltration and preferential flow [18, 19];
- Spatiotemporal distribution of soil water ^{18}O in Hydrohill catchment [9];
- Optimization selection of discharge measuring structures for the application of WHES [20, 21];
- Neutron gauging for vadose water and safety evaluation for users [22];
- Nuclear methods for the monitoring of evapotranspiration from land surface [23].

4.7. Basic research for applications

Preliminary methodological studies for applied hydrological projects

- Unreasonableness of current two-component isotope hydrograph separation [6, 24–26];

- Water tracing using uranium disequilibrium and other tracers for identification of water sources [27, 28, 29];
- Neutron activation for rain water, river water, and groundwater [30, 31];
- Isotope-In-Precipitation Network of China [32];
- Agricultural water demands [33, 34];
- Nonpoint source of agricultural area [35, 36].

4.8. The post graduates dissertations achieved by working in and/or main source from CHL

PhD dissertation

1993: Carol Kendall. Impact of isotopic heterogeneity in shallow systems on stormflow. University of Maryland.

2017: Meng Huang. Using isotopes to evaluate nitrogen sources and transformation processes in Chuzhou hydrological watershed. Hohai University.

Master dissertation

1986: Xue-Wen Wang. Study on the cropland evapotranspiration. Nanjing Institute of Meteorology.

1990: Jing Huang. Application of satellite image for the GIUH confluence model. Wuhan University, Wuhan Institute of Hydraulic and Electric Engineering.

1990: Shu-Qin Xie: Application of satellite imagine for the runoff generation using the SCS Model. Wuhan University, Wuhan Institute of Hydraulic and Electric Engineering.

2016: Niu Wang: Output characteristics of non-point source nitrogen and phosphorus load in Huashan Basin. Hohai University.

2017: Nuo Yang: Study on the tracing of precipitation and runoff in the Hydrohill Experimental Catchment using hydrochemistry and isotopes. Hohai University.

5. Conclusions

The construction and instrumentation of two typical parts of Chuzhou WHES are reviewed, including the Nandadish, a trial target for the CZEB, which is the type of so-called *controlled-natural*, and the Hydrohill, another trial target for CZEB, that is, the type of so-called *artificial-natural*. These trials however all are ongoing, in improving, actually the improving maybe endless as the advancing idea with time seems endless.

Kirkby had warned during 2004 that "There has been a movement away from field work and toward an almost complete dependence on modelling" [37], more than 10 years later, Burt and McDonnell (2015) described how field scientists have posed strong and sometimes outrageous hypotheses – approaches so needed "in an era of largely model-only research", "go further and further down the rabbit hole of model uncertainty estimation" [38]. A more unified and holistic theory as called for by Sivapalan [39] is still on the way depending on experimental efforts. Education also needs to be the antecedence, "field work's primary purpose must be to teach our students to be curious, to look, to collect data, to test existing ideas, to develop new hypotheses, including outrageous ones" [38]. Watershed hydrological experimentation seems in the risk to be marginalization however, "it now is indeed an exciting time for hydrologists/experimentalists to rise up for a new era of scientific hydrology" [9].

Acknowledgements

This review is a commemoration of the 65th anniversary of the Chinese basin study and the 37th anniversary of the establishment of the Chuzhou Hydrology Laboratory (CHL). The senior author (W-Z Gu) has been fortunate to render his services to the both from their beginning, with grateful heart he is indebted to a number of people including colleagues, experts, professors, and local peasants, who have helped, succored, supported make this possible. We are very grateful for those who have made numerous efforts for the reborn renovation of the CHL after long abandonment and, for its leaping forward from manual to digital: Academician Jian-Yun Zhang, Prof. Jiu-Fu Liu, and many other experts in the Chinese Ministry of Water Resources. We also thank many foreign experts who have encouraged, helped, taught and guided our work over the years especially during our difficult, including Jeffrey McDonnell, Keith Beven, Henry Lin, Mebus Geyh, Carol Kendall, Vance Kennedy, Norman Peters, Michael Sklash, А.И. Шикломанов, Joel Gat, Verhagen B and Klaus Froehlich. We also acknowledge many colleagues in the CHL for their hard work in the field.

This work was supported by the grants from the National Key R&D Program of China (No. 2017YFC0403500) and the National Natural Science Foundation of China (No. 91647203, No.91647111, and No.51609145).

Author details

Jiu-Fu Liu*, Ai-Min Liao, Niu Wang, Jin Lin, Hong-Wei Liu, Wen-Zhong Wang, Tao Ma, Zhao Cai, Min-Han Liao, Xue-Gang Li, Peng Zhuo, Na Yang, Jia-Ju Lu and Wei-Zu Gu

*Address all correspondence to: jfliu@nhri.cn

Department of Hydrology and Water Resources, Nanjing Hydraulic Research Institutes, Ministry of Water Resources, China

References

[1] Gu WZ, Wen ZR. Experimental watershed system and the Chuzhou experiments. In: Essays on Hydrological Research. Nanjing Institute of Hydrology; 1982. pp. 214-233. (in Chinese)

[2] Gu W-Z. On the domain and approach of the experimental hydrology. In: Nanjing Hydrology Institute, editor. Treatise on Hydrology and Water Resources. Beijing: Water Resources and Electric Power Press; 1987. pp. 507-515. (in Chinese)

[3] Gu W-Z. Experimental and representative basin study in China: A 35th anniversary review. In: Hooghart JC, Posthumus CWS, Warmerdam PMM, editors. Hydrological Research Basins and Environment. The Hague, The Netherlands; 1990. pp. 175-186

[4] Gu W-Z. A brief decade summary of the Chuzhou Hydrology Laboratory 1981-1991. Institute of Hydrology and Water Resources. Unpublished Internal File;1992. (in Chinese)

[5] Gu W-Z, Liu C-M, Song X-F, Yu J-J, Xia J, Wang Q-J, Lu J-J. Hydrological experimental system and environmental isotope tracing: A review on the occasion of the 50th anniversary of Chinese basin studies and the 20th anniversary of Chuzhou hydrology laboratory. In: Research Basins and Hydrological Planning. A.A. Balkema Publishers; 2004. pp. 11-18

[6] Gu W-Z, Liu J-F, Lu J-J, Frentress J. Current challenges in experimental watershed hydrology. In: Bradley P, editor. Current Perspectives in Contaminant Hydrology and Water Resources Sustainability. Rijeka, Croatia: InTech Publisher; 2013. pp. 299-333. DOI: 10.5772/55087

[7] Kendall C, Wei-Zu G. Development of isotopically heterogeneous infiltration waters in an artificial catchment in Chuzhou, China. In: Isotope Techniques in Water Resources Development 1991. Vienna: IAEA; 1992. pp. 61-72

[8] Kendall C, McDonnell J, Gu W-Z. A look inside 'black box' hydrograph separation models: A study at the Hydrohill catchment. Hydrological Processes. 2001;**15**:1877-1902

[9] Gu W-Z, Liu J-F, Lin H, Lin J, Liu H-W, Liao A-M, Wang N, Wang W-Z, Ma T, Yang N, Li X-G, Zhuo P, Cai Z. Why hydrological maze: The hydropedological trigger? Review of experiments of Chuzhou Hydrology Laboratory. Vadose Zone Journal; **17**:170174. DOI: 10.2136/vzj2017.09.0174. in press

[10] Gu W-Z. Field research on surface water and subsurface water relationships in an artificial experimental catchment. In: Dahlblom P, Lindh G, editor. Interaction between Groundwater and Surface Water, International Symposium. Ystard, Sweden; 1988. pp. 33-41

[11] Gu W-Z. Challenge on some rainfall-runoff conceptions traced by environmental isotopes in experimental catchments. In: Tracer Hydrology. Rotterdam. Netherlands: Balkema Publishers; 1992. pp. 397-403

[12] Hansen DP, Jakeman AJ, Kendall C, Weizu G. Identification of internal dynamics in two experimental catchments. Mathematics and Computers in Simulation. 1997;**43**:367-375

[13] Gu W-Z. Various patterns of basin runoff generation identified by hydrological experiment and water tracing. Journal of Hydraulic Engineering. 1995;**5**:9-17. (in Chinese with English abstract)

[14] Gu W-Z, Freer J. In: Leibundgut Ch, editor. Tracer Technologies for Hydrological Systems. Proceedings of a Boulder Symposium. IAHS Publ.; 1995. Vol. 229. pp. 265-273

[15] Gu W-Z. Experimental research on catchment runoff responses traced by environmental isotopes. Advances in Water Science. 1992;**4**:246-254. (in Chinese with English abstract)

[16] Gu W-Z, Shang M-T, Zhai S-Y, Lu J-J, Jason F, Mcdonnell JJ, et al. Rainfall-runoff paradox from a natural experimental catchment. Advances in Water Science. 2010;**4**:471-478. (in Chinese with English Abstracts)

[17] Gu W-Z, Lu J-J, Zhao X, Peters NE. Responses of hydrochemical inorganic ions in the rainfall-runoff processes of the experimental catchments and its significance for tracing. Advances in Water Science. 2007;**1**:1-7. (in Chinese with English Abstracts)

[18] Gu W-Z, Lu MJ, Chen TY, Lu JJ. Using nuclear method for the distribution of infiltration parameters in experimental watershed. Journal of Nanjing University (Special issue for Natural Science and Geography). 1988:99-107. (in Chinese)

[19] Gu W-Z, Lu M-J, Chen T-Y, Lu J-J. A systematic study of distribution characters of infiltration parameters in an experimental basin by nuclear methods. In: Isotope Applications in Hydrology in Asia and the Pacific. Beijing: IAEA; 1987. pp. 265-276

[20] Gu W-Z. On the measuring structures for watershed discharge monitoring. Hydrology. 1982;**10**:2-9. (in Chinese)

[21] Gu W-Z. On the method of discharge measuring structures for small rivers. In: Nanjing Hydrology Institute, editor. Treatise on Hydrology and Water Resources. Beijing: Water Resources and Electric Power Press; 1987. pp. 507-515. (in Chinese)

[22] Gu W-Z, Lu M-J. Using neutron scattering method for soil moisture research in unsaturated zone of experimental basin. Journal of Nanjing University (Geography). 1983:84-100. (in Chinese)

[23] Gu W-Z, Lu J-J. Nuclear method for the monitoring of evapotranspiration from land surface. In: Nanjing Hydrology Institute (Editor), editor. Treatise on Hydrology and Water Resources. Beijing: Water Resources and Electric Power Press; 1987. pp. 502-506. (in Chinese)

[24] Gu W-Z. On the hydrograph separation traced by environmental isotopes. Advances in Water Science. 1996;**7**:105-111. (in Chinese with English Abstract)

[25] Gu W-Z. Unreasonableness of the current two-component isotopic hydrograph separation for natural basins. In: Isotopes in Water Resources Management 1995. Vienna: IAEA; 1996. pp. 261-264

[26] Gu W-Z, Longinelli. A case study on the hydrological significance of stable isotope data on alpine catchments, Xinjiang, China. In: Snow and Glacier Hydrology. Vol. 218. IAHS Publication; 1993. pp. 371-383

[27] Gu WZ, Lin ZP, Fei GC, et al. The use of environmental sulphur isotopes in the study of the Cambrian-Ordovician aquifer system in the south of Datong. Advances in Water Science. 2000;**V11**(N1):14-20. (in Chinese with English Abstracts)

[28] Gu WZ, Lu JJ, et al. Uranium disequilibrium in the Cambrian-Ordovician aquifer system of Southwest Datong and its application for groundwater recharge. Advances in Water Science. 2001;**V12**(N2):177-184. (in Chinese with English Abstracts)

[29] Gu WZ, Lu JJ, Wu Y. Identification of groundwater recharge sources by using of ^{234}U excess and ^{34}S for the arid Ejina-Badain Jaran Interior Basin of Alaxa plateau in Inner Mongolia. In: Advances in Isotope Hydrology and its Role in Sustainable Water Resources Management. Vienna: IAEA; 2007. pp. 121-131

[30] Gu W-Z, Tudeng-Xiangpei, Daiming L, Danba-Gongjue. A neutron activation study of the geochemistry in the natural waters in Lhasa, Tibet, China. 1998. In: Ecohydrology of High Mountain Areas 1996. Kathmanda, Nepal: ICIMOD; 1998. pp. 573-577

[31] Gu WZ, Lu JJ, Chen TY, Xu QG, Peters NE. Neutron activation analyses on the element composition of natural waters in China. Advances in Water Science. 2003;**V14**(N5):535-541. (in Chinese with English Abstracts)

[32] Zhao KJ, Gu WZ, et al. Isotope-in-precipitation network of China. Hydrology. 1995; **N5**:25-27. (in Chinese)

[33] Jia-Ju L. Analysis and estimation of the water demand and consumptive use of the paddy field in Chuzhou agricultural experimental catchments. In: Essays on Hydrological Research. Nanjing Institute of Hydrology; 1982. pp. 256-269. (in Chinese)

[34] Jiang X-L, Gu W-Z. Field experimental research on the water demand of paddy field. Water Resources Research. 1987;**V8**:N1-2. (in Chinese)

[35] Wang N. Output characteristics of non-point source nitrogen and phosphorus load in Huashan Basin. Master dissertation, Hohai University; 2016. (in Chinese)

[36] Huang M. Using isotopes to evaluate nitrogen sources and transformation processes in Chuzhou hydrological watershed. Dotoral dissertation, Hohai University; 2017

[37] Kirkby MJ, editor. Geomorphology: Critical Concepts in Geography. Vol. II. London: Routledge; 2004. p. 16

[38] Burt TP, Mcdonnell JJ. Whither field hydrology? The need for discovery science and outrageous hydrological hypotheses. Water Resources Research. 2015;**51**:5919-5928. DOI: 10.1002/2014WR016839

[39] Pattern SM. Process and function: Elements of a unified theory of hydrology at the catchment scale. In: Anderson MG, editor. Encyclopedia of Hydrological Science. John Wiley & Sons; 2005. pp. p193-p219